PROGRESS IN ELECTROCHEMISTRY OF ORGANIC COMPOUNDS

Volume 1

Edited by

A. N. Frumkin and A. B. Ershler

Institute of Electrochemistry, Academy of Sciences of the U.S.S.R., Moscow

This new series contains review articles summarising, assessing and assimilating recent research on the electrochemistry of organic compounds, ranging from the kinetics of electrode processes, electrosynthesis, the effect of electrode metal on adsorption processes, electrolytic oxidation and electroreduction of organic substances to polarography.

Problems reviewed include the electrosynthesis of organometallic compounds and electrolytic oxidation of organic substances in fuel cells. Also considered are the possibilities of using electrochemical methods to investigate catalytic processes in solutions and electrochemical generation of free radicals.

Providing an invaluable reference source and guide to recent research, this series will be useful for specialists in the field of theoretical and applied electrochemistry, analytical and organic chemists, and those working in the chemical and fuel industries.

Progress in Electrochemistry of Organic Compounds

1

Progress in Electrochemistry

of Organic Compounds

Edited by A. N. Frumkin and A. B. Érshler

Institute of Electrochemistry,
Academy of Sciences of the U.S.S.R.,
Moscow

Translated from Russian by D. E. Hayler
Translation edited by P. Zuman

Plenum Press London and New York · 1971

Plenum Publishing Company Ltd.,
Donington House,
30 Norfolk Street,
London, W.C.2.

U.S. Edition published by
Plenum Publishing Corporation,
227 West 17th Street,
New York, N.Y. 10011

Library of Congress Catalog Card Number 77-137739
ISBN-13:978-1-4684-3341-8   e-ISBN-13:978-1-4684-3339-5
DOI: 10.1007/978-1-4684-3339-5

The original Russian text was first published by Nauka Press in Moscow in 1969.
The present translation is published under an agreement with Mezhdunarodnaya
Kniga, the Soviet book export agency.

А. Н. ФРУМКИН   *А. Б. ЭРШЛЕР*

ПРОГРЕСС ЭЛЕКТРОХИМИИ ОРГАНИЧЕСКИХ СОЕДИНЕНИЙ

PROGRESS ELEKTROKHIMII ORGANICHESKIKH SOEDINENII

Set in Cold Type by Plenum Publishing Company Ltd.

Foreword

In the Soviet Union, investigations of electrochemical changes in organic substances are being conducted on a comparatively large scale and a large number of specialists are involved. This arises to a large extent from the necessity to solve problems in the applied fields, e.g. in the development of new improved methods for the analysis and synthesis of organic compounds or in the work on fuel cells. The attainment of substantial successes in this field has been linked inseparably with a deeper understanding of the mechanism and kinetics of electrolytic organic reactions and the utilization of modern research methods based on the latest achievements in instrumentation. The theory of organic electrode processes is therefore now developing rapidly. However, the propagation of information on this work has been relatively slow.

The Series of reports on Progress in Electrochemistry of Organic Compounds should stimulate systematic treatment and propagation of information in this field of science. It is proposed that each volume of the series will be compiled on the same lines as the book *Advances in Electrochemistry of Organic Compounds* published by Nauka in 1966. They will form collections of original review articles on the most important aspects of the subject, prepared by competent authorities.

This collection represents the first volume of the series. The reviews published here are based on reports read at the Sixth Conference on Electrochemistry of Organic Compounds held in Moscow in January, 1968. In the review, use is made of the results from the authors' own investigations, the latest information gathered from the world literature, and the contents of reports covering new original investigations in this field which were sent to the Sixth Conference.

In selecting the scope of topics covered by these reviews the Organizing Committee of the Sixth Conference has endeavoured to include problems which interest a large number of Soviet scientists and which have not been considered in sufficient detail at earlier conferences on the electrochemistry of organic compounds.

The collection opens with an examination of the methods, achievements, and potentialities of electron paramagnetic resonance in the electrochemistry of organic compounds. Attention has been paid to the current work on electrosynthesis, investigations with solid electrodes, and investigation of the role of adsorption in electrode processes involving organic substances. Papers on classical polarography are also included. Because the contribution by V. S. Bagotskii and Yu. B. Vasil'ev on the behaviour of organic substances at a platinum electrode was published earlier, another review on this topic, prepared by O. A. Petrii, has been included in the present collection. This departure from the otherwise followed procedure is due both to the need to analyse results obtained since 1965 and to the endeavour to represent as completely as possible the various approaches to this important problem.

The collection can be recommended to the attention of specialists working in the field of theoretical and applied electrochemistry and also to those active in analytical and organic chemistry.

A. N. Frumkin

Contents

Electrochemical Generation of Free Radical Ions and Use of
Electron Paramagnetic Resonance for their Investigation.
Ya. P. Stradyn' and R. A. Gavar   1

Polarographic Reduction of Azomethine Compounds.
Yu. P. Kitaev and T. V. Troepol'skaya 43

Electrode Processes Complicated by Chemical Reactions
and by Adsorption on Mercury Electrodes.
S. G. Mairanovskii 85

Stereochemistry of Electrode Processes.
L. G. Feoktistov 135

Prewaves in Reversible Systems.
*G. A. Tedoradze, E. Yu. Khmel'nitskaya,
and Ya. M. Zolotovitskii* 171

Reduction of Organic Molecules in Region of Low Electrode
Surface Coverages.
A. B. Érshler 203

Electrochemical Synthesis of Organometallic Compounds.
A. P. Tomilov and I. N. Brago 241

Electrolysis of Organic Compounds with Use of
Ion-Exchange Membranes.
M. Ya. Fioshin 287

Electrochemical Oxidation of Organic Compounds at
Metals of the Platinum Group.
O. A. Petrii 319

Electrochemical Methods for Investigation of Catalytic
Hydrogenation Mechanism in Solutions.
D. V. Sokol'skii 377

Electrochemical Reduction Mechanism in Organic Peroxides.
É. S. Levin and A. V. Yamshchikov 411

Electrochemical Generation of Free Radical Ions and Use of Electron Paramagnetic Resonance for Their Investigation

Ya. P. Stradyn' and R. A. Gavar

Among the numerous methods used for the preparation of organic free radicals the most widespread have been the following [1-6]: 1) Thermal decomposition; 2) electric discharge in gases; 3) electron bombardment; 4) radiolysis with γ-rays; 5) photolysis. Alongside these methods in the last 10 years there has also been a rapid development of the electrochemical methods for generation of free radicals, which has a number of special characteristics and advantages over the other methods. Most frequently used for electrolytic generation of free radicals is electrolytic reduction, and more rarely electrolytic oxidation; as a rule radical anions are obtained in the first case and radical cations in the second:

$$R + e \rightarrow R^{\cdot-}; \qquad R - e \rightarrow R^{\cdot+}.$$

Detection of unchanged (neutral) free radicals produced in electrochemical processes is difficult owing to their short life under the conditions of the electrochemical experiment, and the discussion will therefore be mostly concerned with radical ions.

1. METHODS FOR DEMONSTRATING THE FORMATION OF FREE RADICALS AND RADICAL IONS IN ELECTROCHEMICAL PROCESSES

The electrochemical method for generating particles with unpaired spins is based on the appearance of these particles as intermediate or final products in the electrochemical reaction. Transfer of a single electron to a particle (which may be a neutral molecule

1

or an ion) should result in the formation of a free radical as initial product [7, 8].

As a rule free radicals formed during electrochemical processes are short-lived kinetic particles; they determine the course and kinetics of a series of electrochemical reactions but themselves undergo further rapid chemical or electrochemical changes and cannot be isolated from the solution by preparative means.

Methods for their detection and investigation are limited and include the following:

1. Detection indirectly from the nature of the final large-scale electrolysis products [8-10] or by the use of free-radical acceptors, such as unsaturated compounds (styrene, 1,3-butadiene, acrylonitrile, etc.) for example, which readily enter into reaction with the radicals [7]; this also applies to the observation by Bezuglyi and Ponomarev that radical anions formed during electrolysis of acrylic and methacrylic acids can act as initiators of polymerization of the original monomeric depolarizers, which was discovered from the suppression of polarographic maxima [11];

2. Appearance of one-electron steps on polarograms or on potentiometric titration curves [12], as well as the indirect effect of free-radical formation on the polarographic wave parameters (the slope of the wave, the half-wave potential, and others) [13-26];

3. Change in the spectral characteristics of the solution and, in particular, the development of intense coloration due to the free radicals if their life is long enough for them to allow diffusion from the layer adjacent to the electrode into the bulk of the solution [22, 27-30];

4. Direct recording of the EPR spectra of the radical ions; unlike the preceding methods, EPR makes it possible not only to demonstrate the existence of free radicals clearly and unambiguously but also to study the electron-density distribution in these species quantitatively from the hyperfine structure of the spectra, i.e. to identify the nature and structure of the generated radical ions [31, 32].

Historically the earliest indirect evidence for the formation and interaction of free radicals in electrolytic organic chemistry was the establishment of the nature of some electrolysis macro-products, e. g. the products obtained from the classical Kolbe electrosynthesis [10]. The first direct evidence for electrochemical one-electron transfers was provided by the discovery of semiquinones, made independently in 1931 by Michaelis and Friedheim [33] and by Elema [34]. While studying oxidation – reduction potentiometric titration curves for some quinoid systems these investigators noticed the appearance of two one-electron steps and gave the correct interpretation of their observation. Numerous papers by the Michaelis school [35, 36] were devoted to further investigation of semiquinones and their biological role.

Evidence for the formation of free radicals from the shape of potentiometric titration curves can only be obtained for reversible electrochemical systems, the number of which in organic electrochemistry is rather limited. The use of polarography was therefore an important step forward. In the course of a systematic investigation of current – voltage curves for organic compounds at the dropping mercury electrode it was found that in a number of cases, in particular aqueous alkaline solutions and aprotic media, one-electron waves appeared on the polarograms or the analysis of the slope of multielectron waves indicated the transfer of a single electron in the potential-determining step. Evidence for the formation of free radicals during electrolysis is also provided by oscillographic polarography [37] and cyclic current – voltage curves on a suspended mercury drop [38]. By these methods, for example, the formation of free radicals in the electrolytic reduction of benzophenone or nitrobenzene has been demonstrated. The method of cyclic voltammetry was used by Il'yasov, Kargin and co-workers [39] to establish the precedence and stability of radical anions formed from carboxylic esters.

2. ONE-ELECTRON POLAROGRAPHIC WAVES: PRINCIPLES OF ELECTROCHEMICAL GENERATION

Among cathodic reactions which take place within the potential range attainable by means of the dropping mercury electrode, the most important are reactions involving addition of a single electron to multiple π-bonds in conjugated organic molecules, such as hydrocarbons with conjugated double bonds, multinuclear aromatic systems, carbonyl compounds, nitro compounds, quinones, nitriles, and others [20]. The free radicals obtained in electrolytic organic chemistry are therefore π-radicals. The initial products of such electrochemical reactions are either radical anions $R^{\cdot-}$ or (if the molecule has been previously protonated at the electrode surface) uncharged free radicals $RH^{\cdot}$:

$$R + e \to R^{\cdot-} \quad \text{or} \quad RH^+ + e \to RH^{\cdot}.$$

Quantum-mechanical calculations by Hoijtink showed that in the case of alternant hydrocarbons the free radical $RH^{\cdot}$ should possess greater affinity towards the electron than the original depolarizer molecule R, and if there are sufficient AH proton donors a single two-electron wave should appear on the polarogram; on the other hand, under any conditions, a free radical anion $R^{\cdot-}$ is reduced at more negative potentials than the original unchanged molecule R [16–19, 40, 41]. The great affinity of $RH^{\cdot}$ towards the electron is explained by the fact that pairing of the free electron spin is energetically favorable and consequently facilitates electrolytic reduction. In the case of the radical anion $R^{\cdot-}$ it is necessary to overcome electrostatic repulsion. It follows from this that a two-electron wave is formed in those cases where the original molecule is capable of adding a proton or the initial radical anion can be rapidly protonated:

$$R + e \to R^{\cdot-}; \qquad R^{\cdot-} + AH \to RH^{\cdot} + A^-;$$
$$RH^{\cdot} + e \to RH^-; \qquad RH^- + AH \to RH_2 + A^-.$$

If, however, protonation cannot be completed soon enough, a one-electron wave appears on the polarogram in the given potential range, and only at more negative potentials a wave corresponding to the second step appears.

These considerations are valid not only with respect to the electrolytic reduction of hydrocarbons [42, 43], but also to quinones [44–48], carbonyl compounds [49], nitro compounds [22, 24, 31, 50], and other classes of organic depolarizers. Hence the need to prevent or retard protonation of the original R particle or the initial radical anion $R^{\cdot -}$ can be understood. This is achieved in various ways depending on the range of potentials for the electrolytic reduction of the particles, their adsorbability on the electrode, the mechanism of the electrochemical process, etc. Even if the over-all electrochemical process represents a complex combination of the transfer of several electrons and protons, by retarding specific elementary reactions it is possible to check the process at the first stage.

In a number of cases one-electron waves may be obtained when working in aprotic solvents (anhydrous dimethylformamide, dimethyl sulfoxide, acetonitrile, tetrahydrofuran, 1,2-dimethoxyethane, and others), in mixed aqueous organic media [24, 50], or by additions of surface-active substances [13].

A number of authors have traced a transition from a one-electron mechanism of electrolytic reduction to a multielectronic mechanism when the concentration of proton donors in an aprotic medium is progressively increased or, conversely, from a multi-electron wave to a one-electron wave when the pH of the solution is progressively increased or the proportion of the organic component in an aqueous organic medium is increased. In work by Bezuglyi and Ponomarev [11] a transition from a one-electron wave to a two-electron wave was traced during the electrolytic reduction of acrylic and methacrylic esters in anhydrous dimethylformamide with additions of proton donors, and it was shown that the concentration of a given type of proton donor required for the transition to the two-electron wave was lower, the higher its ionization constant ("strength") (Table 1).

In some cases, e.g., in the electrolytic reduction of maleic acid in dimethylformamide in a supporting electrolyte of neutral salts, "self-protonation" of the maleic acid molecules entering into the electrode reaction can evidently take place at the expense of other molecules which dissociate into monobasic anions, and the height of the first wave is limited by the diffusion rate of these molecules, which act as proton donors, as shown in [51]. When weak acids (compared with the first stage of dissociation in maleic acid), such as formic and benzoic acids, are added to the solution the wave height of the protonated form remains practically unchanged.

TABLE 1. Relationship between Autoprotolysis Constants of Proton Donors and Their Concentrations at the Moment when Equilibrium is Attained

Proton donor	Autoprotolysis constant	Proton donor concentration, mole/liter	
		for acrylate	for methacrylate
Water	14.0	9.1	1.0
Propargyl alcohol	13.6	$8 \cdot 10^{-2}$	$1.1 \cdot 10^{-2}$
Phenol	9.9	$5 \cdot 10^{-3}$	$8 \cdot 10^{-3}$

The same type of protonation has also been proposed for the electrolytic reduction of 2-phenylindane-1,3-dione in unbuffered dimethylformamide solutions [52].

However, it is not always possible to obtain one-electron waves even in aprotic solvents; in [53], for example, it was shown that during the electrolytic reduction of Schiff's bases — benzalaniline derivatives — a two-electron wave is formed in nearly all cases, although as a result of large-scale electrolysis at the half-wave potential it was possible to detect the formation of dimeric products (from the nature of the infrared spectra and of cryoscopic measurements). Such polarographic behavior is explained by the instability of the intermediate radical anions.

From polarographic evidence it is possible to detect the formation of free radical anions in the course of an electrochemical process and in some cases also to select in advance conditions for their electrochemical generation (solvent, permissible proton donor concentrations, electrolysis potential range, and so forth).

3. METHODS FOR ELECTROCHEMICAL GENERATION OF FREE RADICAL IONS

The electrochemical method for generation of free radicals involves the formation of free radical ions by electrochemical means and simultaneous or subsequent identification of their nature by means of their EPR spectrum. The free radical ions are generated by large- or small-scale electrolysis in an electrolyzer placed either directly in the resonance cavity of the EPR spectrometer or outside the cavity with provision for sufficiently rapid delivery of the active material into the recording zone. It is a prerequisite for the selection of subjects for investigation and for selection of the experimental conditions that preliminary polarographic or current — voltage measurements (on solid electrodes) should be made.

Electrochemical generation of free organic radical ions was first realized in 1958 by Austen and his co-workers [54]. The subjects for the investigations were anthracene, benzophenone, and anthraquinone, which were reduced at a mercury cathode in dimethylformamide solution at a potential corresponding to the limiting current region on the first polarographic wave. Throughout the electrochemical process samples were taken from the electrolyzer and rapidly frozen in tubes at the temperature of liquid nitrogen. The EPR spectra were then recorded on thawing. Although the first polarographic waves were two-electron waves in the case of anthracene and benzophenone, all the specimens gave single EPR signals, indicating the intermediate formation of free radicals. The spectra did not contain hyperfine structure owing to the polycrystalline state of the specimens, and more detailed information on the nature of the radicals was not obtained.

Austen's work initiated a series of investigations into radical ions by means of electrochemical generation; about 200 articles were published in the period from 1960 to 1967. Since 1963 electrochemical generation of organic radical ions has been accomplished in the U. S. S. R. by various groups of investigators: Ya. K. Syrkin, V. M. Kazakova, A. I. Prokof'ev, and S. P. Solodovnikov in Moscow; A. I. Brodskii, A. S. Degtyarev, and L. L. Gordienko in Kiev; V. D. Bezuglyi and his co-workers in Kharkov; R. Kh. Il'yasov, Yu. P. Kitaev, and others in Kazan'; and R. A. Gavar, Ya. P. Stradyn', and S. A. Giller in Riga.

The principal difficulty in the electrochemical method for generation of radical ions is the very short life of most of the radical ions formed as the initial products of the electrode reaction; as a rule they are immediately subject to further chemical changes such as protonation, disproportionation, dimerization, oxidation, etc., or receive a second electron from the electrode. It is thus necessary to record the EPR spectrum of the radical ions immediately after their formation and to seek possible means of stabilizing them. A sufficient concentration of the radical ions in the resonance cavity of the EPR spectrometer can be created either by increasing the life of the radical ions or by continuously generating them at a steady rate to compensate for their decrease in the specimen. It is fairly complicated to solve these problems experimentally, and it can be achieved by various methods depending on the specific conditions involved.

Two basic methods have been developed — electrochemical generation in the resonance cavity of the spectrometer (internal generation) and electrochemical generation outside the resonance cavity (external generation). These methods differ not only in the positioning of the electrolytic cell but also in a number of other features, which determine whether one or the other method is used in a specific instance.

External Generation of Radical Ions

This method has two variants:

1. Performing the electrolysis in any electrolytic cell, taking and freezing the sample, and recording the EPR spectrum of the frozen specimen by the usual method;

2. Carrying out the electrolysis in a cell outside the resonance cavity with rapid transfer of the solution containing the radical ions into the resonance cavity of the spectrometer and then recording the spectrum.

The first variant was used in the early work by Austen and his co-workers. This method is the simplest and, it would appear, attractive, but it has some extremely important defects:

a) The solid frozen specimens do not give an EPR spectrum with clearly defined hyperfine structure;

b) The unstable radicals rapidly disappear on thawing, and the EPR spectrum can only be recorded for relatively long-lived radicals. This variant has therefore not received widespread use, and later investigators have used one or another variation of the second method.

However, the variant of external generation with the subsequent freezing operation allows rapid preliminary proof of radical-ion formation in the system. In the work by Bezuglyi, Kheifets, and Sobina [30], for example, the formation was detected and the concentration of radical ions from anthraquinone was determined by electrolysis of the original compound at the selected potential in various media (methyl alcohol, isopropyl alcohol, dimethylformamide), rapid freezing of the electrolytic cell in liquid nitrogen, and subsequent recording of the spectrum at 77°K. The external method is of some importance because in designing the cell no limitations are imposed in relation to the positioning of the cell in the spectrometer, and provision can be made for good access to the cell for manipulation and also for observation of changes in the color of the electrolyte, evolution of gas, and other aspects during electrolysis. (Particularly interesting are the observations of Bezuglyi [30] on coloration around the mercury drop during electrolytic reduction of anthraquinone and of 9-vinylanthracene.) Subsequently, nevertheless, investigators have rejected the freezing method, but have retained the idea of external generation itself and proposed alternative procedures.

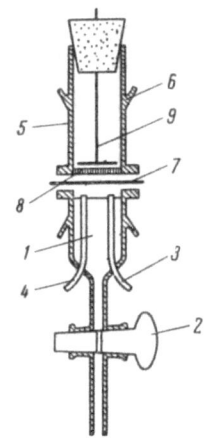

Fig. 1. Diagram of cell for external electrochemical generation [56].

The most convenient cell for work of this type was developed by Fraenkel and his co-workers [55, 56]. The cathode of this cell (Fig. 1) is the mercury surface 1, which is connected through the tap 2 to a mercury reservoir (not shown in the diagram); the height of the mercury in the cell can be regulated by raising or lowering the reservoir. The initial solution is delivered to the cathode compartment through the tube 3, and after electrolysis the solution containing the radical ions is transferred through the tube 4 into a glass ampoule. The anode compartment 5, which is also filled with the investigated solution, is connected to the cathode compartment by means of rubber sockets on the side pieces 6. The anolyte is separated from the catholyte by a thin (0.16 mm) teflon

membrane 7; the glass filter 8 protects the membrane from hydro-
static pressure.   The anode is a platinum disk electrode 9.   Since
diffusion processes do not play a significant part here owing to the
small thickness of the electrolyte layer, the extent of electrolysis
can be roughly estimated from the current drop in the cell.   With
concentration and volume of the electrolyte and the amount of electricity used known it is possible to control the concentration of
radical ions in the electrolytic processes.   Under such controlled
conditions it is possible to make absolute measurements on the
magnitude of the EPR signals and consequently to determine the
yield of the electrolytic process from the concentration of radical
ions.

By separating the cathode section from the anode it is possible
to obtain a better ratio between the working surface of the cathode
and the volume of the catholyte.   The accomplishment of rapid and
complete one-electron reduction of the compound under such conditions makes it possible to create the necessary concentration of
short-lived radical ions.   The complete reduction of the molecules
of radical ions also promotes the development of sharper hyperfine
structure in the EPR spectra.   The time taken for the electrolysis
can be reduced by using a rotating electrode [28].

With these measures an EPR spectrum with hyperfine structure can be easily recorded if the half-life of the free radical is not
less than 10 min.   For less stable radical ions the cell can be used
in a vacuum [55].

It is, nevertheless, difficult or impossible to record the spectra
for radicals with lives less than 5 min unless additional measures
are undertaken.   This can be achieved by generation of a steady concentration of radicals in the resonance cavity.   In principle this can
be realized by the external generation method, if the electrolysis
products are at the same time transferred continuously to the resonance cavity while the spectrum is being recorded.   This can be
carried out in a number of cells designed for the external generation
[57-59].   In one of the cells proposed in [57] intended for investigation of stable radicals an unusual type of nitrogen blower was used
to transfer the catholyte to the resonance cavity.   A similar blower
of simpler design was used in a cell intended for investigating radical

ions obtained in anodic processes [59]. By using a platinum working electrode the authors were able to record both radical cations and radical anions of phenothiazine formed in acetonitrile solution.

According to Nordio, Giacometti, and Favero [58] the cell with a centrifugal pump which they proposed allows shorter-lived radical anions from benzophenone derivatives to be recorded, but this is not in accordance with data on the half-lives of these free radicals.

It is difficult to distinguish between various cell designs with regard to the possibility of recording shorter-lived radical ions, since in the literature on electrochemical generation there is no clear definition of the concepts of "stable" and "unstable" free radicals and as a rule their half-lives were not measured experimentally or at best were only approximately estimated [56].

Internal Generation of Free Radicals

With this method the electrochemical generation of the radical ions is carried out directly in the cavity where the EPR spectrum is recorded. The radical ions are usually generated continuously, which compensates for their loss and creates a steady concentration. This gives two main advantages compared with the external method:

1. The need to transfer the radical ions into the resonance cavity of the spectrometer is eliminated, which saves time and allows shorter-lived radical ions to be recorded;

2. It is possible to follow the electrochemical process at the electrode directly by EPR.

At the same time the method imposes limitations on the cell design owing to the need to place the electrodes in the resonance cavity. These limitations lead to an unfavorable ratio between the volume of the solution and the electrode surface.

In all cells designed to be used for internal generation the rate of the electrochemical process, on which the steady concentration of radical ions depends, is limited by the diffusion rate of the molecules to the electrode. As a result, in such cells it is difficult to record short-lived free radicals, which give a very low steady concentration.

The principle of the internal method was proposed in 1957 by Galkin and his co-workers [60]. An ampoule containing sodium chloride in ammonia was placed in the rectangular resonance cavity (H_{102} type) of a sensitive radiospectrometer. They detected the EPR signal while passing a current through the solution by means of two platinum electrodes sealed into the ampoule.

The next report on this type of experiment appeared recently [61, 62]. Geske and Maki recorded signals from chlorine-containing radical ions by oxidizing a 0.1-mole/liter solution of lithium perchlorate in acetonitrile at a platinum electrode in a special cell adapted so that the electrolysis could be carried out in the resonance cavity. Soon after this, EPR spectra showing complex hyperfine structure due to organic radical ions were recorded for the first time (nitrobenzene radical ions), using the same method [31].

Several more-or-less detailed descriptions of cells for internal generation of free radicals were reported [31, 32, 62-67]. Owing to the limited size of the cavity only one electrode (the working electrode at the surface of which the radical ions are generated) is placed inside it. This electrode and the investigated solution are contained in a glass capillary with an external diameter of 3 mm (for a cylindrical cavity) or in a flat cell of the same thickness (for a rectangular cavity) [62, 66, 68, 69]. Cathodic electrochemical generation is most frequently carried out at a polarized mercury surface [62, 64, 66, 68]. This is usually realized by the surface of a mercury drop which lies freely at the bottom of the capillary introduced into the resonance cavity. By moving the cell in the cavity [31, 64, 66] or by varying the mercury level in the capillary [62, 68], the mercury surface is positioned directly at the cavity center. For the cell in a cylindrical cavity the mercury surface on which the radical ions are generated is equal to approximately $2.5 \cdot 10^{-2}$ cm^2 [31]. A suspended mercury drop is not used as electrode, since it is not convenient when manipulating the cell. It is also not possible to use a dropping mercury electrode, since the radical ion concentration varies periodically with the dropping of the mercury.

In addition to the mercury working cathode, some designs use a cathode made from platinum wire about 0.3-0.5 mm in diameter and ~3 mm long [65, 67, 70] or a fine platinum gauze [65, 66, 69];

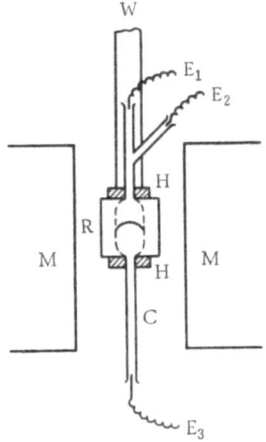

Fig. 2. Diagram of cell for internal electrochemical generation [32]. W is a wave guide; R is the resonant cavity; M is the magnet; C is the cell; H is the cell holder; E_1, E_2, E_3 are the electrodes for electrochemical generation of radical anions. E_3 is connected to the mercury drop, the surface of which forms the cathode.

with such electrodes it is also possible to accomplish oxidative anodic generation [56, 66, 69].

The auxiliary electrode, and also the reference electrode, used by some investigators [63, 66, 67] is placed in the cell and outside the resonance cavity. A calomel electrode is most frequently used as auxiliary electrode [31, 32, 62, 66, 67, 69], but sometimes use is made of mercury electrodes [63], $Ag - AgClO_4$ in acetonitrile [63], platinum wire or foil, and so on [65, 66, 70, 71]. Owing to the considerable separation of the working electrode from the other electrodes it is not always necessary to have artificial division of the investigated solution into anodic and cathodic sections [56, 64]. Such division is only essential in those cases where the auxiliary electrode or the reference electrode can introduce undesirable impurities into the investigated solution. To separate the electrode compartments use is made of an agar-agar bridge [62], plugs of filter paper [69], and filters of fused silicate composition [62, 63]. Before electrochemical generation oxygen is removed from the solution in the cell by introducing (bubbling) nitrogen through a capillary lowered into the solution to the level of the working electrode or by freezing several times in liquid nitrogen followed by evacuation to 10^{-3} mm [64].

In order to work at a specific temperature, e. g., when using liquid ammonia as solvent, the electrochemical generation cell can be immersed in a quartz Dewar vessel and cooled with a stream of cold nitrogen [65].

During electrolysis the cell is supplied from a simple direct-current electrical circuit with a power supply, a potentiometer, a voltmeter, and an ammeter, which make the control of the reduction wave possible. The current passed through the electrochemical generation cell is usually of the order of 10^{-6} to 10^{-5} A. The signals from the radical ions usually appear approximately 10 min after the beginning of electrolysis in nonaqueous solutions. The ratio of the EPR signal to the noise may increase as the electrolysis continues; there are cases where a good EPR spectrum can only be recorded after electrolysis for half an hour [31].

A typical cell for electrochemical generation of radical anions on a mercury surface inside the resonance cavity is the modification of the Geske and Maki cell shown in Fig. 3 [72, 73]. The steady radical anion concentration in the cell is increased by a method in which natural diffusion is supplemented by an artificial supply of the original substance to the region near the cathode. This is achieved by mechanical agitation of the layer near the cathode by means of a capillary, which is introduced into the cell in place of the capillary for blowing in the nitrogen. This prevents prolonged local accumulation of radical anions, which decreases their average life, and also gives more uniform distribution of the radical anions in the recording zone. Agitation of the layer near the cathode in the electrochemical generation cell provides intensive and constant EPR signals required for subsequent selection of experimental conditions for electrochemical generation while recording the spectra for the radical anions, e. g., the potential on the electrodes, the pH of the medium, the composition of the solvent, and the material used for the reactor. These conditions are selected in accordance with the best EPR spectra obtained on the tape of a self-recording instrument with the magnetic field slowly scanned (3 min).

With the internal method the EPR spectra of radical ions with half-lives of 5 min or more can be recorded comparatively easily [74], but by carefully selecting the experimental conditions in order to increase the signal-to-noise ratio (temperature, solvents, EPR apparatus) it is possible to detect radical ions with half-lives of the

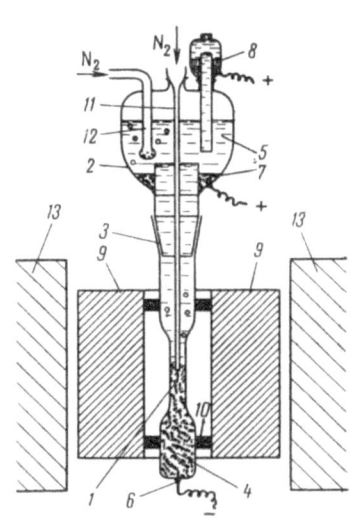

Fig. 3. Cell for electrochemical generation. 1) Reactor, thin-walled tube with external diameter of 2.2 ± 0.1 mm, internal diameter of 1.6 ± 0.1 mm; 2) reservoir; 3) ground-glass joint connecting the reactor to the resorvoir; 4) stationary mercury drop, the surface of which forms the cathode; 5) investigated solution, which fills the reactor over the mercury surface and partly fills the reservoir; 6) platinum wire, 0.5 mm in diameter, sealed into the bottom of the reactor, through which the voltage is applied to the cathode; 7) mercury layer, the surface of which forms the anode in the preliminary tests or in the investigation of nonaqueous solutions; 8) saturated calomel electrode, which forms the anode in the investigation of aqueous solutions; 9) resonance cavity of EPR spectrometer; 10) rubber plungers for securing and centering the electrochemical-generation cell in the resonance cavity; 11) and 12) capillary and branched tube for introduction of inert gas into the investigated solution in the cell to remove oxygen present in the solution; 13) poles of a magnet.

order of a second [75, 76]. There are no systematic comparative data in the literature on the lives of radical ions which would make it possible to compare the possibilities of recording short-lived radicals by internal generation and by external generation with the flow system; indirect assessment of the possibilities and experience, and also independent data on the life of radical ions, indicate the advantage of the internal method for the short-lived radicals [54, 62, 75, 76].

In a paper by Gavar, Grin', and Stradyn' [77] a method was developed for determining the stability of short-lived radical ions in the nitrofuran series, obtained by this method in aqueous and water – alcohol media. This method involves stopping the voltage supply to the cell at a particular moment of time and then recording the fading of the EPR spectrum on the tap of a self-recording spectrometer (Fig. 4). By summing the intensities over all the hyperfine structure

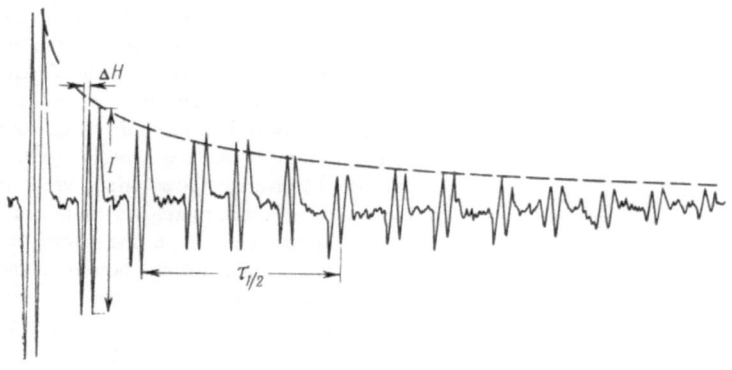

Fig. 4. Fading of EPR signals for radical anions of 5-nitro-2-pyromucic
acid after interruption of electrochemical generation.

lines during fading it is possible to establish the radical-ion concentration at any moment during fading, if a standard solution containing a known and constant amount of free radicals is used. We chose a standard solution of diphenylpicrylhydrazyl in benzene as standard.

By analysis of the data of the dependence of signal on time during electrolysis it is possible by means of equations for homogeneous chemical kinetics to calculate the reaction order, the reaction rate constant, and the half-life for the primary process involved in cleavage of the radical ions [76]. It is thus possible to determine kinetic parameters for the second-order cleavage of radical anions with half-lives not shorter than 1 sec at an initial concentration of $C_0 = 5 \cdot 10^{-5}$ mole/liter.

Sources of error can be the difference in shape between the lines of the investigated radical anions and the line of the standard, nonuniform distribution of the radical anions in the recording zone, cleavage of the radical anions at the electrode surface and the reactor walls, and the effects of mixing. A rather different method for assessing the stability of radical anions has been proposed by Kastening [78].

4. EXPERIMENTAL FACTORS IN ELECTROCHEMICAL GENERATION

Choice of Solvent

The choice of medium in which the radical ions are formed, i. e., the solvent, is determined both by the characteristics of the electrochemical generation method itself and by the properties of the radical ions formed. The requirements for the solvent may be diverse, are frequently difficult to reconcile with each other, and are usually unknown for radical ions which are being produced for the first time. In the latter case the medium for the production of specific radical ions must be found experimentally.

Although it is not possible to predict which solvent is the most suitable for the production of new radical ions, it is nevertheless possible, from experimental results, to formulate various principal requirements that must be fulfilled by the solvent:

1) Sufficiently high solubility of the investigated substance (not less than 10^{-3} to 10^{-2} mole/liter);

2) High solubility of indifferent electrolytes (not less than 10^{-1} mole/liter) and sufficiently high ionizing ability in the solvent;

3) Low dielectric losses of superhigh-frequency electromagnetic field energy due to solvent in a resonance cavity of the EPR spectrometer;

4) Inertness with respect to electrolytic current in the region of potentials required for the electrochemical process;

5) Inertness with respect to the investigated substance and its free radical;

6) Ability to stabilize radical ions to an acceptable lifetime ($\tau_{1/2} \geq 1$ sec);

7) Absence of paramagnetic and other undesirable impurities;

8) Low proton-donating ability.

The first two conditions are consistent with solvents having a sufficiently high dielectric constant ($\varepsilon > 20$); unfortunately such solvents create high dielectric losses of superhigh-frequency field energy in the resonance cavity and thereby complicate the experimental conditions for electrochemical generation.

The suitability of a given polar solvent for a one-electron electrochemical process for a particular substance and the ability of the solvent to stabilize the radical ions are usually found experimentally. For this purpose use is made of both electrochemical generation and data from polarography or large-scale electrolysis.

As a rule it is best to carry out electrochemical generation in media where there is a single one-electron polarographic wave; this occurs under conditions where the protonation of radical anions formed in the first electron transfer is either excluded or retarded, i. e., in polar aprotic solvents (acetonitrile, dimethylformamide, tetrahydrofuran, 1,2-dimethoxyethane, and others).

The principal advantages of the above-mentioned highly polar organic solvents are the following [32, 55, 65, 75, 79, 81]:

1) High pK_{autopr} values compared with water and, consequently, low ability to protonate the radical ions, which considerably prolongs their life;

2) Good dissolving properties for investigated organic compounds and possibility of achieving high concentrations;

3) Comparatively low freezing points, which enables low-temperature stabilization of the radical ions to be used.

Disadvantages of these organic solvents are as follows:

1) Their less common availability when compared with water;

2) The need for purified and dry indifferent electrolytes, the solubilities of which are limited;

3) Relatively greater difficulty of removing oxygen, the solubility of which is extremely high in these solvents;

4) The difficulty of purifying the solvents from various proton donors, especially from water, and also of preventing the acess of moisture during the experimental work.

The last point needs particular consideration; as a result of electrochemical manipulations the concentration of water in dehydrated organic solvents can rise to $1-3 \cdot 10^{-3}$ mole/liter [32], i.e., can even exceed the radical-ion concentration. Consequently during the experiment the concentration of proton donors may increase so much that these solvents are less capable than water of stabilizing the radical ions, since the radical-ion solvates are evidently less stable in them than in water. Although the mechanism of radical-ion stabilization by the solvent is not completely understood, it is necessary to bear in mind that the stability of particles with an unpaired electron can decrease first of all through protonation of radical ions. Solvents with large ε values evidently also stabilize the radical ions by formation of solvates, which prevent them from recombination and disproportionation; in this case temperature reduction should stabilize the solvates and thereby considerably increase the life of the radicals.

The most popular solvent for electrochemical generation is dimethylformamide, since polarographic investigations [42, 43, 47, 49, 82] have shown that the electrochemical reduction of aromatic hydrocarbons [42, 43, 82], carbonyl compounds [47, 49], phosphoryl derivatives [83], nitro compounds [84], nitrogen-containing heterocycles [82], esters, and amides [11, 39] frequently gives a one-electron wave in dimethylformamide solutions. It should, however, be remembered that the ketyl radical ions of benzophenone disappear in dimethylformamide owing to their reaction with the solvent molecules [43, 49] and that the recorded EPR signals correspond not to the primary products but to products from secondary changes.

Other organic solvents, such as acetonitrile, dimethyl sulfoxide and pyridine, have also been used as media for electrochemical generation and even liquid ammonia has been used in some cases for low-temperature work. The use of water as solvent for electrochemical generation is extremely attractive [85, 86]. However, difficulties arise here from protonation of the radical ions and curtailment of their lives. Amongst other disadvantages should be mentioned the high dielectric losses of superhigh-frequency field energy in water, the comparatively high freezing point (which does not permit low-temperature stabilization of radical ions in the liquid phase), and the deformation of the molecular orbital of the unpaired electron owing to hydration of the radical anions (which is expressed by change in the hyperfine coupling constants compared with nonaqueous solvents).

Indifferent Electrolytes

In order to make the solution electrically conductive and to enable the electrolytic process it is necessary to add an indifferent electrolyte to the solution. As a rule salts of alkali metals or alkylammonium bases are used. The concentration of these salts in the investigated solutions amounts to 10^{-1} mole/liter. It is comparatively easy to create such a concentration of indifferent electrolyte in aqueous media by alkali metal salts. In nonaqueous solvents this is not always possible for alkali metal salts; various alkylammonium salts, which have better solubility, are therefore used instead.

The presence of the ions of the indifferent electrolyte in the solution evidently does not substantially change the properties of the radical ions; for example, complexes of alkali metal and radical ion such as those obtained during production of the radical ions by chemical reduction with the alkali metal are not detected in these solutions. This is demonstrated by the fact that when the electrochemical generation method is used, no hyperfine structure arising from splitting at the nuclei of these metals is observed in the EPR spectrum. However, ions of the indifferent salts can evidently change the structure of solvates formed by the radical ion with the solvent molecules; it is possible to observe certain changes in the hyperfine structure of the EPR spectrum depending on the nature of the salts used [87]. A change in the nature of the ions may therefore also lead to some change in the stability of the free radicals and to the formation of ion pairs. Tetraalkylammonium salts are used to avoid the formation of ion pairs.

Concentration of Initial Electrochemically

Active Substance and of Radical Ions

in Solution

It is well known that the EPR spectrometer has limited sensitivity; it would appear to be possible to compensate this by increasing the concentration of the initial substance in the electrolyte solution, thereby increasing the radical ion concentration. However, such increase in concentration does not always assist in producing good EPR spectra, since exchange interactions between the radical ions can cause broadening of the individual components of the spectrum, which leads to loss of hyperfine structure and even to contraction of the whole spectrum [88, 89]. In order to avoid this it is usual to employ the initial substance at a concentration not greater than $5 \cdot 10^{-3}$ to 10^{-2} mole/liter. By reducing the concentration of the initial substance it is also possible to eliminate another effect, which is evidently due to the exchange of an electron between the radical ion and a molecule of the initial substance; as shown in Fig. 5, this manifests itself in a change in the form of the spectrum with change in the concentration of the initial substance [89].

On the other hand, with low concentrations of the initial substance in the solution, it is difficult by electrolysis to create the concentration of unstable radical ions required for detection. In electrochemical generation these contradictory requirements can be reconciled if the time for electrolysis of all the investigated substance is short compared with the life of the radical ions [28].

As a result of the observations mentioned above, the original concentration of the active substance in the solution most frequently amounts to about $5 \cdot 10^{-3}$ mole/liter, but in cases where concentration broadening of the hyperfine structure components appears in the spectrum it is necessary to reduce the concentration to as little as $5 \cdot 10^{-4}$ mole/liter.

Effect of Paramagnetic Impurities

in Solutions

The possibility of recording the spectra of the radical ions is also reduced by their interaction with other paramagnetic particles in the solution [74, 78]. Complications of this sort nearly always arise when organic solvents are used and are related to the presence of dissolved oxygen, the concentration of which amounts to $1-2 \cdot 10^{-3}$ mole/liter in most organic solvents. As a result of magnetic interaction between the unpaired electron of the radical ion and the triplet-state electrons in the oxygen molecule [90] the hyperfine structure components of the EPR spectrum of the free radical become broader, and the effective sensitivity of the spectrometer is thereby reduced. Various methods for removing oxygen from the solutions have been examined [39, 55, 64, 83].

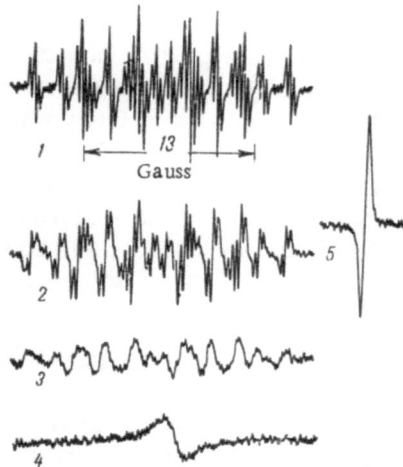

Fig. 5. EPR spectra of radical anions from benzonitrile as functions of the benzonitrile concentration in dimethylformamide. 1) $2.5 \cdot 10^{-3}$ mole/liter; 2) $5.0 \cdot 10^{-3}$ mole /liter; 3) $1.3 \cdot 10^{-2}$ mole/liter; 4) $6.5 \cdot 10^{-1}$ mole/liter; 5) 3.9 mole/liter.

Electrodes and Electrochemical Generation Potentials

A three-electrode system can be used for electrochemical generation. The working electrode is mercury or platinum. A mercury or platinum surface can also be used as auxiliary electrode, or a saturated calomel electrode can be employed. The reference is usually a saturated calomel electrode.

The saturated calomel electrode can be used in aqueous solutions without special precautions. In nonaqueous solutions, however, in spite of the frequent references to the use of such an aqueous electrode in the literature, it can clearly not be recommended since it contaminates the solution with water. In such cases it is better to give up the use of a reference electrode altogether or to use an electrode with a nonaqueous solvent, such as, for example, the $Ag - AgClO_4$ electrode in acetonitrile [63]. Actually, accurate potential

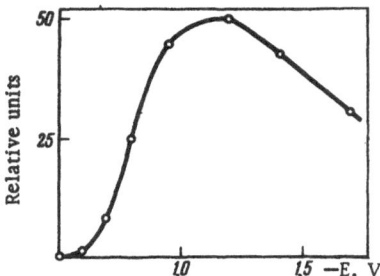

Fig. 6. Variation of EPR signal intensity of radical anions from 5-nitropyromucic acid as function of potential (E, referred to saturated calomel electrode).

determination is not of decisive importance in electrochemical generation, and the reference electrode is therefore frequently omitted. Electrochemical generation is usually performed at potentials corresponding to the limiting-current area of the first one-electron polarographic wave, which is determined from the results of a preliminary polarographic experiment. It should be noticed that it is not essential for the appearance of an EPR signal in steady electrochemical generation that the polarogram in the appropriate medium should contain a one-electron wave; signals can also be recorded in media where the initial substance gives a multielectron polarographic wave and at potentials which lie beyond the potentials of the following waves. With increase in the potential supplied to the electrode, however, the steady radical anion concentration as a rule only increases up to a certain limit. Figure 6 shows the variation of the signal intensity for the radical anion of 5-nitropyromucic acid as a function of the applied potential; the polarogram of this compound in the same medium contains a single four-electron wave with a half-wave potential of ~−0.4 V. The maximum signal intensity is not reached at a potential close to −0.4 V but at ~−1.2 V (referred to the saturated calomel electrode). The form of the spectrum remains constant at all the applied potential values. This shows, on the one hand, that radicals of a different sort are not formed in any appreciable quantities and, on the other, that radicals under the direct influence of a cathodic potential disappear according to a different law at high potentials than at low potentials. Similar relationships were also found during electrochemical generation from 2-nitrofuran, 2-nitrofuryl alcohol, and other derivatives.

In work by Bezuglyi and his co-workers [30] it was shown that in the electrochemical generation of radical ions from anthraquinone in methanol and isopropanol solutions the radical ions can only exist over a limited range of potentials (at potentials more positive than $E_{1/2}$ of the total two-electron wave) and that in dimethylformamide solutions they can be recorded in the range of potentials from the beginning of electrolytic reduction of the investigated substance up to the potentials for decomposition of the supporting electrolyte.

The optimum electrochemical generation potentials should thus be selected empirically, and the polarographic potentials of the one-electron waves can provide tentative preliminary information.

5. Applications and Distinctive Features of Electrochemical Generation Method

The appearance of an EPR signal is evidence for the formation of radical ions in an electrochemical process and thereby gives information on the mechanism of the electrolytic reaction. The absence of signals does not always exclude the possibility of radical ion formation and may be due to their low stability or to the unsuitability of the experimental conditions for the detection of radical ions.

The strength of the EPR signal gives a qualitative idea of the rate of the electrochemical process under given conditions and characterizes the ability of the medium to stabilize the radical ions which are formed, e.g., by solvation or to curtail their lives, e.g., as a result of protonation.

Of the greatest theoretical interest is the production of an EPR spectrum with clearly defined hyperfine structure. The latter arises from isotropic contact spin − spin interaction of the unpaired electron with atoms which have an intrinsic magnetic moment − H^1, H^2, C^{13}, N^{14}, P^{31}, and others.

If an unpaired electron can interact with n equivalent nuclei with spin I, the EPR spectrum will contain instead of a single line a multiplet consisting of 2n +1 separate or partially overlapping lines (components), arranged at equal distances from each other. In interaction with nonequivalent nuclei the form of the spectrum becomes considerably more complicated, but under conditions of good resolution the distribution of the components does yield to analysis even in

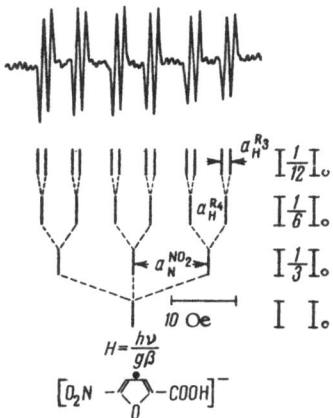

Fig. 7. EPR spectrum of radical ions from 5-nitropyromucic acid and its reconstruction.

this case. The distance between two neighboring hyperfine structure components arising from equivalent nuclei is called the hyperfine coupling constant and is usually expressed in oersteds. The intensity distribution of hyperfine structure components from equivalent nuclei have been examined in a book [4].

Figure 7 shows the EPR spectrum of the radical ion from 5-nitropyromucic acid, obtained by electrochemical generation, and its reconstruction based on a theoretical treatment.

The hyperfine structure of the spectrum gives most valuable information on the structure of individual radical ions — both on the presence of specific atoms in the free radical and on the electron-spin density distribution in the radical. The possibility of obtaining this information theoretically has been considered by a number of investigators [91-93]. In particular, McConnell has shown the existence of direct proportionality between the hyperfine coupling constant A_i, due to i protons in the C — H bonds of various aromatic systems, and the density of the unpaired π-electron at the carbon atom:

$$A_i = Q_i \rho_i.$$

In a number of cases quantum-chemical calculations have confirmed the experimentally determined distribution of unpaired electron density in a radical ion [94]. In addition the hyperfine structure of the spectrum can serve as an experimental basis for verification and confirmation of the validity of complex quantum-chemical calculations applied to the structure of organic molecules and radicals. The hyperfine structure can also serve to identify a given type of radical, and to offer information on the coplanarity and isomerism of radical anions; it can also be used to determine the rates of transitions to various states and the rotational frequencies of individual fragments of the radical anions. Change in the form of the hyperfine structure components can also be used to study the ultrafast reaction of electron transfer between the radical ions and the original molecule By analysis of the hyperfine coupling constants of radical ions it has been possible to establish a relative series of acceptor ability in various groups and molecules and to assess the transmission of the effect of substituents across the nuclei of radical ions.

In the sense of its influence on the initial substance electrochemical generation is "soft"; the structure of the original molecule does not undergo any considerable changes as a result of electrochemical generation; these changes are frequently reversible in their nature, i.e., the reverse electrochemical process can lead to the original substance. As a first approximation it can be assumed that the molecular σ-bonds are not involved in the formation of radicals by electrochemical generation, and that the change in the π-electron system of the molecule, caused by the addition of an electron (formation of radical anions) or by its removal (formation of radical cations), does not affect the geometrical and atomic structures of the molecule. Owing to the "soft" effect on the original molecules and the possibility of precise control on the degree of the effect, the electrochemical generation method makes it possible to obtain radical ions with a specific previously postulated structure.

It is difficult to compare the electrochemical generation method with other methods for producing free radicals — e.g., thermal decomposition, discharge in gases, radiolysis — since with other methods the original molecules undergo substantial changes, as a result of which free radicals with extremely diverse structures are obtained and most frequently represent fragments of the original molecules. The latter may include "molecular ions", which are products of one-electron ionization of the original molecules, although as a rule their concentrations are low [95].

It is more logical to compare electrochemical generation with methods such as the photochemical or chemical methods. The photochemical method for generation of radical anions in combination with EPR spectrometry has not yet received greater attention owing to difficulty in arranging the direct action of light on the specimen in the resonance cavity of the spectrometer [96-105].

The closest method to electrochemical generation with respect to the possibility of producing radical ions of a specific type is the chemical action of sulfuric acid or alkali metal (lithium, sodium, potassium, cesium) on the substance. As with electrochemical generation, this method allows free radicals of a specific type ("molecular ions" or their salts) to be produced. Moreover, it permits production of these radicals in the liquid phase at the concentrations required for the production of well-defined EPR spectra. Reduction with alkali metal has been used to produce a relatively large variety of radical ions [104, 106-108], even including those which have not yet been produced by electrochemical generation, such as the benzene radical anion [109] and its alkyl derivatives [109-111]. The success of the method is at least partly explained by the possibility of stabilizing the free radical at the low temperatures employed ($\sim$ $-70°C$) and the formation of ion pairs with the alkali metal ions. This is confirmed by the fact that radical anions formed by reduction with alkali metal frequently give hyperfine structure components in their EPR spectra arising from the nuclei of these metals [107, 112-114]. On the other hand, hyperfine structure of this type is not observed in the EPR spectra of radical anions produced by electrochemical generation. Compared with electrochemical generation, the chemical method for the production of radical anions is accompanied by the following difficulties:

1) Difficulty or practical impossibility of controlling the degree of conversion of the original substance;

2) Impossibility of obtaining radical anions of compounds which are unstable in alkali medium;

3) Limited choice of solvent (dimethoxyethane, dioxan, tetrahydrofuran);

4) The need to remove the smallest traces of moisture from the solvent;

5) Comparatively complex experimental procedure;

6) Production of "restricted" radical anions — forming ion pairs with the metal cations; nevertheless this feature has been successfully used for investigating the properties of this type of complex [107, 112-115].

A wider selection of solvents can be used in the electrochemical method, including aqueous media, and the purity requirements for the solvents are lower than in the chemical method. In addition it is possible to control the degree of conversion of the original substance by continuous variation of the applied potential.

Thus, it can be deduced that electrochemical generation is a more convenient method than chemical generation for producing well-defined EPR spectra of the same free radical anions.

6. TOPICS WHICH HAVE BEEN INVESTIGATED BY ELECTROCHEMICAL GENERATION

Between 1961 and 1967 the electrochemical generation method has been used to obtain, identify, and investigate free radicals derived from about 400 different organic compounds, such as unsaturated acyclic and alicyclic hydrocarbons, condensed and noncondensed poly-nuclear aromatic hydrocarbons, heterocyclic compounds, quinones, carbonyl compounds, nitriles, nitroso and nitro derivatives, and carboxylic esters.

A. Unsaturated Hyrocarbons and Heterocyclic Compounds

Among this group of compounds spectra with well-defined hyperfine structure have been recorded for radical anions of 1,3-butadiene [65] and 1,3,5-cyclopentatriene [81]. Comparison with the chemical method for producing these radicals shows the advantages of the electrochemical generation method. In particular, reduction of 1,3-butadiene with alkali metal does not give the free radical anions, since 1,3-butadiene readily polymerizes in the presence of the alkali metal. Liquid ammonia was used as solvent for the production of these radicals by electrochemical generation; under these conditions, according to Levy and Myers [65, 81], the compound is reduced not only by a heterogeneous process at the cathode but also by a homogeneous process in the solution, and in the latter case the reducing agent is a solvated electron. The radical ions are stabilized at the temperature of liquid ammonia (−68 to −78°C); however, the life of the radicals is comparatively short under these conditions

(~2 sec), and in order to detect the EPR spectrum it is necessary to carry out the electrochemical generation inside the resonance cavity.

Aromatic Hydrocarbons. There is no information on the production of free radical ions of benzene by the electrochemical method. From noncondensed polynuclear aromatic systems radical anions of diphenyl [62] and of stilbene [116] have been generated.

Radical anions have been obtained electrochemically from condensed aromatic systems by electrolytic reduction of naphthalene [62, 117], 1,5-naphthalenedisulfonate [118], anthracene [54, 55, 62, 117-119], 9,10-diphenylanthracene [69, 120], phenanthrene [62, 117], tetracene [62, 118, 121], pyrene [62], pentacene [62], and perylene [62].

Radical anions have also been obtained from such condensed systems as azulene [122, 123], 4,6,8-trimethylazulene [122], acenaphthylene [124], and sym–dibenzocycloctatetraene [125]. Radical cations of phenanthrene [117] and of 9,10–diphenylanthracene [69] have been obtained by oxidative electrochemical generation at platinum electrodes.

All the above-mentioned radicals were obtained in nonaqueous media.

Heterocyclic Compounds. All the heterocyclic compounds, unsubstituted by electroactive groups, from which radical ions have been obtained by electrochemical generation are nitrogen heterocycles. Among the numerous systems involved radical anions have been obtained from pyrazine [118], pyridazine [126], and tetrazine [126], and radical cations have been obtained from dihydropyrazine [126-128] and dihydrotetramethylpyrazine [128].

Radical anions have also been obtained for such heterocyclic systems as trans-1,2-bis(4-pyridyl)ethylene [116] and 2,2'-dipyridyl [129] in nonaqueous media; a radical cation was obtained from 4,4'-dipyridyl [130] in aqueous medium.

Among condensed heterocyclic compounds formation of radical ions have been proved for the following compounds: radical anions of acridine [117], phthalazine [126], phenazine [70, 126], 1,4,5,8-tetraazanaphthalene [131], 9,10-diazaphenanthrene [129], 2,2'-biisobenzimidazolidine [129], and phenothiazone [108]; radical cations for

5,10-dihydrophenazine [70], phenothiazine [108], and its derivatives substituted with Cl, $(CH_2)_3N(CH_3)_2$ [66], and CH_3 [59] groups in position 10. Unfortunately with the heterocyclic radical ions, as with the radical ions of aromatic hydrocarbons, no data by means of which their stability can be characterized have been given.

B. Compounds Containing Electroactive

Substituents

Many of the electrochemically generated radical ions contain, in addition to the hydrocarbon or heterocyclic frame, reactive functional groups, often as a part of a conjugated system, which substantially affect the stability of the radical ions and also their electron-density distribution. Such radical ions can therefore be treated as a separate group.

Quinones. By electrochemical generation in nonaqueous media radical anions have been obtained from p-benzoquinone [118, 132, 133], p-toluquinone [132], 1,4-naphthaquinone [132, 134], and 9,10-anthraquinone [30, 54, 67, 117, 118, 132, 134].

From o-quinones in nonaqueous solutions radical anions have been obtained for benzocyclobutadienoquinone [135], 9,10-phenanthrenequinone [136] and acenaphthenequinone [136].

In aqueous media radical anions have been obtained for phenolindophenol and its halogen derivatives [137].

The effect of the medium on the stability of 9,10-anthraquinone radical anions was examined in work by Bezuglyi, Kheifets, and Sobina [30].

Aromatic Carbonyl Compounds. Electrolytic reduction of aldehydes and ketones only leads to identifiable radical ions in nonaqueous solutions. Radical anions have been obtained from a number of aromatic aldehydes: the principal representative of the series — benzaldehyde [138, 139] — and its 0- [140] and p-nitro derivatives [138, 140, 141]; p-cyanobenzaldehyde [138]; o- and p-dialdehydes [138, 140].

Radical anions have also been obtained from acetophenone [138, 139], its m- and p-nitro derivatives [138, 141, 142], m-

and p-cyano derivatives [138], and p-fluoro and p-acetyl derivatives [138, 141].

Among other ketones of the aromatic series radical anions have been obtained for triphenylacetophenone [138], benzophenone [54, 58, 68, 138], 4,4'-dichlorobenzophenone, 4,4'-dimethoxybenzophenone [58], 4,4'-diacetyldiphenyl [136] and 4,4'-dimethylbenzophenone [15].

From ketones containing a carbonyl group in the ring radical anions have been obtained in the case of 9-fluorenone [58, 136], 2,7-difluorofluorenone [136], 4,5-phenanthryleneketone [136], and anthrone [58]. Radical anions have also been obtained and investigated for ketones in the heterocyclic series — 4-acetylpyridine [138] and 2-, 3-, and 4-benzoylpyridine [68].

It should be noted that at least in some cases the radical anions of the simplest carbonyl compounds (benzaldehyde, acetophenone) are probably secondary products formed as a result of interaction between the initial radical anion and dimethylformamide according to the following scheme:

$$\text{Ar}\overset{\mid}{\underset{\text{O}^-}{\text{C}}} - \text{R} + \text{HC} \overset{\text{O}}{\underset{\text{N(CH}_3)_2}{\diagup}} \rightarrow \text{ArCR}\overset{\mid}{\underset{\text{O}^-}{}}-\text{CH}\overset{\mid}{\underset{\text{O}^\cdot}{}}-\text{N(CH}_3)_2 \xrightarrow{-\text{N(CH}_3)_2^-} \text{ArC}\overset{\mid}{\underset{\text{O}^-}{}}=\text{CR}\overset{\mid}{\underset{\text{O}^\cdot}{}} + \text{H}^+$$

The inaccuracy of the interpretation proposed by Fraenkel and his co-workers [138] for the structure of the radical ions as primary particles was pointed out by Russell and others [143].

Gazizov, Il'yasov, Razumov, Savicheva, and Sotnikova [83] have described radical anions, obtained by electrolytic reduction in dimethylformamide with 0.05 mole/liter $[\text{N(C}_2\text{H}_5)_4]\text{I}$ as indifferent electrolyte, from arylketonephosphoryl derivatives of the following type:

$$\text{RO}-\overset{\text{R}'}{\underset{\overset{\|}{\text{O}}}{\text{P}}}-\overset{}{\underset{\overset{\|}{\text{O}}}{\text{C}}}-\text{Ar}.$$

Analysis of the spectrum showed that the observed hyperfine doublet splitting with A = 23 Oe arises from interaction between the unpaired electron and the phosphorus nucleus P^{31} (I = $^1/_2$), and further splitting is due to the protons of the benzene rings. Since the observed hyperfine structure is practically independent of the substituents, it must be concluded that the doublet splitting is due mainly to the protons of the benzene ring attached to the carbonyl group [83].

Carboxylic Acid Derivatives. Detailed investigations into the electrolytic reduction of carboxylic acid derivatives (esters, anhydrides, amides) in nonaqueous solutions and the procedures for producing the respective radical anions have been described in the report by Il'yasov and his co-workers [39]. They obtained radical anions from esters of aromatic carboxylic acids (benzoates, phthalates, isophthalates) and from phthalic anhydride and analyzed their EPR spectra. The production of radical anions of acrylates and methacrylates by electrochemical generation and the effect of proton donors on their stability was also described [11].

Amines. Radical cations have been obtained by electrochemical oxidation of monoamino derivatives in the case of 9-amino-10-phenylanthracene [144, 145]; of diamino derivatives in the case of p-phenylenediamine [66, 146, 147], 1,4-diamino-2,3,5,6-tetramethylbenzene [129], tetramethylbenzidine [148, 149], and triethyldiamine [150]; of tetraamino derivatives in the case of tetrakis(dimethylamino)ethylene [151].

Nitriles. In the series of acyclic nitriles radical anions have been obtained from tetracyanoethylene [56, 152, 153], and radical dianions from 1,1,3,3-tetracyanopropylene [56, 152, 153], and 1,1,2,3,3-pentacyanopropylene [56, 152, 153]; in the alicyclic series radical anions have been obtained for dicyanomethylene-2,2,4,4-tetramethylcyclobutanone [63] and bis(dicyanomethylene)-2,2,4,4-tetramethylcyclobutane [63].

Among the aromatic nitriles radical anions have been obtained for benzonitrile [56, 68, 79, 89], its 2-cyano, 3-cyano, 4-cyano, 2,4,5-tricyano, 4-methyl [56, 153], 3-nitro [142], 4-nitro [56, 141, 142], and 3,5-dinitro derivatives [56], for 4-cyanobenzoic acid [56], 4,4'-dicyanodiphenyl [56, 153], and 1,1,8,8-tetracyanoquinodimethane [154].

With nitriles in the heterocyclic series radical anions have only been recorded for 4-cyanopyridine [56, 153].

Nitroso Compounds. Radical anions have been obtained by electrochemical generation for only two nitroso compounds: nitroso-benzene [75] and p-nitrosophenol [155].

Nitro Compounds. The electrochemical method has proved most fruitful for the production of free radicals from nitro compounds; more than half (about 250) of all the radicals which have been obtained by the method are radical nitro anions. This is largely explained by advances made in the polarographic investigation of nitro compounds, which provides the theoretical basis for the electrochemical generation method [156], and also by the fact that these radicals frequently possess high stability even in aqueous media.

The radical anions of acyclic and alicyclic nitro compounds have been investigated in aqueous media. They have been obtained for primary, secondary, and tertiary nitro derivatives:

1) $R - CH_2 - NO_2$: nitroethane, 1-nitropropane, 1-nitrobutane, and 1-nitrohexane [85, 157];

2) $R - \overset{\displaystyle NO_2}{\overset{\displaystyle /}{CH}} - R_1$: 2-nitropropane [79, 85, 157], 2-nitrobutane [157, 158], and 2-nitrocyclohexane [157];

3) $R_1 - \overset{\displaystyle NO_2}{\underset{\displaystyle R_2}{\overset{\displaystyle |}{\underset{\displaystyle |}{C}}}} - R$, where R, R_1, and R_2 represent CH_2OH or CH_3

[157].

The greatest number of radical anions derived from nitrocompounds have been obtained with nitrobenzene derivatives (more than 150). In addition radical anions have been obtained for nitro derivatives of naphthalene; electron-paramagnetic-resonance spectra have been described for 1,8-dinitro- and 1,4,5,8-tetranitronaphthalene [158, 159].

Nitrobenzene radical anions produced by electrochemical generation have been recorded in several solvents with sufficiently high

TABLE 2. Coupling Constants for Radical Ions of Nitrobenzene in Various Solvents, Oe

Coupling constant, Oe				Medium	Reference
A_N	A_{H_o}	A_{H_m}	A_{H_p}		
9.7	3.36	1.07	4.03	Dimethylformamide	[142]
9.79	3.38	1.09	4.06	"	[163]
9.83	3.32	1.09	3.97	"	[79]
10.32	3.39	1.09	3.97	Acetonitrile	[31]
10.48	3.36	1.09	3.82	"	[79]
10.62	3.30	1.03	3.82	"	[134]
10.51	3.48	1.09	3.86	Pyridine	[79]
11.46	3.42	1.11	3.89	Liquid ammonia	[75]
13.87	3.30	1.12	3.52	Water	[157]

dielectric constants. The hyperfine coupling constants (Table 2) show
that the spin of the unpaired electron is mainly localized at the nitro
group, but the spectrum shows partial delocalization on the benzene
ring. The highest unpaired electron spin density in the benzene ring
occurs at the p- and o-positions in relation to the nitro group. The
density at the m-position is less than that at the p- and o-positions by
a factor of approximately 3. Variation of the unpaired electron spin
density in the benzene ring with the solvent used is slight; in the investigated cases the hyperfine coupling constant due to interaction
with the proton in the p- and o-positions of the benzene ring has
changed with variation of the solvent by not more than about 10%, and
that due to interaction with the m-position not more than 5%. The
constant at the nitrogen of the nitro group is more sensitive to the
effect of the medium. Fraenkel and his co-workers [142, 160] give
the following expression for this constant:

$$A_N = 99.0\, \rho_N^\pi - 71.6\, \rho_O^\pi,$$

where ρ^π is the unpaired electron spin density at the nitrogen and
oxygen atoms, respectively.

Since the corresponding values of $Q_N^{NO_2}$ and $Q_O^{NO_2}$ are large and
have opposite signs, even a slight redistribution of spin density in the
nitro group should cause comparatively large changes in the coupling

constant at the nitrogen of the nitro group. It is possible that such changes may be due to solvation equilibria of radical anions by solvent molecules [75, 79, 80, 87, 161].

Electrochemical generation has also been used to investigate interaction between the unpaired electron of the nitrobenzene radical anion and an O^{17} nucleus in the nitro group [162].

Among mononitro derivatives of the benzene series radical anions have been obtained by electrochemical generation for mono- [118, 141, 142, 164], di- [142, 164], tri- [79, 157, 164, 165], and tetramethyl derivatives [142, 164-166]; mono-, di-, and tri-terbutyl derivatives [164]; monomethoxy derivatives [79, 134, 141]; SO_3CH_3- [141] and amino derivatives [79, 118, 141, 142, 157, 164]; mono- [79, 155, 157] and dicarboxy derivatives [157]; monoacetoxy derivatives [141]; carbamyl [141] and monoformyl derivatives [139-141]; monoacetyl [138, 141, 142], cyano [56, 141, 142], fluoro [141], chloro [79, 134, 141, 157, 163, 167], bromo [134, 141, 163, 168], and iodo [163] derivatives; p-phenyl derivatives [141, 142, 169]; p-diphenyl derivatives [170]; p'- derivatives of various p-diphenyl derivatives [57, 141, 142, 169, 171]; p-substituted bridge compounds linked to the nitrobenzene nucleus through a bridge of the $-CH_2-$, $-CH_2CH_2-$, $-S-$, $-O-$, or $-CH=CH-$ types [57, 64, 69, 171]; p-substituted diphenylhydrazines [172].

Radical anions have also been investigated in the case of dinitrobenzenes [74, 79, 141, 142, 160, 164, 166, 173], their methyl [142, 164-166, 174], tertbutyl [164, 166], hydroxy [142, 166], amino [142, 166], carboxy [155], formyl [140], acetyl [166], and cyano [56] derivatives.

Radical anions have also been obtained from trinitrobenzene and its methyl, methoxy, amino, carboxy [175], and diphenylhydrazine [172] derivatives.

In work by Kitaev and his co-workers [84] EPR spectra are described for hydrazones and phenylhydrazones of nitrobenzaldehydes, and it was found that the spectra could only be recorded for compounds containing the nitro group in the m-position; this is evidently explained by the instability of the radicals from the p- and o-isomers due to mobility of the aniline and aldehyde protons.

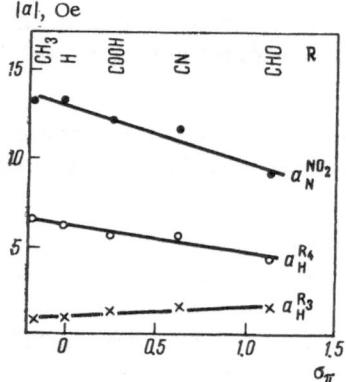

Fig. 8. Dependence of hyperfine coupling constants in EPR spectra of radical anions of 2-R-substituted 5-nitrofuran in aqueous solution of the Hammett σ_π-constant of the substituent R.

Among nitro derivatives of heterocyclic compounds radical anions have been obtained electrochemically for 5-nitropyridine [79] and its N-oxide [118] and also for 2-nitrothiophene [176].

The electrochemical method was also [86] used to obtain radical anions from 2-nitrofuran and eight of its derivatives containing substituents in the 3-, 4-, and 5-positions; the EPR spectra of these compounds had well-resolved hyperfine structure and contained nearly all the theoretically derived constants. These radicals were obtained in aqueous and in aqueous ethanolic solutions, and a method was developed for studying the kinetics of their disappearance [76, 77]. It was shown that the hyperfine coupling constants in the spectra for these radicals correlated linearly with data from quantum-chemical calculations.

The list which we have given of compounds investigated by the electrochemical generation method is far from exhaustive. We have restricted ourselves to discussion of main objectives and some of the problems which arise in their investigation. The ever-increasing number of publications on this section of electrochemistry and the manufacture by certain foreign firms of special electrochemical generation attachments for electron-paramagnetic-spectrometers bear witness to the urgency of researches in this field.

LITERATURE CITED

1. D. Ingram, Free Radicals as Studied by Electron Spin Resonance, Academic Press, New York (1958).
2. G. Minkoff, Frozen Free Radicals, Wiley, New York (1960).
3. A. Bass and H. Broida eds., Formation and Trapping of Free Radicals, Academic Press, New York (1960).
4. L. A. Blyumenfel'd, V. V. Voevodskii, and A. G. Semenov, Use of Electron

Paramagnetic Resonance in Chemistry [in Russian], Izd. Sibirsk. Otd. Akad. Nauk SSSR, Novosibirsk (1962).

5. A. L. Buchachenko, Stable Free Radicals [in Russian], Izd. Akad. Nauk SSSR, Moscow (1963).

6. M. S. Blois et al. eds., Symposium on Free Radicals in Biological Systems, Academic Press, New York (1961).

7. Á. P. Tomilov and M. Ya. Fioshin, Usp. Khim., 32:60 (1963).

8. M. J. Allen, Organic Electrode Processes, Reinhold, New York (1958).

9. R. Pasternak, Helv. Chim. Acta, 31:735 (1948).

10. B. C. L. Weedon, Quart. Rev., 6:380 (1952).

11. V. D. Bezuglyi and Yu. P. Ponomarev, Advances in Electrochemistry of Organic Compounds [in Russian], Nauka, Moscow (1968), p. 56.

12. W. M. Clark, Oxidation – Reduction Potentials of Organic Systems, Baltimore (1960).

13. L. Holleck and H. Exner, Z. Elektrochem., 56:416 (1952).

14. B. Kastening and L. Holleck, Z. Elektrochem., 63:166 (1959).

15. L. Holleck and B. Kastening, Z. Elektrochem., 63:177 (1959).

16. G. J. Hoijtink, Rec. Trav. Chim., 71:1089 (1952).

17. G. J. Hoijtink, Rec. Trav. Chim., 72:903 (1953).

18. G. J. Hoijtink, Rec. Trav. Chim., 74:1525 (1955).

19. G. J. Hoijtink, Rec. Trav. Chim., 76:860 (1957).

20. S. Wawzonek, Talanta, 12:1229 (1965).

21. S. Wawzonek and A. Gundersen, J. Electrochem. Soc., 111:324 (1964).

22. W. Kemula and R. Sioda, Bul. Acad. Polon., Ser. Sci. Chim., 10:107 (1962).

23. J. Stradiñ and S. Hillers, Tetrahedron Supplement: Proceedings of the International Symposium on Nitro Compounds, Warsaw, 1963 (1964), p. 409.

24. Ya. P. Stradyn', G. O. Reikhmanis, and R. A. Gavar, Elektrokhimiya, 1:955 (1965).

25. O. H. Müller, Ann. New York Acad. Sci., 91:11 (1940)

26. S. G. Mairanovskii, Izv. Akad. Nauk SSSR, Otd. Khim. Nauk, 2140 (1961).

27. D. Paul, A. Lipkin, and S. Weisman, J. Am. Chem. Soc., 78:116 (1956).

28. B. Kastening, Electrochim. Acta, 9:241 (1964).

29. B. Kastening, Coll. Czechosl. Chem. Communs., 30:4033 (1965).

30. V. D. Bezuglyi, L. Ya. Kheifets, and N. A. Sobina, in: Advances in Electrochemistry of Organic Compounds [in Russian], Nauka, Moscow (1968), p. 49.

31. D. H. Geske and A. H. Maki, J. Am. Chem. Soc., 82:2671 (1960).

32. R. N. Adams, J. Electroanalyt. Chem., 8:151 (1964).

33. E. Friedheim and L. Michaelis, J. Biol. Chem., 91:355 (1931).

34. B. Elema, Rec. Trav. Chim., 50:807 (1931).

35. L. Michaelis, Naturwiss., 22:461 (1931); J. Biol. Chem., 96:703 (1932); Chem. Rev., 16:243 (1935); J. Am. Chem. Soc., 63:2446 (1941).

36. L. Michaelis and S. Granick, J. Am. Chem. Soc., 63:1636 (1941); 70, 624 (1948).

37. G. K. Budnikov, Usp. Khim., 33:590 (1964).

38. W Kemula and Z. Kublik, Bull. Acad. Polon. Sci., Sér. Sci. Chim., Géol.
     Géogr., 6:653(1958); Nature, 182:793 (1958); Roczniki Chem., 32:941 (1958).
39. A. V. Il'yasov, Yu. M. Kargin, Ya. A. Levin, I. D. Morozova, N. N.
     Sotnikova, and V. Kh. Ivanova, in: Advances in Electrochemistry of Organic
     Compounds [in Russian], Nauka, Moscow (1968), p. 53.
40   G. Hoijtink and J. van Schooten, Rec. Trav. Chim., 72:691 (1953).
41. G. J. Hoijtink, J. van Schooten, and W. I. Aalbersberg, Rec. Trav. Chim.,
     73:895 (1954).
42. S. Wawzonek, E. Blaha, R. Berkey, and M. Runner, J. Electrochem. Soc., 102:235
     (1955).
43. S. Wawzonek and D. Wearrig, J. Am. Chem. Soc., 86:2067 (1959).
44. P. H. Given and M. E. Peover, Coll. Czechosl. Chem. Communs., 25:3195 (1960).
45. S. Wawzonek, R. Berkey, E. W. Blaha, and M. E. Runner, J. Electrochem. Soc.,
     103:546 (1956).
46. M. E. Runner, J. Electrochem. Soc., 103:456 (1956).
47. P. H. Given, M. E. Peover, and J. Schoen, J. Chem. Soc., 2674 (1958).
48. I. M. Kolthoff and T. B. Reddy, J. Electrochem. Soc., 108:980 (1961).
49. S. Wawzonek and A. Gunderson, J. Electrochem. Soc., 107:537 (1960); 111:324
     (1964).
50. Ya. P. Stradyn' and I. Ya. Kravis, in: Electrochemical Processes Involving Or-
     ganic Compounds [in Russian], Nauka, Moscow (1969).
51. V. P. Gul'tai, S. G. Mairanovskii, and N. K. Lisitsina, in: Advances in
     Electrochemistry of Organic Compounds [in Russian], Nauka, Moscow (1968),
     p. 37.
52. Ya. P. Stradyn' I. K. Tumane, O. Ya. Neiland, and G. Ya. Vanag, Dokl. Akad.
     Nauk SSSR, 166:631 (1966).
53. L. V. Kononenko, V. D. Bezuglyi, and V. N. Dmitrieva, in: Advances in Electro-
     chemistry of Organic Compounds [in Russian], Nauka, Moscow (1968), p. 57.
54. D. E. Austen, P. H. Given, D. S. Ingram, and M. E. Peover, Nature, 182:1784
     (1958).
55. J. K. Bolton and G. K. Fraenkel, J. Chem. Phys., 40:3307 (1964).
56. Ph. H. Rieger, I. Bernal, W. H. Reinmuth, and G. F. Fraenkel, J. Am. Chem.
     Soc., 85:683 (1963).
57. J. E. Harriman and A. H. Maki, J. Chem. Phys. 39:778 (1963).
58. P. L. Nordio, G. Giacometti, and P. Favero, Ric. Sci., 33, Ser. 2, Pt. II-A3,
     No. 4, 107 (1963).
59   J. P. Billon, G. Cauquis, and J. Combrisson, J. Chim. Phys., 61:374 (1964).
60. A. A. Galkin, Ya. L. Shanfarov, and A. V. Stefanishina, Zh. Éksper. Teor.
     Fiz., 32:1581 (1957).
61. Information Discussion of Electroanalytical Chemistry, Analyt. Chem., 31:1450
     (1959).
62. K. Möbius, Z. Naturforsch., 20A:1093, 1117 (1965).
63. M. T. Jones, E. A. La Lancette, and R. E. Benson, J. Chem. Phys., 41:401
     (1964).

64. B. I. Shapiro, V. M. Kazakova, and Ya. K. Syrkin, Zh. Strukt. Khim.,
    6:540 (1965).
65. D. H. Levy and R. J. Myers, J. Chem. Phys., 41:1062 (1964).
66. L. H. Piette, P. Ludvig, and R. N. Adams, Analyt. Chem., 34:916 (1962).
67. H. Rinkel and W. Windsch, Z. Phys. Chem. (Leipzig), 227:281 (1964).
68. P. Favero, G. Giacometti, P. L. Nordio, and G. Rigatti, Nuovo Cim., 24:21
    (1962).
69. R. E. Sioda and W. S. Koski, J. Am. Chem. Soc., 87:5573 (1965).
70. K. H. Hauser, A. Häbich, and V. Franzen, Z. Naturforsch., 16a:836 (1961).
71. J. M. Hirschson and G. K. Fraenkel, Rev. Sci. Instr., 26:34 (1965).
72. R. A. Gavar, Ya. P. Stradyn' and S. A. Giller, Izv. Akad. Nauk Latv. SSR, Ser.
    Khim., 381 (1964).
73. R. A. Gavar, Ya. P. Stradyn', and S. A. Giller, Zavod. Lab., 1:41 (1965).
74. A. H. Maki and D. H. Geske, J. Chem. Phys., 33:825 (1960).
75. D. H. Levy and R. J. Myers, J. Chem. Phys., 42:3731 (1965).
76. Ya. P. Stradyn', R. A. Gavar, V. K. Grin', and S. A. Giller, Teor. Éksper.
    Khim., 4:774 (1968).
77. R. A. Gavar, V. K. Grin', and Ya. P. Stradyn', in: Advances in Electrochemistry
    of Organic Compounds [in Russian], Nauka, Moscow (1968), p. 52.
78. B. Kastening, Z. Analyt. Chem., 224:196 (1967).
79. P. Ludvig, Th. Layloff, and R. N. Adams, J. Am. Chem. Soc., 86:4568 (1964).
80. J. Q. Chambers, Th. Layloff, and R. N. Adams, J. Phys. Chem., 68:661 (1964).
81. D. H. Levy and R. J. Myders, J. Chem. Phys., 43:3063 (1965).
82. P. H. Given, J. Chem. Soc., 2684 (1958).
83. M. B. Gazizov, A. V. Il'yasov, A. I. Razumov, G. A. Savicheva, and N. N.
    Sotnikova, in: Advances in Electrochemistry of Organic Compounds [in Russian],
    Nauka, Moscow (1968), p. 54.
84. Yu. P. Kitaev, A. V. Il'yasov, I. M. Skrebkova, and I. D. Morozova, in:
    Advances in Electrochemistry of Organic Compounds [in Russian], Nauka, Moscow
    (1968), p. 55.
85. L. H. Piette, P. Ludvig, and R. N. Adams, J. Am. Chem. Soc., 83:3909 (1961).
86. R. A. Gavar, Ya. P, Stradyn' and S. A. Giller, in: Advances in Electrochemistry
    of Organic Compounds [in Russian], Nauka, Moscow (1968), p. 50.
87. T. Kitagawa, Th. Layloff, and R. Adams, Analyt. Chem., 36:925 (1964).
88. J. G. Powles and M. H. Mosley, Proc. Phys. Soc., 78:370 (1961).
89. P. Ludvig and R. N. Adams, J. Chem. Phys., 37:828 (1962).
90. J. Deguchi, Bull. Chem. Soc., Japan, 35:598 (1962).
91. S. J. Weissman, J. Chem. Phys., 25:890 (1956).
92. S. Aono, Progr. Theoret. Phys. (Kyoto), 21:779 (1959).
93. H. McConnell, J. Chem. Phys., 25:890 (1956).
94. I. D. Morozova, Usp. Khim., 31:1231 (1962).
95. J. H. Beynon, Mass Spectrometry and Its Use in Organic Chemistry, Elsevier,
    London — New York (1960).
96. L. G. Stoodley, J. Electron. Control, 14:531 (1963).

97. C. Lagercrantz and M. Yhland, Acta Chem. Scand., 16:1043, 1799, 1808 (1962).
98. C. S. Johnson, Jr. and H. S. Gutowsky, J. Chem. Phys., 39:58 (1963).
99. E. E. Beasley and R. S. Anderson, J. Chem. Phys., 40:2565 (1964).
100. H. Berg, H. Schweiss, and D. Tresselt, Exper. Techn. Phys., 12:116 (1964).
101. H. Mauser and H. Heitzek, Z. Naturforsch., 20b:200 (1965).
102. R. Livingston and H. Zoldes, J. Am. Chem. Soc., 88:4333 (1966).
103. T. E. Gough and C. R. Symons, Trans. Faraday Soc., 22:269, 279 (1966).
104. F. Schneider, K. Möbius, and M. Plato, Angew. Chem., 77:888 (1965).
105. W. M. Fox, J. M. Gross, and M. C. R. Symons, J. Chem. Soc., A13:448 (1966).
106. A. Carrington, Quart. Rev. Chem. Soc., 17:67 (1963).
107. B. J. McClelland, J. Chem. Rev., 64:301 (1964).
108. B. I. Shapiro, V. M. Kazakova, and Ya. K. Syrkin, Dokl. Akad. Nauk SSSR, 171:156 (1966).
109. T. R. Tuttle and S. J. Weissman, J. Am. Chem. Soc., 80:5342 (1958).
110. J. R. Bolton and A. Carrington, Mol. Phys., 4:497 (1961).
111. J. R. Tuttle, J. Am. Chem. Soc., 84:2839 (1942).
112. N. M. Atherton and S. J . Weissman, J. Chem. Soc., 83:1330 (1961).
113. S. Aono and K. Oohashi, Progr. Theor. Phys. (Kyoto), 30:162 (1963).
114. E. De'Boer, Rec. Trav. Chim., 84:609 (1965).
115. N. Hirota and R. Krellick, J. Am. Chem. Soc., 83:614 (1966).
116. Ch. S. Johnson, Jr., and R. Chang, J. Chem. Phys., 43:3183 (1965).
117. R. Pointeau, J. Favéde, and P. Delhaés, Proc. XII Coloque Ampére, Amsterdam (1964), p. 276.
118. J. M. Fritsch, Th. P. Layloff, and R. N. Adams, J. Am. Chem. Soc., 87:1724 (1965).
119. J. R. Bolton and G. K. Fraenkel, J. Chem. Phys., 41:944 (1964).
120. P. A. Malachesky, L. S. Marcoux, and R. N. Adams, J. Phys. Chem., 70:2064 (1966).
121. K. Möbius and M. Plato, Z. Naturforsch, 19a:1240 (1964).
122. J. Bernal, P. H. Rieger, and G. K. Fraenkel, J. Chem. Phys., 37:1489 (1962).
123. P. L. Nordio, M. V. Pavan, and G. Rigatti, Ric. Sci., Ser. 2r, A3:851 (1963).
124. M. Iwaizumi and T. Isobe, Bull. Chem. Soc. Japan, 37:1651 (1964).
125. Th. Katz, M. Yoshida, and L. C. Siew, J. Am. Chem. Soc., 87:4516 (1965).
126. E. W. Stone and A. H. Maki, J. Chem. Phys., 39:1635 (1963).
127. B. L. Barton and G. K. Fraenkel, J. Chem. Phys., 41:695 (1964).
128. M. R. Das and G. K. Fraenkel, J. Chem. Phys., 42:792 (1965).
129. D. H. Geske and G. R. Padmanbhan, J. Am. Chem. Soc., 87:1651 (1965).
130. C. S. Johnston, Jr., R. E. Visco, H. S. Gutowsky, and A. M. Hartley, J. Chem. Phys., 37:1580 (1962).
131. F. Gerson and W. L. F. Armarego, Helv. Chim. Acta, 48:112 (1965).
132. E. W. Stone and A. H. Maki, J. Chem. Phys., 36:1944 (1962).
133. J. Gendel, J. H. Freed, and G. K. Fraenkel, J. Chem. Phys., 41:949 (1964).
134. T. Fujinaga, J. Deguchi, and K. Umemoto, Bull. Chem. Soc. Japan, 37:822 (1964).

135. D. H. Geske and A. L. Balch, J. Phys. Chem., 68:3423 (1964).
136. R. Dehl and G. K. Fraenkel, J. Chem. Phys., 39:1793 (1963).
137. P. L. Kolker and W. A. Waters, Chem. and Ind., 1205 (1963).
138. P. H. Rieger and G. K. Fraenkel, J. Chem. Phys., 37:2811 (1962).
139. N. Steinberger and G. K. Fraenkel, J. Chem. Phys., 40:723 (1964).
140. E. W. Stone and A. H. Maki, J. Chem. Phys., 38:1999 (1963).
141. A. H. Maki and D. H. Geske, J. Am. Chem. Soc., 83:1852 (1960).
142. P. H. Rieger and G. K. Fraenkel, J. Chem. Phys., 39:609 (1963).
143. G. A. Russell, E. Th. Strom, E. R. Talaty, and St. A. Weiner, J. Am. Chem. Soc., 88:1998 (1966).
144. F. Tonard, Compt. Rend., 280:2793 (1965).
145. F. Tonard and S. Odiot, J. Chim. Phys., 63:227 (1966).
146. M. T. Melchior and A. H. Maki, J. Chem. Phys., 34:471 (1961).
147. H. J. Lee and R. N. Adams, Analyt. Chem., 34:1587 (1962).
148. Z. Galus and R. N. Adams, J. Chem. Phys., 36:2814 (1962).
149. J. M. Fritsch and R. N. Adams, J. Chem. Phys., 43:1887 (1965).
150. T. M. McKinney and D. H. Geske, J. Am. Chem. Soc., 87:3013 (1965).
151. K. Kuwata and D. H. Geske, J. Am. Chem. Soc., 86:2101 (1964).
152. P. H. Rieger, J. Bernal, and G. K. Fraenkel, J. Am. Chem. Soc., 83:3918 (1961).
153. P. H. Rieger and G. K. Fraenkel, J. Chem. Phys., 37:2795 (1962).
154. P. H. Fischer and C. A. McDowell, J. Am. Chem. Soc., 85:1694 (1963).
155. P. L. Nordio, M. V. Pavan, and C. Corvaja, Trans. Faraday Soc., 60:1985 (1964).
156. Ya. P. Stradyn', Polarography of Organic Nitro Compounds [in Russian], Izd. Akad. Nauk LatvSSR, Riga (1961).
157. L. H. Piette, P. Ludvig, and R. N. Adams, J. Am. Chem. Soc., 84:4212 (1962).
158. F. Gerson and R. N. Adams, Helv. Chim. Acta, 48:1539 (1965).
159. E. Brunner, R. Mücke, and F. Dörr, Z. Phys. Chem., Neue Folge, 60:30 (1966).
160. J. H. Freed, P. H. Rieger, and G. K. Fraenkel, J. Chem. Phys., 37:1881 (1962).
161. D. H. Geske and J. L. Rangle, J. Am. Chem. Soc., 83:3532 (1961).
162. W. M. Gulick and D. H. Geske, J. Am. Chem. Soc., 87:4049 (1965).
163. T. Kitagawa, Th. P. Layloff, and R. N. Adams, Analyt. Chem., 35:1086 (1963).
164. D. H. Geske, J. L. Rangle, M. A. Bambenek, and A. L. Balch, J. Am. Chem. Soc., 86:987 (1964).
165. J. Bernal and G. K. Fraenkel, J. Am. Chem. Soc., 86:1671 (1964).
166. J. H. Freed and G. K. Fraenkel, J. Chem. Phys., 41:699 (1964).
167. N. N. Vylegzhanina, A. V. Il'yasov, and Yu. P. Kitaev, Zh. Strukt. Khim., 6:153 (1965).
168. J. P. Colpa and J. R. Bolton, Mol. Phys., 6:273 (1963).
169. A. S. Degtyarev, L. N. Ganyuk, A. M. Golubenkova, and A. I. Brodskii, Dokl. Akad. Nauk SSSR, 157:1406 (1964).

170. R. L. Hansen, R. H. Young, and P. E. Toren, J. Phys. Chem., 70:1657 (1966).
171. A. I. Brodskii, L. L. Gordienko, and A. S. Degtyarev, Zh. Vses. Khim. Obshch.
     im. D. I. Mendeleeva, 11:196 (1966).
172. F. G. Valitova, A. V. Il'yasov, N. N. Sotnikova, and S. Yu. Baigildina, Zh.
     Strukt. Khim., 5:777 (1965).
173. J. H. Freed and G. K. Fraenkel, J. Chem. Phys., 40:1815 (1964).
174. J. H. Freed and G. K. Fraenkel, J. Chem. Phys., 37:1156 (1962).
175. S. H. Glarum and J. H. Marshall, J. Chem. Phys., 41:2182 (1964).
176. E. A. C. Lucken, J. Chem. Soc., A, 991 (1966).

Polarographic Reduction of Azomethine Compounds

Yu. P. Kitaev and T. V. Troepol'skaya

Compounds with molecules containing the azomethine grouping have been known for a very long time and continually attract attention owing to the great possibilities they offer in syntheses and for practical purposes. In spite of this, problems concerning the molecular structure of the azomethines have been insufficiently investigated and remain largely controversial. A number of chemical and physical properties of such compounds were unexpected owing to the unique way in which electronic and steric factors are mutually influenced. The study of these factors is one of the most important problems in modern theoretical organic chemistry [1].

The first polarographic investigations of azomethines were carried out in the early forties [2, 3]. It was then found that they represented a convenient analytical form for the polarographic determination of various carbonyl compounds, and for a long time the investigations were concerned with their analytical applications. At that time very little information was obtained on the mechanism of the reduction of the azomethine group at the dropping mercury electrode.

The polarographic method, as is now known, makes it possible to ascertain the differences in properties and the structural characteristics of the C = N bond, and can be used for comparative investigation of carbonyl compound derivatives. Polarographic data hence allow characterization of reactivity of azomethine compounds in such simple reactions as nucleophilic additions of an electron. However, in the course of application of polarography it was possible to find certain limitations, arising either from chemical reactions of the azomethines in the investigated solutions or from their adsorbability at the dropping mercury electrode.

As yet no attempt has been made to generalize the vast amount of information on the polarography of azomethine compounds. The recently published review of polarographic reduction mechanisms in organic compounds only contains a brief account on this matter [4]. In the present paper information on electrode reactions involving participation of azomethines with various structural groups is treated systematically, and the peculiarities of their behavior at the dropping mercury electrode, arising from their molecular structure, are discussed. Compounds in which the azomethine grouping forms a part of a heterocyclic system have been omitted from consideration, as such substances possess properties which differ from those of the more common azomethines [5].

1. ALDIMINES AND KETIMINES

Derivatives of Ammonia and Aliphatic Amines

A polarogram of a ketimine was first obtained by Zuman [6], who showed that the imine was reduced when acetone was dissolved in ammonium − ammonia buffer solution. The limiting current here corresponded to the equilibrium concentration in the body of the solution. Zuman then proposed the use of the less volatile primary amines in place of ammonia. He also discovered the first limitations of the method − benzophenone and acetoacetic ester could not be determined in this way [7]. There thus was available a very simple and convenient method for determining carbonyl compounds, and its details are still being developed to this day [8–13]. The method can also be used to detect aliphatic amines [12, 14, 15].

The behavior of the imines was found to be analogous with the behavior of carbonyl compounds, and the following scheme was proposed for their reduction:

$$\underset{R}{\overset{R}{>}}C = NR + H^+ \rightleftarrows \underset{R}{\overset{R}{>}}C = \overset{+}{N}RH;$$

$$\underset{R}{\overset{R}{>}}C = N^+RH + 2e + 2H^+ \rightarrow \underset{R}{\overset{R}{>}}CH - \overset{+}{N}RH_2.$$

According to this scheme, the active form is only the conjugate acid.

Zuman explained the extremely significant shift in the half-wave potential $E_{1/2}$ with change in pH (90–100 mV/pH unit) by participation of three protons and two electrons in the reaction. However, he did not take account of the specific role of the ammonium salts as proton donors. It may be that the shift arises from the nature of the buffer solution, i.e., from greater variation in the concentration of NH_4^+ cations possessing proton-donor properties with pH-change than in the normal buffer solutions, where water or undissociated acid serve as proton-donors. In fact the slope of the $E_{1/2}$ – pH curve for the derivative from cyclohexanone and hexamethylenediamine (in a mixture of the diamine and hydrochloric acid) amounts to ~60 mV/pH unit [10], i.e., is close to the theoretical for a two-electron two-proton process. The value of $dE_{1/2}/d(pH)$, determined from data on $CH_2 = N(CH_2)_5COOH$ in [13], also amounts to 60 mV/pH unit.

Zuman first demonstrated that the reduction of azomethine aldehyde derivatives proceeds more readily than that of ketone derivatives [7] and described the appearance of peaks on their waves [16]. Acetone [17], cyclic ketones [18, 19], pyruvic acid [20], oxidation products from ascorbic acid [21], certain amino acids [16], and keto steroids [22] were determined in the form of imines.

A two-electron reduction mechanism was also established for these imines (by comparing the limiting current i_{lim} with the wave heights of other compounds). In the same period the oscillopolarographic behavior of pyruvic acid in ammonium – ammoniacal solutions was described [23]. There is, however, some doubt about the assignment of the peak or wave with $E_{1/2} = -1.0$ V (referred to the saturated calomel electrode) at pH 9.0 to the reduction of the $CH_3C(NH)CO_2^-$ anion. Norton and Furman [24] determined amino acids from the decrease in the phthalic aldehyde wave without mentioning the wave of the imine, although there is a poorly defined wave on the polarograms in the paper (near the current of the supporting electrolyte) which may belong to the imine.

In most of the investigations the imines were obtained by pouring the initial reagents directly into the polarographic cell without isolating the individual compounds. Only in the case of cyclopentanone was the imine obtained by reaction of the ketone and ammonia in ethereal solution [25] and then investigated polarographically.

Nevertheless it did not differ in its behavior on the dropping mercury electrode from the behavior of cyclopentanone in the ammoniacal supporting electrolyte. This provides evidence for the fact that, in the polarography of carbonyl compounds in ammoniacal solutions, we are in fact concerned with the imines.

However, as suggested again by Zuman [7], the imines in these solutions are in equilibrium both with the original compounds and with the intermediate product:

$$\begin{array}{c}R \\ \diagdown \\ \diagup \\ R\end{array} C \begin{array}{c} OH \\ \diagdown \\ \diagup \\ NHR,\end{array}$$

dehydration of which gives the imine. In fact it has been shown [26], in an investigation into the kinetics of the condensation of aliphatic aldehydes with ammonia, that not the imine but an α-amino alcohol is formed exclusively in aqueous solutions. It can therefore be expected that its dehydration kinetics would affect the wave heights, decreasing them in the pH region where the amino alcohol was most stable (and this is in fact observed in practice).

The imines obtained by the reduction of 2-chlorocyclohexanone in ammoniacal buffer solution were investigated in more detail [27]. The wave, which evidently belongs to a mixture of cyclohexanone imine and 2-iminocyclohexanol, is kinetic in nature and, according to coulometric data, corresponds to a two-electron reduction process. Cyclohexylamine, identified in the form of its derivative with phenylthiourea, was detected in the solution after large-scale electrolysis. Similar results were later obtained by Lund [28], who showed that two electrons are used in the reduction of cyclohexanone in a solution of methylamine and its hydrochloride, and isolated the expected N-methylcyclohexylamine.

Derivatives of Aromatic Amines

Whereas in most cases polarographic determination of aliphatic imines is carried out in solutions containing ammonia or amines (excesses of which suppress hydrolysis) as supporting electrolyte, such a possibility is limited for aniline derivatives due to the poor solubility of aromatic amines in water – alcohol mixtures. Hydrolysis of the anils, which occurs readily, particularly in acidic solutions, also complicates their investigation by polarography.

In 1952 Zuman [29] showed that benzylideneaniline is polarographically active and is reduced more readily than the original benzaldehyde. A number of o-nitrobenzalazomethines were submitted to polarography in 1954 [30], but the reduction potentials of the azomethine group were not even quoted. It can therefore be assumed that aromatic anils were first investigated by Hollek and Kastening [31], who obtained N-benzylamine by reduction of benzylideneaniline and found that this anil is reduced like the aliphatic imines but that the reaction takes place in two one-electron steps:

$$C_6H_5CH{=}NC_6H_5 + e + H^+ \rightarrow \begin{bmatrix} C_6H_5\dot{C}H{-}NH{-}C_6H_5 \\ \uparrow\downarrow \\ C_6H_5CH_2{-}\dot{N}{-}C_6H_5 \end{bmatrix} \xrightarrow{+e}$$

$$\rightarrow \begin{bmatrix} C_6H_5\bar{C}H{-}NH{-}C_6H_5 \\ \uparrow\downarrow \\ C_6H_5CH_2{-}\bar{N}\cdot C_6H_5 \end{bmatrix} \xrightarrow{+H^+} C_6H_5CH_2NHC_6H_5.$$

The kinetics of hydrolysis and formation of this anil under various conditions was also investigated in detail [32].

Lund [28], indicating the ready hydrolyzability of the anils of benzaldehyde and salicylaldehyde in acidic and alkaline media, investigated the anil of benzophenone, which is fairly stable at pH > 5.0, and also discovered two waves of almost equal height, which merged into one at pH > 9.0. He later [33] attributed them to two one-electron stages. Two waves also appear in strongly alkaline solution. The polarographic behavior of the anil observed by Lund does not differ from that described in [31], but he rigorously demonstrated an over-all two-electron reaction mechanism, by isolating the product of controlled-potential electrolysis in neutral and alkaline solutions; attempts to carry out electrolysis at the first wave at pH 6.0 were unsuccessful.

In the work by Ono and Uehara [34], which appeared almost simultaneously, such an attempt proved more successful, and electrolysis at the first wave at pH 6.0 gave a product of a one-electron reduction – dianilinodibenzyl, and furthermore electrolysis at the second wave gave benzylaniline. These authors observed a single two-electron wave in nonaqueous medium. They postulated the intermediate formation of a free radical, which could dimerize or be further reduced. Two-step reduction was also described for the condensation products of pyrrole-2-aldehyde with various amines in neutral and alkaline media [35], whereas these anils hydrolyze extraordinarily quickly in acidic solutions.

Bezuglyi and co-workers [37] used polarography to investigate the kinetics of azomethine formation and also to investigate the polar effect of substituents in the molecules on the kinetics of this reaction and on the half-wave potential of the aromatic anils obtained. They related the nature of this effect to the structural characteristics of the molecule — its noncoplanarity and the existence with the −CH=N− group of two possibilities for transmission of the effect, i.e., through the π-electrons of the double bond and through the electron pair on the nitrogen [36-44].

The effects of noncoplanarity of Schiff's bases were also observed by Vainshtein, Davydovskaya, and others while interpreting the results of a polarographic investigation of salicylaldehyde and o-tosylaminobenzaldehyde anils [45-49]. They developed a procedure for polarography of the anils, suppressing their hydrolysis by a large excess of the amine.

Recently nonaqueous solutions have also been used to eliminate hydrolysis in the studies of Schiff's bases [42-44]. Vainshtein and co-workers [49] consider that in anhydrous dimethylformamide and in aqueous ethanolic buffer solutions, the C=N group in all the salicylalanils and o-tosylaminobenzalanils bearing various substituents in the aniline residue gives two one-electron waves.

Somewhat different conclusions about the mechanism of the electrode reaction were reached in recent papers by Bezuglyi and his co-workers [43, 44]. They observed a single two-electron wave with a large number of benzalanilines bearing different substituents both in the aldehyde and in the amine residue. Exceptions are benzal-p-aminobenzenesulfonamide and derivatives of aminophenols and p-hydroxybenzaldehyde, which give two or three reduction waves; their sum corresponds to the addition of two electrons to the azomethine bond. It was shown that the first compound is reduced in two consecutive one-electron steps with the intermediate formation of a radical anion, which is capable of further reduction to a dianion (secondary amine) or of dimerization. A similar mechanism, analogous to that for aromatic carbonyl compounds is assumed to exist in the majority of Schiff's bases which are reduced in a single two-electron step, although it has not been possible to detect the intermediate radical. It was also shown that the products of the electrode reaction are more strongly adsorbed than the original azomethines. This is manifested in the appearance, in numerous cases, of an adsorption prewave. This fact, and not only the irreversibility of the electrode reaction at the first wave, explains the shift of the half-wave potential towards less negative potentials with increase in the drop time.

The position is more complex with derivatives of o- and p-aminophenol. The three reduction waves of o-aminophenol derivatives were attributed to one-electron reduction of quinoid (first wave) and benzenoid (second wave) tautomeric forms of these compounds and to subsequent one-electron reaction of intermediate radical anions (third, total wave). With the p-aminophenol derivatives, reduction of which results in the appearance of two waves on the polarograms, it was assumed that the first wave of the benzenoid form merges with the wave for subsequent reduction of both tautomers. Unfortunately the authors cite no evidence for this tautomerism from other methods. In the case of the o-derivatives the intramolecular hydrogen bond could stabilize the benzenoid structure. Two reduction waves were also observed with p-hydroxybenzalaniline, for which no similar quinoid structure can be proposed.

The appearance of two waves is most likely due to the ability of the OH group to dissociate, so that in solution the Schiff's base exists in two forms reducible at different potentials — anion and undissociated acid. The proton which splits off can also participate in the reduction, leading to the formation of a secondary amine.

2. AMIDINES

The polarography of amidines is only represented by a single paper [50], in which it was shown that compounds in this group behave extremely diversely at the mercury electrode. Aliphatic and aliphatic – aromatic amidines are polarographically inactive (to − 1.8 V, referred to the saturated calomel electrode). Substituted phenoxy-acetamidines are reduced to phenol and acetamidine in the following manner:

$$PhOCH_2C\underset{NH_2}{\overset{\overset{+}{N}H_2}{{\Large\diagup}\atop{\Large\diagdown}}} + 2e \rightarrow Ph-\bar{O} + CH_3C\underset{NH_2}{\overset{NH}{{\Large\diagup}\atop{\Large\diagdown}}}$$

For aromatic amidines the most likely course of the electrode reaction,

$$Ar-C\underset{NH_2}{\overset{NH}{{\Large\diagup}\atop{\Large\diagdown}}} + 4e \rightarrow Ar\bar{C}H-\bar{N}H + NH_3$$

is also consistent with the results from preparative electrolysis at controlled potential. For compounds with two amidine groups the number of electrons, determined by comparison with the wave heights of other compounds, is eight.

The half-wave potentials of all the amidines do not depend on the acidity of the investigated solution between pH 7 and 9. This was explained by the fact that hydrogen ions do not participate in the electrode reaction and their addition to the primary electrolysis products occurs at such a low rate that protonation has no effect on $E_{1/2}$.

At pH 9.5 the amidines are hydrolyzed very quickly, and polarographic measurements are almost impossible. There is, however, a report that half-wave potentials have been determined for some aromatic amidines at pH 12.5 with tetraethylammonium hydroxide as supporting electrolyte [51].

3. OXIMES

Among the azomethine compounds the greatest attention from electrochemists has for a long time been attracted to oximes. In 1940 Kolthoff and Langer [3] obtained a polarogram for dimethyl-glyoxime in ammonium – ammoniacal solution during amperometric titration of nickel and found that its wave height was proportional to the concentration of the compound. During amperometric titration of copper Langer [52] recorded a polarogram for the oxime of α-benzoin. The polarogram of salicylaldoxime was also first obtained in the determination of copper [53]. These papers were therefore closely related to inorganic polarography.

The first investigations of oximes in buffer solutions were carried out by Stone and Furman [54], Sartori and Gaudiano [55], and Hartnell and Bricker [56]. Polarograms for the oxime

$$HON{=}CH{-}\underset{\underset{O}{\|}}{C}{-}CH{=}CH{-}\underset{\underset{O}{\|}}{C}{-}OCH_3$$

have shown a four-electron wave which decreased in alkaline solutions, accompanied by other waves, the nature of which was not elucidated [56]. A single cathodic wave was observed by Calzolari [57] for oximes of β-isatin, α-benzoin, and salicylaldehyde; for the last two compounds in alkaline media it disappeared. Calzolari and Furlani [58] showed that the oximes of acetophenone and benzophenone are reduced with the consumption of four-electrons per molecule, i.e., to amines. In their opinion [59], the presence of two two-electron waves, one of which was observed on polarograms for acidic solutions and decreased in neutral solutions, and the other which appeared in alkaline solutions, but then also disappeared with further increase in pH, indicated that the following two electrode processes took place:

a) In acidic medium:

$$\underset{\diagup}{\diagdown}C{=}NOH \xrightarrow{2e,\ 2H^+} \underset{\diagup}{\diagdown}CH{-}NHOH \xrightarrow{2e,\ 2H^+} \underset{\diagup}{\diagdown}CH{-}NH_2 + H_2O;$$

b) In alkaline solution:

$$\underset{\diagup}{\diagdown}C{=}NOH \xrightarrow{2e,\ H_2O} \underset{\diagup}{\diagdown}C{=}NH \xrightarrow{2e,\ H_2O} \underset{\diagup}{\diagdown}CH{-}NH_2.$$

With p-phenylacetophenone oxime, however, they found a two-electron reduction process [60].

Among papers published during the fifties the most interesting is the detailed investigation by Souchay and Ser [61] on simple oximes, dioximes, and oximes where the C = N bond is coordinated with the carbonyl group. They found that for some oximes, e. g., benzophenone and acetophenone derivatives, the wave height did not reach its maximum four-electron value, and this was explained by the formation of difficultly reduced secondary products in the electrode reaction. This paper also described current maxima and decreases which Lund later observed for other azomethines [28]. Souchay and Ser interpreted these decreases by a change in the structure of the double layer at the electrode surface with increasingly negative potential, resulting in a decrease in the recombination rate of the protons with the oxime molecules. The waves in acidic solution were attributed to reduction of protonated molecules and those in alkaline solution to reduction of the $RR'C = N - O^-$ anion, the formation of which was detected from the appearance of color in the solution (in contrast to [59]). The decrease in the four-electron wave for the oximes of aryl alkyl and diaryl ketones to almost zero at pH > 6.0 was also explained [62] by reduction of the protonated form of the oxime, while the four-electron wave in alkaline medium was attributed to the neutral oxime and its decrease at very high pH values was attributed to dissociation of the molecules:

$$R_2C = NOH \rightleftarrows R_2C = NO^- + H^+.$$

Thus, whereas all the authors consider that the wave in acidic medium corresponds to reduction of the protonated oxime, there is some controversy about the nature of the wave in alkaline solutions. Souchay and Ser [61] investigated the division of the wave for some oximes in strongly acidic solutions into two sections of almost equal height. Lund [28] also observed two two-electron waves with the oxime of testosterone propionate; electrolysis at the first wave gave testosterone propionate, formed by hydrolysis of the intermediate ketimine, and electrolysis at the second wave gave the corresponding amine.

Lund [28] proposed the following scheme for the reduction of oximes in acidic and, possibly, in neutral solutions:

$$RR'C{=}NOH + H^+ \rightleftarrows (RR'C{=}NOH)\,H^+ \xrightarrow{2e,\ 2H^+} (RR'C{=}NH)\,H^+ +$$

$$+ H_2O \xrightarrow{2e,\ 2H^+} RR'CH{-}NH_3^+.$$

In a subsequent paper [63] he was able to isolate, instead of hydroxylamine, a salt of the ketimine formed in the first stage of the electrode reaction, i. e. , he was able to show that the $N - O$ bond was cleaved first.

For the majority of oximes which form a single overall wave. Lund interpreted the reduction mechanism in the following manner:

$$RR'C{=}NOH + H^+ \rightleftarrows (RR'C{=}NOH) H^+ \xrightarrow{4e, \; 4H^+} RR'CHNH_3^+ + H_2O.$$

The decrease of the four-electron wave of oximes of α, β-unsaturated compounds at very high pH values was attributed to the formation of anion RNO^-.

Zuman and Exner [64] investigated O- and N-substituted oximes and showed that aliphatic O-substituted oximes are polarographically inactive. Aliphatic N-substituted and aromatic O-substituted oximes in acidic medium behave like aromatic nitrones and are reduced to secondary amines with the intermediate formation of the respective Schiff's bases; unlike the latter they are not reduced in alkaline solutions. The study of such problems has been continued recently [65–67].

Lund [28] proposed an empirical rule which, nevertheless, has only been verified for a limited number of examples, namely that compounds with the $C = N - Y$ grouping are reduced under polarographic conditions with consumption of two electrons to form a saturated compound if Y is a carbon atom; if Y is nitrogen or oxygen, four electrons are consumed in the reaction, the $N - Y$ bond splits, and the double bond is then saturated. However, as shown by later investigations, the Lund rule is not always valid, and even related compounds can be reduced by both a two-electron and a four-electron mechanism depending on the pH of the medium or on the nature of substituents (amidines for example). It would be better to restrict the use of this relationship to the reduction of azomethines in acidic solutions.

Stradyn' and co-workers [68] investigated the polarographic behavior of certain oximes and phenylhydrazones (derivatives of phenyl- and 2-furylglyoxal), which in buffered neutral and alkaline solutions revealed a four-electron wave for the reduction of the oxime of hydrazone grouping to amines and also a wave for the carbonyl group. At pH <5.0, however, an increase was observed in the wave height to a six–electron level, which was shown to be due to split-ting of the activated $>C=\overset{+}{N}H$ bond, formed in the preceding four-electron process.

Certain differences in the polarographic behavior of syn and anti isomers of oximes were observed by Lund [28], particularly in alkaline medium where the waves of the syn isomers were greatly reduced and the waves of the anti isomers had almost the same height as in acidic solutions. Bulgarian investigators [69–78] have been engaged in the polarographic investigation of geometrical oxime iso-mers. It was shown that with quaternary alkyl ammonium salts as supporting electrolyte the syn form of an aldoxime (α) gave two waves, while the anti form (β) gave one wave, the half-wave potential of which almost corresponded to that of the first wave in the syn form. The first wave of the syn form had kinetic character and the second wave had diffusion character. The ratio of wave heights de-pended on a number of factors which influence the kinetics of the nitrone $\rightleftharpoons$ oxime tautomerization. With the anti isomers the equilib-rium is evidently shifted completely to the side of the oxime form. However, no other evidence for tautomerism was given. Data on the mechanism of the electrode reaction are also incomplete and uncertain, in particular since the reduction was carried out in unbuffered solu-tions at very negative potentials.

By comparing the wave heights of oximes with the wave of the nitro group Tyutyulkov and co–workers [76] represented the reduction mechanism in the oximes as follows:

$$>C=N-OH + 6H^+ + 6e \rightarrow >CH_2 + NH_3 + H_2O.$$

Rather different conclusions were reached recently by Pas-paleev [79], who suggested that the nature of the first wave, which is observed both with the syn (α)-form and the anti (β)-form, is de-termined not only by the oxime – nitrone tautomerization but also by isomerization of the α- and β-forms in the bulk of the solution.

According to results obtained by Japanese investigators [80], the syn forms of aldoximes (derivatives of aromatic and heterocyclic aldehydes) with $(CH_3)_4NBr$ as supporting electrolyte give one wave, and the anti forms give two. In buffered acidic solutions each isomer gives one wave, the syn isomer being reduced at more negative potentials.

With the syn and anti isomers of 5-nitrofuraldoxime Kemula and Zawadowska [81] only found differences in neutral medium, where the syn isomer gave an additional cathodic wave. In alkaline solution of the anti isomer an additional wave was observed, which was attributed to interaction between the reduction products from the nitro and oxime groups.

Several papers have been devoted to the oximes of heterocyclic carbonyl compounds [82-84, 86-89], e. g. , to the oximes of nitropyrrole-2-aldehydes [82], and formylpyridines [83, 84]. The three isomeric oximes of formylpyridines [84] give one irreversible four-electron wave for the protonated form which decreases at pH > 10. The oximes of 2- and 4-formylpyridine can give further a wave of the unprotonated compound in alkaline media and another two-electron wave, corresponding to aminomethylpyridine. However, the half-wave potentials of aminomethylpyridine formed in this reaction and aminomethylpyridine added to the solution were reported to differ, which was explained by a difference in the orientation of the molecules at the electrode. This effect, which has also been observed with other compounds, was later called the inheritance effect [85].

The inflection of the $E_{1/2}$ – pH plot was attributed [84] to the existence of two forms of these oximes protonated in different ways, and this conclusion was confirmed by isolation of the products from electrolysis at controlled potential.

A similar pattern is characteristic [86] for six pyridinealdoximes, which form primary amines both in acidic and in alkaline solution according to the Lund scheme:

$$pH < 2: \quad {>}C{=}N{-} + 5H^+ + 4e \rightarrow {-}CH_2{-}NH_3^+;$$

$$pH > pK_a': \quad {>}C{=}N{-} + 4H_2O + 4e \rightarrow {-}CH_2{-}NH_2 + 4OH^-.$$

Laviron, Person, and Fournari [87] confirmed this form of reduction for the oximes of neutral aromatic and heterocyclic derivatives (furan, pyrrole), whereas basic heterocyclic derivatives (pyridine) only gave a two-electron wave for the unprotonated compound. Four-

electron reduction was, however, found with the methyl iodide derivative of 2-pyridinealdoxime [88], accompanied by another wave at pH
4-8, attributed to further reduction of the amine.   Manoušek and
Zuman [89] have also reported on the four-electron reduction of
pyridoxinealdoxime to amine at pH < 9.0*.

The sugar oximes exhibit a number of peculiarities in their
polarographic behavior. Haas and co-workers [90] obtained well-defined diffusion waves for irreversible transfer of four electrons
for most sugars, but with the glucose and galactose derivatives
the wave heights were considerably lower since only the acyclic form
possesses polarographic activity. The slow transformation of the inactive cyclic oxime into the acyclic form results in a small kinetical-
by controlled limiting current. Polarography was also used [91] to
determine the formation rate of sugar oximes and other azomethine
sugar derivatives. The existence of active and inactive forms was
also observed for the oxime of formaldehyde [92], where the "active"
form (which is only stable at pH ~7) forms colored complexes with
nickel, manganese, and cobalt salts and is converted into the inactive
form in acidic and alkaline solutions. The nature of these forms was
not investigated, and their existence seems doubtful. It is possible
that the oxime is hydrolyzed in acidic solution and exists in alkaline
solution as a polarographically inactive anion which does not form
complexes.

The reduction of dioximes, examined in [61, 93-98], presents
a more complex pattern. Cyclohexanedione dioxime was investigated for analytical purposes [93]. In acidic solutions two waves wer
observed, the first of which disappeared at pH > 5.0 and the second
at pH > 9.0. A further wave was observed at pH > 4.0, but a detailed
interpretation of the fairly complex behavior at the dropping mercury
electrode was not undertaken. The number of electrons added to this
dioxime in ammoniacal solution was found to be eight, which is consistent with the formation of 1,2-diaminocyclohexane.

The three waves far cyclohexanedione dioxime were attributed [93] to the syn form, the anti form, and an unsettled equilibrium between the syn and possibly the anti form. Since the wave

---

* No difference between the half-wave potential of the subsequent two-electron wave and
  that of pyridoxamine has been observed [89] over wide pH-range -- Editor.

attributed to anti form largely decreased when nickel salts were added, it was suggested that the syn form does not react with nickel ions. The authors did not consider the possibility that the wave of the "syn form" (which was only observed in acidic medium) may belong to a monoxime, obtained by hydrolysis of the dioxime; this interpretation does not contradict the reported [93] experimental material. The small wave of the "syn form" has kinetic nature and increases with increased temperature, whereas the second wave decreases. Its half-wave potential lies in the same region as that of other monoximes and is about 300 mV more positive than the half-wave potential of the second wave, which is too much for geometrical isomers.

Osterud and Prytz [98] explained the decrease in height of the waves for dioximes and monoximes with increase in pH by kinetically controlled antecedent protonation. They gave a proof for the irreversible nature of the reduction of oximes and noticed splitting of the waves at certain pH values (which is particularly characteristic for dioximes) without interpreting it either by isomerization or tautomerization, as was done by investigators mentioned earlier. They were not able to determine the number of electrons and, on the basis of previous work [56], took it as four for the monoximes.

Eight-electron reduction waves have also been described for dimethylglyoxime [95] and α-furyldioxime [96]. With the latter the wave was divided into two, corresponding to the addition of six and of two electrons, but the second wave was less than a two-electron wave and was absent in strongly acidic solutions; this was attributed to partial conversion of the product from the six-electron reaction in the solution before it could be further reduced. However, only the product resulting in the eight-electron transfer was formed in electrolysis at a large mercury cathode.

The dioximes (and also monoximes) of benzoquinone, naphthaquinone, and anthraquinone were investigated by Elofson and Atkinson [97], who observed one wave in acidic and alkaline solutions and two waves in neutral solution; the smallest wave, of kinetic character, was explained by nitrone – oxime isomerization at the electrode:

$$>C{=}NOH \rightleftarrows >C{=}N(H) \rightarrow O.$$

They established a six-electron mechanism for the reduction of these compounds and isolated the corresponding diamines from the solution after electrolysis:

$$\text{Dioxime} \xrightarrow{\text{4H}^+, \text{ 4e}} \text{diimine} \xrightarrow{\text{2H}^+, \text{ 2e}} \text{diamine.}$$

The monoximes of the quinones are reduced irreversibly in one wave in acidic and in two waves in alkaline solutions.

Nitrones

The behavior of nitrones or isooximes at the dropping mercury electrode, resembles that of the oximes. Zuman and Exner [64] found that aromatic nitrones are reduced both in acidic and in alkaline solution in a four-electron wave to give a secondary amine and an intermediate Schiff's base; aliphatic nitrones are only polarographically active in acidic solution.

The probable mechanism for the reduction of these compounds is the following*:

$$\underset{\overset{|}{\text{O}}\text{H}}{C_6H_5CH=\overset{+}{N}R} \xrightarrow{\text{2e, H}^+} C_6H_5 + OH \cdots CH=\overset{+}{N}HR \xrightarrow{\text{2e, 2H}^+} C_6H_5CH_2\overset{+}{N}H_2R.$$

If R is an alkyl group, a single four-electron wave is observed at all pH values; if R is a phenyl group, the wave is divided into two waves at low pH values (a three- and a one-electron wave). This is due to the relative stability of the intermediately formed free radical $C_6H_5\overset{\cdot}{C}H - NHC_6H_5$.

Hydroxamic Acids

The hydroxamic acids, which are related to the oximes, have been investigated in less detail. Aceto- and propionohydroxamic acids are reduced from aqueous ethanolic buffer solutions containing

---

* In the opinion of the translation Editor the following scheme is preferable:

$$\underset{\overset{|}{\text{O}}\text{H}}{C_6H_5CH=\overset{+}{N}R} \xrightarrow{\text{2e, H}^+} OH^- + C_6H_5CH=\overset{+}{N}HR \xrightarrow{\text{2e, 2H}^+} C_6H_5CH_2\overset{+}{N}H_2R.$$

a tetraalkylammonium salt and form two irreversible waves at negative potentials [99]; their half-wave potentials are not dependent on
pH. These compounds are not reduced in the usual buffer solutions.
As proposed by Prytz and Osterud [99], the electrode reaction involves reduction of the C = N bond, but its mechanism has not been
investigated. The more negative reduction of hydroxamic acids (as
compared with oximes) is explained by their more highly acidic characteristics. A method has also been developed for polarographic
determination of hydroxamic acids in the presence of hydroxylamine
[100].

Polarographic reduction of high-molecular hydroxamic [101]
and aliphatic aminohydroxamic acids [102] has been reported, but
their reduction mechanisms have not been examined.

Amidoximes

The amidoximes differ from other oximes in their behavior at
the dropping mercury electrode. Already Lund [28] noticed that
benzamidoxime was reduced under the consumption of two electrons
per molecule, and benzamidine was isolated from the solution after
electrolysis in acidic medium. This corresponds to the first stage
in the reduction of normal oximes:

$$C_6H_5C\,(NH_2) = NOH + H^+ \rightleftarrows [C_6H_5C\,(NH_2) = NOH]\,H^+ \rightarrow$$

$$\xrightarrow{2e,\ 2H^+} [C_6H_5C\,(NH_2) = NH]\,H^+ + H_2O.$$

In addition to the diffusion two-electron reduction waves, Mollin
and Kašparek [103, 104] have described catalytic hydrogen evolution
waves (except for acetamidoxime) and, in some cases, formation of
another diffusion wave at more negative potentials. In their opinion
a protonated form of the compound (present in the bulk of the solution)
is reduced at low pH values, and a form formed by recombination
with protons at the electrode surface is reduced at high pH values.
Unfortunately it was not possible to establish the group at which protonation occurred. The pK_B values, determined from potentiometric
and polarographic measurements, proved identical. Generally speak-
ing this is uncharacteristic of azomethine compounds, which exhibit
considerable adsorption on the electrode.

Reduction of compounds containing two amidoxime groups gives a four-electron wave. Similar behavior was described earlier [105] for bis-amidoximes of oxalic acid, but the nature of the waves was not discussed.

4. AZINES

Until recently only benzalazine [28, 106, 107] and naphthalazine [107] had been investigated. Lund found that in acidic medium benzalazine gave a six-electron wave for reduction to benzylamine, and this wave decreased at pH > 7.0 owing to kinetic control of the limiting current resulting in protonation:

$$C_6H_5CH=N-N=CHC_6H_5 + 6e + 8H^+ \rightarrow 2C_6H_5CH_2NH_3^+.$$

Benzaldehyde benzylhydrazone was formed in alkaline solutions:

$$C_6H_5CH=N-N=CHC_6H_5 + 2e + 2H_2O \rightarrow C_6H_5CH_2NH-N=CHC_6H_5 + 2OH^-,$$

and the reduction wave split into two at pH values between 9.5 and 12; with increase in pH the second wave became larger at the expense of the first.

Other investigators have also observed similar waves in buffered solutions of benzalazine [106]. With quaternary ammonium salts as supporting electrolyte in unbuffered solutions benzalazine and naphthalazine each form two waves of approximately equal height [106, 107], and according to the Ilkovič equation each wave corresponds to the addition of two electrons, i. e., reduction probably gives 1, 2-dibenzylhydrazine, like reduction with sodium amalgam in alkaline solution.

Benzalazine and other aromatic azines were also investigated recently by Bezuglyi and Shimanskaya [108] both in aqueous buffered solutions and in methanol with $(C_2H_5)_4NI$ as supporting electrolyte. Unlike the results obtained in [28], not one wave but two waves were obtained with benzalazine at pH 4-7. Only one wave was obtained in alkaline solution (with one exception).

It was shown that the two protonated C = N bonds in the investigated azines give two two-electron waves (with similar half-wave potentials) or one four-electron wave:

$$RCH{=}N{-}N{=}CHR \xrightarrow{2e,\ 2H^+} RCH_2NH{-}N{=}CHR \xrightarrow{2e,\ 2H^+}$$

$$\rightarrow RCH_2{-}NH{-}NH{-}CH_2R.$$

The number of electrons participating in the electrode reaction during reduction in methanol, with $(C_2H_5)_4NI$ as supporting electrolyte, was found to be two. Bezuglyi and co-workers [41] explained the more negative reduction of azines as compared with the related aldimines by the effect of the second nitrogen atom, bearing an unshared electron pair. In ring systems with similar structure to the azines — oxadiazoles — reduction of the bond takes place at even more negative potentials [41].

5. HYDRAZONES

The first hydrazones submitted to polarography were products of the reaction between ketosteroids and the Girard T reagent [2, 109]. A number of references to the early papers on the polarography of these hydrazones are cited in [110]. Among the most interesting is the investigation by Prelog and Häfliger [111], who observed in readily hydrolyzed cyclic ketone derivatives two waves of almost equal height, merging into one and disappearing with increase in the pH value. The maximum limiting current at pH 8.0 corresponded to a two-electron conversion of the betainyl hydrazones into hydrazines. The scheme proposed for the electrode process is similar to that involved in the reduction of ketones:

1) $>C{=}NNHCOCH_2\overset{+}{N}(CH_3)_3 \underset{-e}{\overset{+e}{\rightleftarrows}} >\overset{.}{C}{-}\overset{-}{N}NHCOCH_2\overset{+}{N}(CH_3)_3,$

   $+\,H^+ \downarrow\uparrow - H^+$

   $>\overset{.}{C}{-}N\overset{}{H}NHCOCH_2\overset{+}{N}(CH_3)_3 \underset{-e}{\overset{+e}{\rightleftarrows}} >\overset{.}{C}{-}NHNHCOCH_2\overset{+}{N}(CH_3)_3.$

2) $>\overset{.}{C}{-}NHNHCOCH_2\overset{+}{N}(CH_3)_3 + e + H^+ \rightarrow\ >CHNHNHCOCH_2\overset{+}{N}(CH_3)_3.$

However, Prelog and Häfliger noticed that this mechanism did not explain all the observed facts, particularly the high value of $\Delta E_{1/2}/\Delta pH$ shift (85 mV/pH unit). Results consistent with these data were obtained by Young [112], who described splitting of waves for the Girard hydrazones in acidic medium. The reduction currents which he observed corresponded to a transfer of ~0.6–1.4 electron. These papers [111, 112] have been frequently quoted, although the reaction mechanism was not proved. The hydrazones are reduced [110] irreversibly in protonated form by a four-electron mechanism:

$$> C = NHNHR^+ + 4e + 4H^+ \rightarrow > CHNH_3^+ + H_2NR.$$

The first wave for diphenylphosphorylhydrazones at pH < 2.5 was attributed to reduction of the protonated form of the hydrazone to a free radical, and the second wave was attributed to its further reduction to hydrazine [113] based on unconfirmed conclusions reached by previous authors [111, 112]. These conclusions have recently been reconsidered by Japanese investigators [114], who showed convincingly that reduction of betainylhydrazones takes place with the consumption of four electrons at all pH values with cyclopentanone derivatives and only in acidic solutions with benzophenone derivatives; the electrode reaction mechanism changes with increasing pH value, and gradually becomes a two-electron process. They also investigated the hydrolysis kinetics for these compounds by polarography [115].

According to data obtained by Lupton and Lynch [116, 117], unsubstituted hydrazones are also reduced in acidic solution and form two waves on the polarograms, but the nature of the waves was not determined. Other investigators found only one wave on polarograms for acetone hydrazone and certain other hydrazones [14]. A single wave is characteristic of benzaldehyde dimethylhydrazone, and in alkaline solution its height corresponds to an addition of approximately two electrons [106] and formation of 1,1-dimethyl-2-benzyl-hydrazine. The currents were approximately twice as large in acidic solution as in alkaline solution, but this fact was ignored and only the difficulty of accurate measurements of the wave height at low pH values was mentioned.

Like other azomethine sugar derivatives, the hydrazones of aldopentoses reveal (under identical conditions) waves with different heights, since their reduction currents arise not only from diffusion of the polarographically active hydrazone form to the electrode but also are limited by the kinetics of conversion of the inactive cyclic form into the hydrazone form. It has been assumed that in these compounds a two-electron reduction of the C = N bond takes place [118].

Arylhydrazones, which are interesting not only as analytical forms of carbonyl compounds and intermediate products for organic syntheses but also from the standpoint of their structure and their changes in solution, have been examined in more detail. In the first investigations on the polarography of arylhydrazones it was found that, on storage in ethanol – water solutions, they undergo changes leading to the appearance of new waves; this was first attributed to tautomeric changes [119–121]. Here it was necessary to assume that, in the free state, phenylhydrazones of aliphatic and alicyclic ketones exist in the enhydrazine form, while the derivatives of the aromatic series and aliphatic aldehydes are hydrazones. Such an assumption did not contradict available data on the chemical differences between these groups of arylhydrazones, since at that time there were no sufficiently reliable physicochemical data on their structure.

The experimental data obtained subsequently for other arylhydrazones by the method employed [119] also agreed fairly well with the proposed scheme of tautomeric changes [122, 123]:

$$\text{Enhydrazine} \rightleftarrows \text{hydrazone} \rightleftarrows \text{benzeneazoalkane.}$$

The reason for this remained obscure [124], since it had already been established [124–126] that in the free state the arylhydrazones had the hydrazone and not the enhydrazine structure. A detailed investigation into the polarographic behavior and electrode reaction mechanism for arylhydrazones [127] showed (in combination with spectroscopic data) the erroneous nature of the conclusion about tautomeric changes in ethanol – water solutions of these compounds and proved that the observed changes were caused by oxidation of the hydrazones by atmospheric oxygen to form hydroperoxide compounds with azo structures:

$$C_6H_5N = N-C\underset{HO-O}{\overset{R'}{\underset{|}{\big\langle}}}{}^{R'}_{R''}$$

It has thus not been possible to demonstrate the existence of an enhydrazine tautomer by polarography.

The reduction of five aromatic and unsaturated arylhydrazones was examined by Lund [28]. In an investigation of the behavior of carvone phenylhydrazone in a solution of mineral acid Lund discovered two waves of almost equal height; at pH > 3.0 they merged into one, which decreased with further rise in pH value. By preparative electrolysis in acidic solution it was possible to isolate aniline and show that four electrons were used in the reduction and, what was most important, that the initial stage was the cleavage of the $N-N$ bond of the hydrazone. The other phenylhydrazones behaved similarly but only formed one four-electron wave, and the overall electrode reaction was therefore represented in the following form:

$$RR'C = NNHC_6H_5 + 4e + 6H^+ \rightarrow RR'CH-NH_3^+ + H_3\overset{+}{N}-C_6H_5.$$

Lund considers that here too a cleavage of the $N-N$ bond, activated by protonation, takes place first and is then followed by reduction of the aldimine:

$$C_6H_5CH = N-NHC_6H_5 + H^+ \rightleftarrows (C_6H_5CH = N-NHC_6H_5)\overset{+}{H} \rightarrow$$

$$\xrightarrow{2e,\ 3H^+} (C_6H_5CH = NH)\overset{+}{H} + H_3\overset{+}{N}C_6H_5 \xrightarrow{2e,\ 2H^+} C_6H_5CH_2\overset{+}{N}H_3.$$

The four-electron reduction of phenylhydrazones, in particular of those derived from pyridine, was further confirmed [87]. Kitaev and his co-workers [127, 128] also found that in buffered acidic solution a number of arylhydrazones from various carbonyl derivatives were reduced in the protonated form by the four-electron mechanism according to Lund. They consider, however, that in some cases the mechanism can change, with increase in the pH of the solution, to two-electron reduction of the $C = N$ bond. In addition it was shown that reduction of arylhydrazones is complicated by adsorption phenomena, which affect their polarographic behavior. In the remaining few papers concerned with the polarography of phenylhydrazones the reduction mechanism was not discussed [129-132]. Nitrophenylhydrazones have also been submitted to polarographic electrolysis, but attention was only restricted to the wave of the nitro group, which is used for analytical purposes. The reduction of the hydrazone group in mono-, di-, and trinitrophenylhydrazones was first studied by Kitaev and Skrebkova [128, 133-135], who showed that this process takes place in acidic medium and involves two two-electron stages according to the Lund scheme.

A fairly complex picture is presented by the polarographic behavior of isonicotinylhydrazones, which were first investigated by Kolusheva and co-workers [136]. For the derivatives of aliphatic carbonyl compounds and of fural they described a single wave, which split into two with increase in concentration. With aromatic derivatives two waves were also observed at low concentrations over a narrow pH range. These diffusion waves were attributed to two-electron reduction of the C = N group, but the reason for the splitting was not established. In the study of reduction of aromatic acylhydrazones [137], the authors proposed that reduction will proceed in a single wave, if the phenyl radical of the aldehyde or ketone residue is not attached to the C = N group and that splitting of the wave will occur at certain pH values, if the phenyl radical is directly attached. The coulometrically determined number of electrons involved in the reduction was found at pH 2.0 to be equal to two [137].

Two waves for pyrrole-2-aldehyde isonicotinylhydrazone were suggested [35] to arise from the existence of tautomerism or resonance between the following structures:

$$\underset{\overset{\|}{\text{C}}}{\text{O}}-\text{NH}-\text{N}=\text{CH}- \rightleftarrows -\underset{\overset{|}{\text{C}}}{\overset{\text{OH}}{}}=\text{N}-\text{N}=\text{CH}-.$$

Numerous recent investigations on acylhydrazones [138, 141] described the polarographic behavior and discussed the reduction mechanism. The glucosazones, which are similar in structure to the hydrazones and are reduced in a single four-electron wave [142, 143] in alkaline medium containing boric acid gave another wave corresponding to a special case of catalytic evolution of chelate-bound hydrogen [143]. A wave for catalytic hydrogen evolution has also been described by Mairanovskii and his co-workers [130] for 1,3,5-trimethyl-4-piperidone phenylhydrazone.

In nonaqueous solutions, however, reduction of hydrazones may take place in a different way. As shown by Kitaev and Skrebkova [144], phenylhydrazones, nitrophenylhydrazones, semi- and thiosemicarbazones, nicotinyl- and isonicotinylhydrazones, benzoylhydrazines, and also the products from the coupling of diazonium salts with diphenyldiketopyrazolidine and barbituric acid, which are derivatives of aromatic aldehydes, aliphatic aromatic ketones, or di- and tricarbonyl compounds, are reduced in dimethylformamide, where the first irreversible one-electron wave corresponds to the formation of an unstable radical anion. In a number of cases the second wave is also a one-electron wave, i. e., the reduction apparently only affects the C = N bond. However, with some hydrazones a large number of electrons (two or three) are added at the potential of the second wave, and this seems to be connected with a cleavage of dimeric molecules. Under the conditions used the $E_{1/2}$ values correlate linearly with the polar constants of substituents in the benzene ring of the hydrazone.

6. SEMI- AND THIOSEMICARBAZONES

From a polarographic point of view semi- and thiosemicarbazones resemble the structurally related arylhydrazones. Earlier papers reported only isolated data on the reduction of aldehyde and ketone semicarbazones [145-147] and p-acetaminobenzaldehyde thiosemicarbazone [148, 149] at the dropping mercury electrode. The latter compound has been studied in more detail by Dušinský [150], who observed a four-electron wave on the polarograms in acidic solutions; at pH 7.0 this wave decreased almost to zero. A two-electron wave, the limiting current of which did not depend on pH, appeared in alkaline solution. These results were interpreted on the assumption that tautomeric equilibrium existed between the ionic form in acidic solutions and the thiol form in alkaline solutions. The opinions of the various authors differed on the question of the number of electrons involved in the electrode reactions with semi- and thiosemicarbazones.

Souchay and Graizon [151] considered that two electrons were involved in the reduction of benzophenone semicarbazone. Lund [28] demonstrated a four-electron mechanism for the electrode reaction with initial cleavage of the N − N bond in benzaldehyde semicarbazone, as with benzophenone semicarbazone in mineral acid solution. In the reduction of benzophenone semicarbazone in a solution buffered at pH 4.0 1-benzhydrylsemicarbazide, i. e., the product from the addition of two electrons, was isolated with 30% yield. The semicarbazones of cinnamaldehyde and benzalacetone can be reduced both in the protonated and in the unprotonated form; four electrons are required in alkaline solution, but it has not been possible to determine

the number of electrons required in acidic solution owing to hydrolysis of the compounds.

Beginning in 1959 Kitaev and his co-workers [152-156] have
published a series of articles on the polarography of semi- and thiosemicarbazones, and these contain a large amount of factual material
about the behavior of more than 50 aliphatic, alicyclic, aromatic, and
aliphatic – aromatic compounds at the dropping mercury electrode.
Like the other azomethines they are reduced in the form of protonated
complexes formed both in the electrolyte near the electrode and at the
electrode surface. With increase in the pH value the limiting currents
are gradually governed more by the preceding protonation. In solutions at pH ~6.0 the number of electrons required for reduction was
found to be close to two with aliphatic and alicyclic derivatives. A
two-electron mechanism was also proposed for the reduction of
aromatic and aliphatic – aromatic semi- and thiosemicarbazones.

In a further investigation of the structure and mechanism of the
electrode reaction in semi- and thiosemicarbazones it was shown
that in acidic solution the protonated carbazone form is reduced under
consumption of four electrons. The mechanism can change with increase in pH value [128, 157], and adsorption at the dropping mercury
electrode, described by the S-shaped isotherm of Frumkin [157],
affects the polarographic behavior and the form of the waves for
these compounds.

Laviron and his co-workers [87] investigated the properties of
semi- and thiosemicarbazones in heterocyclic series. As in the case
of oximes, they found two-electron reduction with pyridine derivatives and four-electron reduction with benzene derivatives and neutral
heterocyclic compounds. In alkaline solutions, however, the latter
require two electrons and are converted into the corresponding substituted hydrazines. According to data obtained by Haas and co-
workers [90], like the other azomethine sugar derivatives, the limiting currents for aldose semi- and thiosemicarbazones depend largely
on the structure of the sugar residue, and the number of electrons
calculated from the wave height varies between 4 and 0.48, because
part of the material is present in the non-reducible form. Other
authors have also indicated that the aldose semicarbazones are cyclic
but rapidly isomerize in aqueous solution to the active acyclic form

[158]. A polarographic investigation of 5-nitro-2-furaldehyde semi-carbazone [159] and substituted benzaldehyde semi- and thiosemi-carbazones [160] may also be mentioned. Several methods were described for the determination of carbonyl compounds as their semi-carbazones [151, 161-163].

A critical review of polarographic researches on semicarba-zones was given in the recently published article by Fleet and Zuman [164]. They showed that the four-electron reduction of semicarba-zones is preceded by protonation. They examined also quantitatively the effect of polar substituents on the reduction of aromatic aldehyde derivatives and found that in this series of compounds the $E_{1/2}$ values are relatively little sensitive to the effects of substituents ($\rho_\pi = 0.10$ V).

7. TETRAZOLIUM SALTS AND FORMAZANES

The formazanes are the hydrazine analogs of the amidines and have a genetic relationship to the tetrazolium salts and the reductions of both groups of compounds here are therefore considered together. Polarography of tetrazolium salts is fairly complex owing to the large number of stages involved in the electrode process and the instability of some of the intermediate products. Tetrazolium salts were first investigated polarographically by Doskočil and Doskočilova [165]. With reference to the stepwise chemical hydrogenation of tetrazolium salts proposed by Jerchel and Kuhn [166], Ried and Wilk [30] proposed a similar mechanism for their polarographic reduction:

formazane   hydrazidine

amidrazone

All the current — voltage curves given in this paper [30] show extreme-ly poorly defined waves, which is evidently due to unfortunate choice of experimental conditions, and it is therefore difficult to make any conclusions about the mechanism of the process.

An extensive investigation on the polarography of tetrazolium salts was accomplished by Jambor and co-workers [167-173]. They found that at pH > 6.0 triphenyltetrazolium chloride is reduced to triphenylformazane and then to benzhydrazidine (see the Ried and Wilk scheme), and that at pH < 6.0 the wave corresponds to two reactions:

$$\underset{+}{PhC \Big\langle \begin{matrix} N-NPh \\ N=NPh \end{matrix}} \xrightarrow{2e,\ H^+} PhC \Big\langle \begin{matrix} N- \quad NPh \\ NH-NPh \end{matrix} \xrightarrow{2e,\ 2H^+} PhC \Big\langle \begin{matrix} N-NHPh \\ NHNHPh \end{matrix}$$

The electrode reaction of galactodiphenyltetrazolium salts is affected by the simultaneous occurrence of two processes in which formazane and a dihydrotetrazolium derivative (two-electron stage) are formed. The latter is then converted into hydrazidine (also with the use of two electrons); the contributions of the two mechanisms depend on the pH value [168]. After irradiation phenyltetrazolium chloride was converted to a free radical by accepting four electrons [169]. Ditetrazolium compounds [169, 170], formazanes, and tetrazolium derivatives of sugars [171] have also been investigated and the results were summarized [172, 173].

With some tetrazolium salts Jambor discovered anomalies which he attributed to adsorption prewaves and catalytic hydrogen evolution waves [170].

Campbell and Kane [174] studied the polarographic behavior of tetrazolium bromide and some of its derivatives and found overall eight-electron reduction. Two waves were observed on the polarogram at pH < 4.0, and a third appeared when the pH was increased. The first wave corresponded to the transfer of four electrons, and to the isolated reaction product was assigned the hydrazidine structure. In alkaline solution the first wave split into two two-electron waves. Campbell and Kane suggested that the reduction product at the second wave was $PhCH(NHNHPh)_2$ and that the third wave corresponded to further splitting of this molecule:

$$\underset{+}{PhC \Big\langle \begin{matrix} N-NPh \\ N=NPh \end{matrix}} \xrightarrow{2e} PhC \Big\langle \begin{matrix} N-NHPh \\ N=NPh \end{matrix} \xrightarrow{2e} PhC \Big\langle \begin{matrix} NNHPh \\ NHNHPh \end{matrix} \xrightarrow{2e}$$

$$\longrightarrow PhHC \Big\langle \begin{matrix} NHNHPh \\ NHNHPh \end{matrix} \xrightarrow{2e} PhHC \Big\langle \begin{matrix} NH_2 \\ NHNHPh \end{matrix} + PhNH_2 .$$

Only the first two stages were observed at low pH values. By analogy with phenylhydrazones, however, we can describe the further reduction of the hydrazidine by the following mechanism:

$$PhC\begin{subarray}{l} \diagup NNHPh \\ \diagdown NHNHPh \end{subarray} \xrightarrow{2e} PhC\begin{subarray}{l} \diagup NH \\ \diagdown NHNHPh \end{subarray} + PhNH_2 \xrightarrow{2e} PhCH\begin{subarray}{l} \diagup NH_2 \\ \diagdown NHNHPh \end{subarray}$$

The analogy with phenylhydrazones is supported by the fact that the $E_{1/2}$ values for the second and third waves are very close to $E_{1/2}$ for the corresponding waves in arylhydrazones.

The results given by Campbell and Kane are not entirely consistent with the conclusions reached by Jambor, who proposed the formation of 1,2-dihydro-2,3,5-triphenyltetrazolium,

$$PhC\begin{subarray}{l} \diagup N-NPh \\ \quad\ \ | \\ \diagdown NH-NPh \end{subarray}$$

in acidic solutions. The polarography of tetrazolium salts and formazanes has also been investigated by others [175–180].

POLAROGRAPHY AND STRUCTURE
OF AZOMETHINES

Here we will deal briefly with the mechanism of electrode reactions involving various hydrazones and the factors affecting the rates of these reactions. The more detailed description given to the mechanism for the reduction of hydrazones arises from the fact that in them the azomethine group in some cases exhibits special characteristics owing to interaction with the neighboring heteroatoms.

As shown above, all the hydrazine derivatives are reduced at the dropping mercury electrode under consumption of four electrons per molecule. In buffered solutions hydrazones are, as a rule, first protonated at the electrode surface, and the primary step in the electrochemical reaction is the irreversible transfer of an electron, leading to cleavage of the single nitrogen — nitrogen bond by activated conjugation and protonation. The reduction of the imines formed — derivatives of aromatic aldehydes and aliphatic — aromatic ketones —

takes place at the same potentials as cleavage of the nitrogen–nitrogen bond. As a result, one four-electron wave is observed on the polarograms; in the case of derivatives in the aliphatic and alicyclic series in acidic solutions these two reduction steps appear as separate waves, which merge into one wave in weakly acidic solutions. The validity of the reduction mechanism depicted below has been confirmed by investigation of model compounds, by analysis of the products from preparative microelectrolysis, and by other methods:

$$\frac{R}{R'}\!\!>\!\!C = N\!\!-\!\!NH\!\!-\!\!\begin{bmatrix} C_6H_4X \\ CONH_2 \\ CSNH_2 \\ COC_6H_4X \\ COC_5H_4N \end{bmatrix} + H^+ \rightleftarrows A + 2e + 3H^+ \rightarrow \frac{R}{R'}\!\!>\!\!C\!\!=\!\!\overset{+}{N}H_2 +$$

$$+ H_3\overset{+}{N}\!\!-\!\!\begin{bmatrix} C_6H_4X \\ CONH_2 \\ CSNH_2 \\ COC_6H_4X \\ COC_5H_4N \end{bmatrix}$$

$$\frac{R}{R'}\!\!>\!\!C\!\!=\!\!\overset{+}{N}H_2 + 2e + 2H^+ \rightarrow \frac{R}{R'}\!\!>\!\!CH\!\!-\!\!\overset{+}{N}H_3.$$

(In this scheme A denotes a protonated hydrazone complex.) Since the imine is formed at the electrode in protonated form, its half-wave potential does not depend on the pH value

The derivatives of aromatic aldehydes and aliphatic – aromatic ketones are also reduced in nonaqueous solutions in dimethylformamide. In this case two waves appear which correspond to two successive one-electron irreversible transfers or to a two-electron process, respectively. The azomethine group may be assumed to be the acceptor of the first electron. It had already been noticed earlier that amalgams and metals reduce hydrazones by different mechanisms: the former lead at first to imine – amine cleavage, whereas the latter first of all hydrogenate the azomethine bond. Probably a similar pattern is also observed in the electrochemical reaction, where the direction of attack is determined by the potential at which the process begins. Of great importance is the question as to whether the free molecule or the protonated molecule is reduced, since protonation undoubtedly changes the polarizability of the bonds. Substituents in the aldehyde residue have been shown to have a stronger effect on the potential of the first wave in dimethylformamide solutions than those in the hydrazine residue.

The polarographic behavior and the form of the polarograms with many hydrazones are largely affected by the adsorbability of the molecules at the surface of the dropping mercury electrode. At the limiting currents frequently more or less sharply defined decreases are observed, the magnitude of which depends on the concentration of depolarizer and methanol and also on the pH value of the solution. Furthermore it depends on the drop time and on the buffer capacity. As the clearest current decreases are most pronounced on polarograms of semi- and thiosemicarbazones and phenylhydrazones, the nature of this effect was therefore investigated in solutions of these substances.

Calculation of the thickness of the reaction layer in which the protonation occurred from the value of the protonation constant gave a value in the order of several angstrøms or fractions of an angstrøm, which showed that the depolarizer molecules were adsorbed. As a rule the slope of the waves does not depend on the pH value but depends on the region of reduction potentials. Most frequently the $\Delta E_{1/2}/\Delta pH$ value differs from the slope of the logarithmic analysis of the wave, which is usually due to adsorption of an inactive form of the depolarizer.

To describe the change in the observed reduction current with increased potential use was made of the Mairanovskii equation [85]:

$$\log[i_{\lim}/(i_d - i_{\lim})] = A - 0.43a\,(E - E_M)^2,$$

in which the factor a characterizes the value and slope of the current decrease; the quantity A is related to the rate constant of the interaction between the adsorbed particles and the proton donors.

In the case of semicarbazones, thiosemicarbazones, and benzoylhydrazones the a value does not depend on the pH and is a characteristic of the compound, which expresses the adsorbability and orientation of its molecules at the dropping mercury electrode [157]. This parameter depends little on the concentration of the indifferent electrolyte, if the latter does not retard the electrode process. Investigation of the dependence of the polarographic wave parameters on the drop time showed that when the drop time was shortened the slope of the current decreased and the a value also decreased. The effect of the drop time on the relationship between $E_{1/2}$ and the pH value, demonstrating the substantial role of adsorption in the kinetics of electrolytic reduction, was here noticed for the first time.

Detailed investigation of the adsorbability and polarographic behavior has made it possible to ascertain that the observed effect of many factors on the polarographic parameters, including a, is due to the great sensitivity of adsorbability to variation of the electrode potential. Adsorption isotherms which we plotted for some compounds were S-shaped. The attraction constant, $\gamma = 0.8-1.3$, and adsorption constant, $\beta_0 = (6-9) \cdot 10^3$ liter/mole, of the Frumkin isotherm were determined for some substances.

It has already been said above that, in addition to minima, there are also maxima on the polarograms for many phenylhydrazones, semi- and thiosemicarbazones, isohydrazones, and isonicotinyl-hydrazones. It has been shown that the appearance of maxima on the waves for the hydrazones is due to tangential movements of the electrode surface. Investigation of the electrocapillary curves has shown that the maxima appear in the region of potentials where desorption of the larger part of the depolarizer occurs.

The polarographic investigation of hydrazones had the aim of characterizing the effect of conjugation on the half-wave potential for the reduction of $C = N$ and thereby of assessing the intramolecular interactions in the various series of compounds. It was found, however, that polarography gives the characteristics not of the azomethine bond but of the ordinary hydrazine nitrogen – nitrogen bond. It is seen from Table 1 that this bond is most easily reduced in the derivatives of aromatic carbonyl compounds. At the same time, as was shown in [152, 181], conjugation increases when aromatic rings and carbonyl groups are introduced into the hydrazine residue, and this is also reflected in the polarographic behavior. The sequence of reducibility of investigated types of hydrazones at the dropping mercury electrode is the same as deduced from spectroscopic data. Consequently, intramolecular interactions do affect the course of the electrode reactions; it is true they are only one of the factors which determine the kinetics of the electron transfer and their contribution

TABLE 1. Half-wave Potentials for Various Azomethines at Certain pH Values

Compound	$E_{1/2}$, V (sat. cal. elect.)				
	pH				in DMFA
	1,05	3,55	5,20	7,40	
Phenylhydrazones					
Acetaldehyde	0.74	0.98	1.26	1.36	—
Acetone	—	1.14	1.42	1.56	—
Benzaldehyde	0.75	1.00	1.11	—	1.79
p-Nitrophenylhydrazones					
Acetaldehyde	0,71	1.05	1,13	—	—
Acetone	0.85	1.11	1.28	—	—
Benzaldehyde	0.80	0.98	1.10	1.24	—
Semicarbazones					
Acetaldehyde	0.87	1.03	1.26	1,39	—
Acetone	0.93	1.06	1,37	1.50	—
Benzaldehyde	0.69	0.91	1,05	1.24	1.66
Thiosemicarbazones					
Acetaldehyde	0.86	0.99	1.27	1.38	—
Acetone	0.91	1.04	1.38	1.49	—
Benzaldehyde	0.75	0.96	1.06	1.19	1.48
Benzoylhydrazones					
Acetaldehyde	—	—	—	—	—
Acetone	1.02	1.15	1,37	1.38	—
Benzaldehyde	0.72	0.92	1,05	1,23	1.50
p-Nitrobenzoylhydrazones					
Acetaldehyde	0.71	1,05	1.13	—	—
Acetone	0.85	1.11	1.28	—	—
Benzaldehyde	0.80	0.98	1.10	1,24	—
Isonicotinylhydrazones					
Acetaldehyde	—	—	—	—	—
Acetone	0.56	0.76	0.88	1.05	—
Benzaldehyde	0.65	0.82	1.03	1.18	1.21
Oximes					
Acetaldehyde	—	—	—	—	—
Acetone	—	—	—	—	—
Benzaldehyde	0.70	0.92	1.08	1.20	—
Imines					
Acetaldehyde	0.94	1,10	1.20	1,25	—
Acetone	—	1.35	1.39	1.45	—
Benzaldehyde	—	—	—	—	1.74

is far from always noticeable. In the case of the hydrazones, therefore, it has only been possible to reveal differences in structure between the series, whereas it has not been possible to characterize by means of polarography the more subtle effects of substituents within the series.

The half-wave potential serves as an indication of the effect of various substituents on the reactivity. In the reaction series investigated by Kitaev and co-workers the effect of substituents on the kinetics of the electrode process was found to be insignificant owing possibly to the substantial role of adsorption. It was found that, although the linear correlation was retained between $E_{1/2}$ and σ for the substituent, the ρ value was as a rule small. In the case of arylhydrazones $\rho = +0.025$ for a series of different hydrazones derived from the same carbonyl compound and somewhat larger (+0.10) for hydrazones substituted at the aniline nitrogen. In the case of isonicotinylhydrazones the correlation analysis was carried out for two waves. Here the cleavage of the $N - N$ bond was found practically independent of the Taft constants ($\rho = +0.045$), and $E_{1/2}$ for the reduction of the imine formed in the first stage of the process correlated with the Hammett constants at $\rho = +0.22$. Conditions in which the same course for the electrode process was most probable (pH, depolarizer concentration, composition and concentration of buffer solution) were selected for the correlation analysis.

However, the effect of the substituent on the half-wave potential can be reduced or eliminated not only by adsorption but also by complete or partial exclusion of the polarographically active group from the conjugated chain. In all probability there is a competition between the effects of conjugation and steric interactions in the molecules of the compounds examined, leading to noncoplanarity. It was therefore necessary to check the ability of the hydrazone group to conduct electronic effects.

For this purpose we investigated the polarographic behavior of the following series:

I. $NO_2-\langle\ \rangle-NH-N=CH-\langle\ \rangle_X$

II. $X-\langle\bigcirc\rangle-NH-N=CH-\langle\bigcirc\rangle-NO_2$

III. $\langle\bigcirc\rangle-NH-N=CH-\langle\bigcirc\rangle$ with X below left ring and O_2N below right ring

IV. $\langle\bigcirc\rangle-NH-N=CH-\langle\bigcirc\rangle-X$ with NO_2 below left ring

The correlation analysis was performed at pH 3.0 and showed that in all the series the half-wave potential did not depend on the nature of the substituent. This indicates that the hydrazone group did not conduct electronic displacements from one aromatic ring to another, although spectral data indicated some decrease in absorbance. This can be explained by the fact that rearrangement of the electronic structure takes place in the adsorbed molecule under the influence of the electrode field, leading to uncoupling of the aromatic rings.

Less ambiguous conclusions can be reached if the polarographic characteristics obtained under conditions which exclude adsorption and the effect of protolytic equilibria are compared. We therefore carried out such an analysis for the half-wave potential of the first wave formed by these compounds in the polarography of dimethyl-formamide solutions. In fact, under these conditions, we were able to observe the transmission of the substituent effect in the first series, which in accordance with the views expressed earlier indicated decrease of conjugation ($\rho = +0.1$). The polarographic data can thus be explained if a nonplanar conformation of the hydrazone molecules is adopted.

Correlations of $E_{1/2}$ values and the polar substituent constants, which allow certain information to be obtained on the spatial distribution of the molecules, have also been obtained recently for other series of azomethines. Thus, in the work by Davydovskaya and Vainshtein [49], the effects of substituents in the m- and p-positions of the phenyl group bound to nitrogen in Schiff's bases on the half-wave potentials in dimethylformamide solution were expressed by the Hammett equation, and the points lay close to a straight line although the ρ_π values were small (0.02-0.3); this was one of the arguments in favor of the conclusion about the noncoplanarity of the molecules in these compounds. The ρ_π value for o-tosylaminobenzalanilines is considerably lower than that for salicylanilines. This is explained by the fact that cleavage of the intramolecular hydrogen bond takes place in dimethylformamide solution, and as a result the noncoplanarity of the molecules increases. (The unshared electron pair of the nitrogen is not included in the quasi-aromatic ring.)

A satisfactory linear relationship between $E_{1/2}$ and the Hammett constant was also obtained by Bezuglyi and co-workers [42, 43] for various groups of Schiff's bases in dimethylformamide solution. By comparing the results from optical and polarographic investigations they showed [42] that the presence of an intramolecular hydrogen bond in the azomethines, like increased conjugation in these molecules, causes a bathochromic effect and increased absorbance. These facts cause a decrease of electron density at the azomethine group and facilitate its reduction.

The relationship between $E_{1/2}$ for the anils and their optical characteristics has also been investigated by Davydovskaya, Vainshtein, and Bolotin [182]. In order to make wider use of polarography for the solution of structural problems in organic chemistry it is necessary above all to make detailed investigations of the elementary events in the electrode process, where the structural peculiarities of the depolarizer molecules are most completely revealed. Only in this way is it possible to reach a quantitative assessment of the reactivity of compounds by the polarographic method.

LITERATURE CITED

1. Azomethines: Structure, Properties, and Use [in Russian], Izd. RGU, Rostov (1967).
2. J. K. Wolfe, E. B. Hershberg, and L. F. Fieser, J. Biol. Chem., 136:653 (1940); Chem. Abs., 35:1434 (1941).
3. J. M. Kolthoff and A. Langer, J. Am. Chem. Soc., 62:211 (1940).
4. C. L. Perrin, Progress in Physical Organic Chemistry, Vol. 3, Interscience Publishers, New York (1965), p. 246.
5. J. Volke, Talanta, 12:1081 (1965).
6. P. Zuman, Nature, 165:485 (1950); Coll. Czechosl. Chem. Communs., 15:839 (1950).
7. P. Zuman, Sborník I Mezinarodního Polarografického Sjezdu v Praze, Praha, 1:704 (1951).
8. R. E. Van Atta and D. A. Jamieson, Analyt. Chem., 31:1217 (1959).
9. D. V. Sokol'skii, V. P. Shmonina, and G. V. Taneeva, Zavod. Lab., 30:793 (1964).

10. M. E. Hall, Analyt. Chem., 31:2007 (1959).
11. M. U. Yanotovskii, V. G. Mairanovskii, and G. I. Samokhvalov, Zh. Fiz. Khim., 38:2995 (1964).
12. N. F. Alekseev, Zavod. Lab., 24:684 (1958).
13. Ya. I. Tur'yan and B. P. Zhantalai, Zavod. Lab., 27:1211 (1961).
14. D. K. Banerjee, G. C. Riechmann, and C. C. Budke, Analyt. Chem., 36:2220 (1964).
15. Yu. S. Ignat'ev, Zavod. Lab., 31:1188 (1965).
16. P. Zuman, Chem. Listy, 46:516 (1952).
17. P. Zuman and M. Březina, Chem. Listy, 46:599 (1952).
18. M. Březina, Analyt. Chem., 24:916 (1952).
19. M. Březina and P. Zuman, Chem. Listy, 47:975 (1953).
20. P. Zuman, Chem. Listy, 45:65 (1951).
21. P. Zuman, Chem. Listy, 46:521 (1952).
22. M. Březina, V. Volková, and I. Volke, Chem. Listy, 48:194 (1954).
23. P. Favero, Atti Accad. Naz. Lincei. Rend., Cl. Sci. Fis., Mat. e Natur, 14:433, 520 (1953).
24. D. R. Norton and N. H. Furman, Analyt. Chem., 26:1116 (1954).
25. M. T. Falqui, Rend. Seminar, Fac. Sci. Univ. Cagliari, 26:186 (1956).
26. Y. Ogata and A. Kawasaku, Tetrahedron, 20:855, 1573 (1964).
27. P. J. Elving and R. E. Van Atta, J. Electrochem. Soc., 103:676 (1956).
28. H. Lund, Acta Chem. Scand., 13:249 (1959).
29. P. Zuman, Chem. Listy, 46:11, 668 (1952).
30. W. Ried and M. Wilk, Ann., 590:91 (1954).
31. L. Holleck and B. Kastening, Z. Electrochem., 60:127 (1956).
32. B. Kastening, L. Holleck, and G. Melkonian, Z. Electrochem., 60:130 (1956).
33. H. Lund, Studier Over Electrodereaktioner i Organisk Polarografi og Voltammetri, Aarhus (1961), p. 45.
34. S. Ono and M. Uehara, J. Electrochem. Soc. Japan, Overseas Ed., 27:25 (1959); Bull. Univ. Osaka Prefecture, Ser. A, 7:193 (1959); Chem. Abs., 54:7376i (1958).
35. M. Deželič, A. Lackovič, and M. Trkovnik, Croat. Chem. Acta, 32:31 (1960).
36. V. D. Bezuglyi, in: Methods for Analysis of Chemical Solutions and Preparations [in Russian], No. 3, IREA, Moscow (1962), p. 7.
37. V. D. Bezuglyi, L. V. Konenenko, V. N. Dmitrieva, and E. V. Titov, in: Polarography and Kinetics of Chemical Reactions [in Russian], No. 6, IREA, Moscow (1964), p. 96.
38. V. D. Bezuglyi, V. N. Dmitrieva, and L. V. Kononenko, Papers of Fifth Conference on Electrochemistry of Organic Compounds [in Russian], Nauka, Moscow (1964), p. 9.
39. V. N. Dmitrieva, L. V. Kononenko, and V. D. Bezuglyi, Teor. Eksper. Khim., 1:456 (1965).
40. V. D. Bezuglyi, V. N. Dmitrieva, and L. V. Skvortsova, Kinetika i Kataliz, 6:737 (1965).

41. V. N. Dmitrieva, N. P. Shimanskaya, and V. D. Bezuglyi, in: Polarography and pH Measurement [in Russian], Kharkov State University, Kharkov (1966), p. 32.
42. N. F. Levchenko, L. Sh. Afanasiadi, and V. D. Bezuglyi, Zh. Obshch. Khim., 37:666 (1967).
43. L. V. Kononenko, V. D. Bezuglyi, and V. N. Dmitrieva, in: Advances in Electrochemistry of Organic Compounds [in Russian], Nauka, Moscow (1968), p. 57.
44. I. A. Rozanel'skaya, V. N. Dmitrieva, B. I. Stepanov, and V. D. Bezuglyi, in: Advances in Electrochemistry of Organic Compounds [in Russian], Nauka, Moscow (1968), p. 59.
45. Yu. I. Vainshtein, Yu. A. Davydovskaya, and B. M. Bolotin, Papers of Fifth Conference on Electrochemistry of Organic Compounds [in Russian], Nauka, Moscow (1964), p. 8.
46. B. M. Bolotin, Yu. A. Davydovskaya, Yu. I. Vainshtein, and V. G. Brudz', in: Chemical Reagents and Preparations [in Russian], No. 27, IREA, Moscow (1965), p. 283.
47. Yu. I. Vainshtein and Yu. A. Davydovskaya, in: Chemical Reagents and Preparations [in Russian], No. 27, IREA, Moscow (1965), p. 317.
48. Yu. A. Davydovskaya, Yu. I. Vainshtein, and B. M. Bolotin, in: Chemical Reagents and Preparations [in Russian], No. 30, IREA, Moscow (1967), p. 281.
49. Yu. A. Davydovskaya and Yu. I. Vainshtein, in: Advances in Electrochemistry of Organic Compounds [in Russian], Nauka, Moscow (1968), p. 61.
50. P. O. Kane, Z. Analyt. Chem., 173:50 (1960).
51. M. Kuna and M. I. Kopac, Ann. N. Y. Acad. Sci., New York, 58:261 (1954); Chem. Abs., 49:6176 (1955).
52. A. Langer, Ind. Eng. Chem., Analyt. Ed., 14:283 (1942).
53. C. Carruthers, Ind. Eng. Chem., Analyt. Ed., 17:398 (1945).
54. K. G. Stone and N. H. Furman, J. Am. Chem. Soc., 70:3062 (1948).
55. G. Sartori and A. Gaudiano, Gazz. Chim. Ital., 78:77 (1948).
56. E. D. Hartnell and C. E. Bricker, J. Am. Chem. Soc., 70:3385 (1948).
57. C. Calzolari, Bull. Soc. Adriat. Sci. Nat. Trieste, 45:109, 116, 171 (1949-1950); Pubbl. Facolta Sci. Ing-Univ. Trieste, Ser. B, 81:1; 82:3; 83:3 (1950).
58. C. Calzolari and C. Furlani, Ann. Triest. Cura Univ. Trieste, Ser. 2, 21:47 (1951); Chem. Abs. 47:4220i (1953).
59. C. Calzolari and C. Furlani, Bull. Sci. Fac. Chim. Ind. (Bologna), 12:50 (1954); Chem. Abs., 47:420i (1953).
60. C. Calzolari and C. Furlani, Ann. Triest. Cura Univ. Trieste, Ser. 2, 22-23:43 (1953); Chem. Abs., 49:938e (1955).
61. P. Souchay and S. Ser, J. Chim. Phys., 49:172 (1952).
62. H. J. Gardner and W. P. Georgans, J. Chem. Soc., 4180 (1956).
63. H. Lund, Acta Chem. Scand., 18:563 (1964).
64. P. Zuman and O. Exner, Coll. Czechosl. Chem. Communs., 30:1832 (1965).
65. C. A. Benassi, E. Fornasari, L. Griggio, and F. M. Veronese, Gazz. Chim., Ital., 94:467 (1964).
66. K. S. Tewari, Indian J. Chem., 3:204 (1965).

67. V. M. Chursina, E. F. Litvin, and L. Kh. Freidlin, Zh. Analit. Khim.,
     21:365 (1966).
68. Ya. P. Stradyn', I. Ya. Kravis, and N. O. Saldabol, Zh. Obshch. Khim.,
     37:977 (1967).
69. N. Tütülkoff (Tyutyulkov) and St. Buduroff (Budurov), Dokl. Bolg. Akad. Nauk,
     6:5 (1953).
70. N. Tyutyulkov and St. Budurov, Izv. Khim. Inst. Bulgar. Akad. Nauk, 3:361
     (1955).
71. N. Tütülkoff (Tyutyulkov) and L. Stefanova, Dokl. Bolg. Akad. Nauk, 9:49
     (1956).
72. N. Tütülkoff (Tyutyulkov) Zh. Fiz. Khim., 32:1389 (1958).
73. N. Tütülkoff (Tyutyulkov) and B. Panayotova, Dokl. Bolg. Akad. Nauk, 11:201
     (1958).
74. N. Tütülkoff (Tyutyulkov) and E. Paspaleev, Bulg. Akad. Nauk. Izvest. Khim.
     Inst., 6:389 (1958); Chem. Abs., 54:5295g (1960).
75. N. Tütülkoff (Tyutyulkov) and I. Bakardziev, Dokl. Bolg. Akad. Nauk, 12:133
     (1959).
76. N. Tütülkoff (Tyutyulkov) and E. Paspaleev, Dokl. Bolg. Akad. Nauk, 14:159
     (1961).
77. N. Tütülkoff (Tyutyulkov) and V. Palikarov, Dokl. Bolg. Akad. Nauk, 15:399
     (1962).
78. N. Tütülkoff (Tyutyulkov), K. Tsvetanov, and V. Stamatova, Dokl. Bolg. Akad.
     Nauk, 16:389 (1963).
79. E. Paspaleev, Monatsh. Chem., 97:230 (1966).
80. J. Nakaya, K. Mori, H. Kinoshita, and S. Ono, Nippon Kagaku Zasshi, 80:1212
     (1959).
81. W. Kemula and J. Zawadowska, Roczn. Chem., 40:1303 (1966).
82. J. Tirouflet and P. Fournari, C. R. Acad. Sci., 248:1182 (1959).
83. J. Tirouflet, P. Fournari, and J. P. Chane, Compt. Rend., 242:1799 (1956); Chem.
     Abs., 50: 10562e (1956).
84. J. Volke, R. Kubiček, and F. Šantavý,Coll. Czechosl. Chem. Communs., 25:871
     (1960).
85. S. G. Mairanovskii, Papers of Fifth Conference on Electrochemistry of Organic Com-
     pounds [in Russian], Nauka, Moscow (1964), p. 5.
86. N. G. Lordi and E. M. Gohen, Analyt. Chem. Acta, 25:281 (1961).
87. E. Laviron, M. Person, and P. Fournari, Compt. Rend., 255:2440 (1962).
88. J. Vachek, Českosl. Farmac., 11:469 (1962).
89. O. Manousek and P. Zuman, Coll. Czechosl. Chem. Communs., 29:1432 (1964).
90. J. W. Haas, I. D. Storey, and C. C. Lynch, Analyt. Chem., 34:145 (1962).
91. J. W. Haas and R. E. Kadunce, J. Am. Chem. Soc., 84:4910 (1962).
92. M. Bartusek and A. Okáč, Coll. Czechosl. Chem. Communs., 26:52 (1961).
93. M. Ishibashi, T. Fujinaga, and K. Kawamura, Bull. Chem. Soc., Japan, 26:513
     (1953).

94. A. D. Monnier and W. Haerdi, Helv. Chim. Acta, 41:2205 (1958).
95. M. Spritzer and L. Meites, Analyt. Chim. Acta, 26:58 (1962).
96. R. I. Gelb and L. Meites, J. Phys. Chem., 68:2599 (1964).
97. R. M. Elofson and J. G. Atkinson, Can. J. Chem., 34:4 (1956).
98. Th. Osterud and M. Prytz, Acta Chem. Scand., 10:451 (1956).
99. M. Prytz and Th. Osterud, Acta Chem. Scand., 11:1530 (1957).
100. Th. Osterud and M. Prytz, Acta Chem. Scand., 13:2114 (1959).
101. N. K. Yusupova and A. L. Markman, Uzb. Khim. Zh., 36 (1963).
102. B. V. Matveev and G. G. Tsybaeva, Zh. Obshch. Khim., 34:2491 (1964).
103. J. Mollin and F. Kašpárek, Coll. Czechosl. Chem. Communs., 25:451 (1960).
104. J. Mollin and F. Kašpárek, Coll. Czechosl. Chem. Communs., 26:2438 (1961).
105. M. Kuraš and J. Mollin, Chem. Listy, 52:344 (1958).
106. G. C. Whitnack, J. E. Young, H. H. Sisler, and E. St. C. Gantz, Analyt. Chem., 28:833 (1956).
107. N. P. Shimanskaya, L. Ya. Malkes, and V. D. Bezuglyi, Zh. Obshch. Khim., 33:2094 (1963).
108. B. D. Bezuglyi and N. P. Shimanskaya, Zh. Obshch. Khim., 35:17 (1965).
109. J. Barnett and C. J. O. Morris, Biochem. J., 40:450 (1946).
110. M. Březina, V. Volková, and J. Volkova, and J. Volke, Coll., 19:894 (1954).
111. V. Prelog and O. Häfliger, Helv. Chim. Acta, 32:2088 (1949).
112. J. R. Young, J. Chem. Soc., 1516 (1955).
113. B. A. Arbuzov and E. A. Berdnikov, Izv. Akad. Nauk SSSR, Otd. Khim. Nauk, 165 (1962).
114. M. Masui and H. Ohmori, Chem. Pharm. Bull., 12:877 (1964).
115. M. Masui and H. Ohmori, J. Chem. Soc., 3951 (1964).
116. J. M. Lupton and C. C. Lynch, J. Am. Chem. Soc., 62:211 (1940).
117. J. M. Lupton and C. C. Lynch, J. Am. Chem. Soc., 66:697 (1944).
118. J. W. Haas and C. C. Lynch, Analyt. Chem., 29:479 (1957).
119. Yu. P. Kitaev and A. E. Arbuzov, Izv. Akad. Nauk SSSR, Otd. Khim. Nauk, 1037 (1957).
120. Yu. P. Kitaev and A. E. Arbuzov, Izv. Akad. Nauk SSSR, Otd. Khim. Nauk, 1405 (1960).
121. Yu. P. Kitaev, in: Theory and Practice of Polarographic Analysis [in Russian], Kishinev State University, Kishinev (1962), p. 88.
122. Yu. P. Kitaev and T. V. Troepol'skaya, Izv. Akad. Nauk SSSR, Otd. Khim. Nauk, 454 (1963).
123. Yu. P. Kitaev and T. V. Troepol'skaya, Izv. Akad. Nauk SSSR, Otd. Khim. Nauk, 465 (1963).
124. R. O'Connor, J. Org. Chem., 26:4375 (1961).
125. R. R. Shagidullin, F. K. Sattarova, T. V. Troepol'skaya, and Yu. P. Kitaev, Izv. Akad. Nauk SSSR, Otd. Khim. Nauk, 473 (1963).
126. R. R. Shagidullin, F. K. Sattarova, N. V. Semenova, T. V. Troepol'skaya, and Yu. P. Kitaev, Izv. Akad. Nauk SSSR, Otd. Khim. Nauk, 633 (1963).

127. Yu. P. Kitaev and T. V. Troepol'skaya, Izv. Akad. Nauk SSSR, Ser. Khim.,
     1903 (1967).
128. Yu. P. Kitaev, G. K. Budnikov, T. V. Troepol'skaya, and I. M. Skrebkova,
     Zh. Obsch. Khim., 37:1437 (1967).
129. Z. V. Pushkareva and M. K. Murshtein, Dokl. Akad. Nauk SSSR, 156:389 (1964).
130. S. G. Mairanovskii, D. I. Dzhaparidze, and O. I. Sorokin, Izv. Akad. Nauk
     SSSR, Otd. Khim. Nauk, 795 (1964).
131. V. F. Lavrushin, V. D. Bezuglyi, G. G. Belous, and V. G. Tishchenko, Zh.
     Obshch. Khim., 24:7 (1964).
132. V. G. Tishchenko, G. G. Belous, V. D. Bezuglyi, and V. F. Lavrushin, in: Polaro-
     graphy and Kinetics of Chemical Reactions [in Russian], IREA, Moscow (1964),
     p. 70.
133. Yu. P. Kitaev and I. M. Skrebkova, Papers of Fifth Conference on Electrochemistry
     of Organic Compounds [in Russian], Nauka Moscow (1964), p. 13.
134. Yu. P. Kitaev and I. M. Skrebkova, Zh. Obshch. Khim., 27:1204 (1967).
135. Yu. P. Kitaev and I. M. Skrebkova, Zh. Obshch. Khim., 27:1198 (1967).
136. A. Kolusheva, N. Nin'o, and V. Koen, Izv. Khim. Inst. Bulg. Akad. Nauk, 7:27
     (1960).
137. A. Kolusheva and N. Nin'o, Farmatsiya (Bulgarian), 16:(9) 25 (1966).
138. Yu. P. Kitaev and G. K. Budnikov, Izv. Akad. Nauk SSSR, Ser. Khim., 554
     (1967).
139. Yu. P. Kitaev and G. K. Budnikov, Izv. Akad. Nauk SSSR, Ser. Khim., 562
     (1967).
140. G. Alessandro and E. Mecarelli, Bull, Chim. Farmac., 103:427 (1964).
141. B. Bobarevič, M. Deželič, and V. Javanovič, Glasnik Hem. i Tehnol., Bi. H,
     47, (1964-1965).
142. B. Jámbor and L. Mester, Acta Chim. Hungar., 9:485 (1956).
143. B. Jámbor and K. Kisban, Acta Chim. Hungar., 9:493 (1956).
144. Yu. P. Kitaev and I. M. Skrebkova, in: Advances in Electrochemistry of Organic
     Compounds [in Russian], Nauka, Moscow (1968), p. 62.
145. I. Tate, Japan. J. Pharm. Chem., 20, 38 (1948); Chem. Abs., 45:3257a (1949).
146. R. W. Brockman and D. E. Pearson, J. Am. Chem. Soc., 74:4128 (1952).
147. T. Sasaki, Pharm. Bull., 2:99 (1954); Chem. Abs., 49:11460g.
148. K. Mňouček and E. Knobloch, Ceskosl. Farmac., 2:306 (1953).
149. F. Icha, Českosl. Farmac., 2:308 (1953).
150. G. Dušinský, Pharmazie, 8:897 (1953).
151. P. Souchay and M. Graizon, Chim. Analyt., 36:85 (1954).
152. Yu. P. Kitaev, G. K. Budnikov, and A. E. Arbuzov, Izv. Akad. Nauk SSSR, Otd.
     Khim. Nauk, 824 (1961).
153. Yu. P. Kitaev, G. K. Budnikov, and A. E. Arbuzov, Izv. Akad. Nauk SSSR, Otd.
     Khim. Nauk, 1772 (1961).
154. Yu. P. Kitaev, G. K. Budnikov, and I. M. Skrebkova, Izv. Akad. Nauk SSSR, Otd.
     Khim. Nauk, 244 (1962).

155. Yu. P. Kitaev and G. K. Budnikov, Izv. Akad. Nauk SSSR, Otd. Khim. Nauk, 978 (1964).
156. Yu. P. Kitaev and G. K. Budnikov, Zh. Obshch. Khim., 33:1396 (1963).
157. G. K. Budnikov, S. Yu. Baigil'dina, and Yu. P. Kitaev, Élektrokhimiya, 2:1263 (1966).
158. J. M. Van der Velen, J. Org. Chem., 28:564 (1963).
159. L. Vignoli, B. Cristau, F. Gouezo, and C. Fabre, Chim. Analyt., 45:439 (1963).
160. J. Asahi, Chem. Pharm. Bull. 11:930 (1963); Chem. Abs., 59:10989e.
161. D. M. Coulson, Analyt. Chim. Acta, 19:284 (1958).
162. A. V. Khoroshin, Zavod. Lab., 28:420 (1961).
163. B. Fleet, Analyt. Chim. Acta, 36:304 (1966).
164. B. Fleet and P. Zuman, Coll. Czechols. Chem. Communs., 32:2066 (1967).
165. J. Doskočil and D. Doskočilová, Rozpr. M. Tridy Česke Akad., 60:12 (1950).
166. D. Jerchel and R. Kuhn, Ann. 568:185 (1950).
167. B. Jámbor, Acta, Chim. Hungar., 4:55 (1954); Chem. Abs. 48:12580e (1952).
168. B. Jámbor and L. Mester, Acta Chim. Hungr., 6:263 (1955).
169. B. Jámbor, Nature, 176:603 (1955).
170. B. Jámbor and E. Bajucz, Agrokémia es Talajtan, 5:127 (1956); Chem. Abs., 50:15283i (1956).
171. B. Jámbor and L. Mester, Agrokémia es Talajtan, 5:127 (1956).
172. B. Jámbor, Tetrazoliumsalze in der Biologie, Gustav-Fischer-Verlag, Jena (1960).
173. B. Jámbor, in: Die Polarographie in der Chemotherapie, Biochemie und Biologie, I. Jeanaer Symposium, Berlin (1964), p. 49.
174. H. Campbell and P. O. Kane, J. Chem. Soc., 3130 (1956).
175. P. Kivalo and U. K. Mustakallio, Suomen Kem., 29:13154 (1956).
176. R. Ralea and M. A. Petrovanu, Rev. Chem., 1:69 (1956); Chem. Abs., 52: 133d (1958).
177. R. Ralea, M. A. Petrovanu, and A. Papp. Stud. Cerc. Sti. Chim. Acad. R. P. Rum., 7:131 (1956).
178. G. Giacometti, Ric. Sci., 26:2167 (1956); Chem. Abs., 51:48:844a (1957).
179. T. Arkawa and Y. Oguro, J. Exper. Med., 3:107 (1956).
180. V. M. Ostrovskaya, Yu. A. Davydovskaya, A. A. Pryanishnikov, Yu. I. Vain-shtein, and V. M. Dziomko, Zh. Obshch. Khim., 35:230 (1965).
181. A. E. Arbuzov, Yu. P. Kitaev, and R. R. Shagidullin, Izv. Akad. Nauk. SSSR, Ser. Khim., 1198 (1967).
182. Yu. A. Davydovskaya, Yu. I. Vainshtein, and B. M. Bolotin, Trudy IREA, No. 30, 281 (1967).

Electrode Processes Complicated by
Chemical Reactions and by Adsorption
Mercury Electrodes

S. G. Mairanovskii

The number of research papers in which electrode processes involving chemical steps are studied have increased significantly in recent years. Whereas after publication of the first papers on kinetic currents by Brdička, Wiesner, and co-workers [1-7] electrode processes involving chemical stages appeared to be rare exceptions, we now know that in the majority of cases electron transfers, particularly to organic molecules, are accompanied by chemical reactions taking place near the electrode. As a result of the appreciable adsorbability of many organic compounds at the electrode – solution interface, adsorption phenomena greatly affect both the electrochemical and the chemical reactions.

In the present review a number of new examples are given of electrode processes involving chemical steps and adsorption of components in electrode and chemical reactions, some recently discovered types of chemical reactions in the vicinity of the electrodes are examined, and new views are put forward on the nature of some complex electrode processes. In order to complete the picture, in addition to data from the researches of recent years, use is also made of the results from some of the earlier investigations.

1. ELECTRODE PROCESSES DETERMINED BY PROTONATION REACTIONS

Electron transfers preceded by protonation are observed in the electrolytic reduction of various classes of organic compounds (see for example [8, 9]), and the protonation rate frequently determines the nature of the whole electrode process. This effect arises, for example, in the polarography of nitro compounds.

Protonation in Electrochemical
Reduction of Nitro Compounds

It has been shown in a number of papers [10-13] that when surface-active substances are added to the polarographed solution, the
four-electron diffusion waves for the reduction of a number of nitro
compounds in weakly alkaline aqueous solution decrease to a one-
electron level and that simultaneously, at a more negative potential,
a new three-electron wave appears, so that the overall process
corresponds to the reduction of the nitro compound to the respective
hydroxylamine derivative. The decrease of the four-electron wave
to the one-electron step has also been observed when switching from
aqueous solutions to solutions in aprotic solvents — acetonitrile, dimethylformamide [14-16]. This phenomenon is due to retarded protonation of the radical anion derived from the nitro compound in the
first electron transfer; the surface-active substances added to the
aqueous solution displace the radical anions from the electrode surface [12, 13, 17], thereby preventing their protonation at the surface
[18].

It was recently found [19] that with the reduction of 2-nitrofuran
and its derivatives, splitting of the four-electron wave is observed
when ethanol is added to neutral buffer solutions. In an investigation
of the effect of ethanol on the polarographic behavior of aromatic
nitro compounds it was found [20] that the waves for nitrobenzene
derivatives capable of undergoing acidic dissociation (i. e., containing hydroxyl or carboxyl substituents in the ring) split in neutral or
alkaline solutions even when they contain 10% ethanol (Fig. 1).

With increase in the pH value the wave height does not fall to
zero but to a quarter of its initial value, i. e., to the level corresponding to a one-electron diffusion current. As emphasized in [20],
this fact contradicts the point of view of some authors [21, 22] who
previously observed splitting of the wave for nitrobenzoic acids. In
the opinion of these authors, the first wave corresponds to reduction
of the undissociated nitrobenzoic acid and the second to its anion,
formed as a result of dissociation of the carboxyl group. With aqueous
solutions, however, such splitting is only observed in the presence of
surface-active substances — e. g., in the Prideau-Ward buffer [23],
containing the strongly adsorbed phenylacetic acid. Person [24]

suggested that the observed splitting of the wave for nitrobenzoic acids is explained in the same way as the splitting of the waves for other aromatic nitro compounds, i. e., by retarded protonation of the radical anion formed during the first electron transfer. During dissociation of the carboxyl group in nitrobenzoic acid (or the radical anion formed from it) its adsorbability on the cathode surface decreases sharply [20], and it is clear moreover that, with other conditions equal, the rate of a cathodic process involving a particle which has increased its negative charge becomes lower.

Both these factors promote splitting of the waves for nitrobenzoic acids and nitrophenols in alkaline solution. Similarly dissociation of nitrobenzene derivatives facilitates separation of the waves, but such a process does not necessarily take place; when the ethanol content of the solution was increased to 20% it was possible [20] to observe splitting of the waves for nitrobenzene, p-nitrotoluene, and m-nitroaniline. On the other hand, in the case of halonitrobenzenes there was no separation of the waves observed even with high concentrations of ethanol. As already mentioned, an increase in the ethanol concentration leads to a considerable decrease in the adsorbability of the radical anions and consequently causes a sharp decrease in their protonation rate at the surface.

The relationship between $E_{1/2}$ for the nitro compound reduction waves and the pH of the solution gives an S-shaped curve (see Fig. 1, for example, and also the numerous published data [25]), and consequently in weakly acidic and neutral solutions the electron transfers are preceded by protonation of the nitro group; this is also demonstrated by the dependence of $E_{1/2}$ for the nitro compound reduction waves on the buffer capacity of the solution and the nature of the buffer components [27]. In alkaline solution reduction of unprotonated molecules takes place and is followed by protonation of the radical anion formed. Under conditions involving splitting of the four-electron wave at pH values where the height of the first wave has not fallen to the level of the one-electron diffusion current, the limiting current of the first wave corresponds to the sum of the one-electron diffusion current for formation of the radical anion and the kinetic current limited by the protonation rate of the radical anions and the subsequent three-electron reduction of the radicals formed. Protonation of both nitro group [27] and radical anions [11-13, 20] takes place to

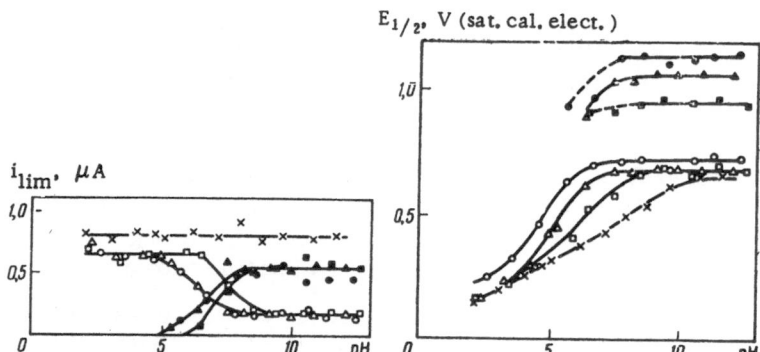

Fig. 1. Dependence of limiting current (i$_{lim}$) and half-wave potential (E$_{1/2}$) of nitro group reduction waves for p-nitrobenzoic acid on pH value of buffer (Britton — Robinson) solutions with various ethanol contents (vol. %)[20]. × × represents 0; □■ represents 10 vol.%; △▲ represents 20 vol.%; ○● represents 40 vol.%. Light points represent the first wave and full points the second wave.

a considerable extent at the electrode surface and involves participation of adsorbed particles.

Splitting of waves which results in separation of a one-electron step has also been observed for aliphatic nitro compounds [28]. Thus, for example, a one-electron polarographic wave was observed for the potassium salt of dinitromethane in an unbuffered solution of lithium perchlorate (Fig. 2). To enable observation of a one-electron wave in solutions of nitroform it is sufficient to add 4% (by volume) of dimethylformamide or acetonitrile to its salt in lithium perchlorate solution as supporting electrolyte (Fig. 2).

Interesting phenomena are observed in the reduction of nitroform (trinitromethane) at the dropping mercury electrode [29]. On polarograms in unbuffered aqueous solutions the first wave corresponds to one-step reduction of the undissociated $(NO_2)_3CH$ molecule, and the second and subsequent waves correspond to multistep reduction of its anion $(NO_2)_3C^-$. Accordingly the polarograms for an unbuffered solution of the potassium salt of nitroform (nitroform is a strong acid with pK$_A$ = 0.15) only contain the multistep wave for the reduction of the anion (Curve 1, Fig. 3), and this remains practically

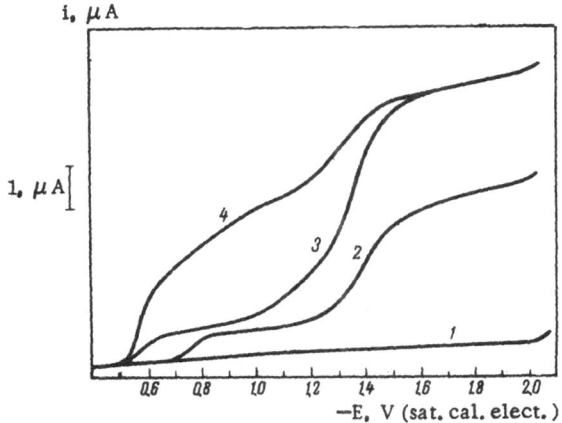

Fig. 2. Polarograms for potassium salts $(5 \cdot 10^{-4}$ mole
/liter) of dinitromethane (2) and nitroform (3 and 4)
with aqueous 0.1 N lithium perchlorate solution as
supporting electrolyte.   3) With addition of 4% dimethylformamide; 1) supporting electrolyte.

unchanged even when the solution is made considerably alkaline
(Curves 8 and 9, Fig. 3).   However, even with very low concentrations of sulfuric acid added to a neutral solution of the potassium
salt of nitroform at constant concentration, a more positive wave
corresponding to the reduction of undissociated nitroform appears on
the polarogram (Curve 2); with further increase in the sulfuric acid
concentration this wave increases and the waves for the reduction of
the anion decrease (Curves 2-6, Fig. 3).   It is interesting that the
limiting current of the reduction wave for undissociated nitroform
under these conditions has a purely diffusion character at any sulfuric acid concentration.   However, if the equilibrium amount of undissociated nitroform is determined at a given sulfuric acid concentration and the diffusion current corresponding to the reduction of
only undissociated nitroform is calculated, the observed wave is
found to be much larger than the calculated.   Comparison of the wave
height for nitroform at a given sulfuric acid concentration with the
wave for discharge of hydrogen ions in the same solution but without
the nitroform (Curves 2'-6', Fig. 3) shows that these waves are equal

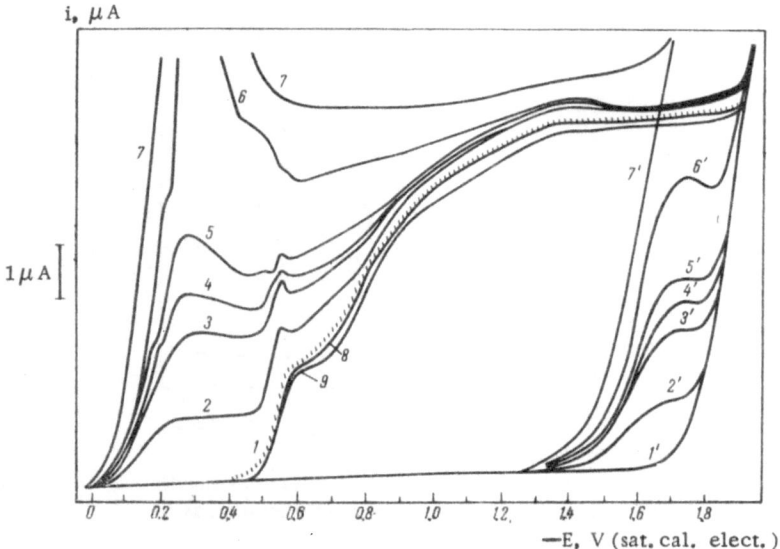

Fig. 3. Polarograms for nitroform (c = $5 \cdot 10^{-4}$ mole/liter). 1) In aqueous 0.1 N potassium sulfate solution; 2)-7) in presence of sulfuric acid at concentrations of $9.2 \cdot 10^{-4}$, $10.4 \cdot 10^{-4}$, $17 \cdot 10^{-4}$, and $90 \cdot 10^{-4}$ g-equiv/liter respectively; 8) and 9) in presence of 0.02 and 0.16 N potassium hydroxide; 2')-7') waves corresponding to reduction of hydrogen ions in absence of nitroform and at the same sulfuric acid concentrations as 2)-7); 1') current for supporting electrolyte in aqueous 0.1 N potassium sulfate solution.

in height. With comparatively large sulfuric acid concentrations ($c_{H_2SO_4}$) a polarographic maximum develops on the nitroform wave (which distorts the wave), and in this case the current after the decrease of the maximum should be taken as the i_{lim} value of the wave.

The equality of the wave heights for the reduction of nitroform and hydrogen ions shows that the protonation rate of nitroform anions by hydrogen ions (undissociated nitroform is formed here) is very high and does not affect the height of the first wave. In other words, the first wave on the polarograms for the nitroform salt in the presence of small quantities of strong acid has a quasidiffusion character [9], and with a relative excess of nitroform ions its limiting current is practically equal to the diffusion current of the hydrogen ions. With considerable increase in $c_{H_2SO_4}$ the anion reduction waves disappear, and the one-step wave for the reduction of nitroform reaches a certain maximum and then undergoes no further change with increase in $c_{H_2SO_4}$; after the nitroform wave, however, a wave for the discharge of the superfluous hydrogen ions appears and increases. Under these conditions the first wave is also quasidiffusion in nature (the acidity of the solution is still insufficient for a significant shift of equilibrium towards the undissociated form, so that in this case too protonation of the anions takes place in the layer near the electrode), but its height is determined by diffusion of the nitroform anions.

The quasidiffusion character of the first wave observed both at hydrogen ion concentrations lower and higher than equivalent results, in accordance with Eqs.(71) or (125) in [9], with the shift of the wave to more positive potentials both with increase in $c_{H_2SO_4}$ (with constant concentration of the nitroform salt) and with increase in the content of nitroform anions in the solution at $c_{H_2SO_4}$ = constant. Here, when the current is quasidiffusion in nature with respect to hydrogen ions (i.e., under conditions with a relative excess of anions), the ratio of the concentration of the electrochemically inactive to the active form takes the form $\sigma = [H^+]/[HA]$, whereas with an excess of hydrogen ions it takes the form $\sigma = [A^-]/[HA]$.

The half-wave potential for reduction of nitro compounds as a function of the pH value of the solution is determined to a greater or lesser degree by the rates of the preceding protonation and electron transfer, and both of these processes are affected substantially by adsorption of nitro compounds at the electrode surface. This is why, as shown by Stradyn' and his co-workers [30-31], the Hammett reaction constant ρ_π for the half-wave potentials, which expresses the susceptibility of the nitro group to the electronic effects of the substituents in the 2-nitrofuran and nitrobenzene rings, depends to a considerable degree on the pH of the medium and the concentration of ethanol in the solution. The ρ_π values are comparatively small in strongly acidic and alkaline solutions and increase appreciably in neutral solutions, particularly when the concentration of ethanol in the solution is increased (Fig. 4). The variation of ρ_π with the pH of the medium is explained by the different effects of the substituents, on the one hand, on the protonation rate of the nitro group and radical

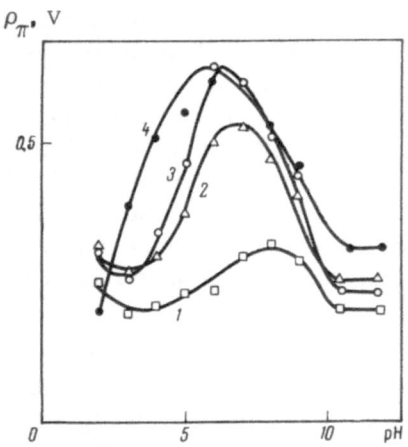

Fig. 4. Dependence of Hammett reaction
constant ρ_π on pH for $E_{1/2}$ values of first reduction wave of series of nitrobenzene derivatives in solutions with various contents of
ethanol [31]: 1) 0; 2) 10 vol. %; 3) 20 vol. %;
4) 40 vol. %.

anion and, on the other, on the electron transfer rate [30, 31]. Increase in the ethanol concentration clearly affects the adsorbability
and also the kinetics of protonation and electron transfer to different
degrees in different representatives of the reaction series — 2-nitro-
furan and nitrobenzene derivatives. The increase in the ρ_π values
with increase in the ethanol content of the solution may also arise
from the greater effect of substituents in the unadsorbed molecules [31].

It should be noticed that the hydroxylamine derivatives formed
as a result of the four-electron reduction of nitro compounds are
further reduced (to amines) only after protonation, which also has a
surface character [20].

Chemical Reactions in Reduction of

Aldehydes and Ketones

A few years ago it was shown that under certain conditions it
was possible, in the reduction of acetophenone, to observe a kinetic

wave, the height of which was limited by its protonation rate at the electrode surface [32]. In the case of cyclohexenone (where the carbonyl group is conjugated with the double bond of the ring) it was possible to trace a relationship between the height of the kinetic wave, limited by its protonation rate, and the pH value of the solution; as with classical examples of waves for weak acids, the form of this relationship is reminiscent of a dissociation curve [33]. A kinetic wave due to retarded protonation was observed in the polarography of 2-acetylfuran [34], and also in the case of benzaldehyde, acetophenone, and ferrocenealdehyde in anhydrous ethanolic buffer solutions [35].

Decrease in the limiting current and the acquisition of kinetic character with increase in the pH of the solution between 4.5 and 5.5 was observed for the first wave of propiophenone and p-hydroxypropiophenone [36]. In the case of p-hydroxypropiophenone the first wave disappears completely even at pH 6, and there are no waves at all on the polarograms right up to pH 7.0 since the reduction wave for the unprotonated p-hydroxypropiophenone is overlapped by the current of the supporting electrolyte. (Introduction of a OH group in the p-position of propiophenone shifts its wave by 150-160 mV towards more positive potentials [36].) With increase in the pH value of the solution above 7.0 the total heights of the waves for o- and p-hydroxypropiophenones fall on account of dissociation at the hydroxyl group. Owing to the formation of an intramolecular hydrogen bond the wave of o-hydroxypropiophenone can be observed in considerably more alkaline solutions than the wave of the p-isomer, but nevertheless the wave of the o-hydroxy derivative in alkaline solutions is much lower than the wave of unsubstituted propiophenone [36].

The variation in height of the first wave with increase in the pH value, reminiscent of a dissociation curve in form, with the simultaneous appearance and growth of a second wave also occurs in the reduction of β-dicarbonyl compounds, particularly with 2-arylindane-1,3-diones [37-38]. This phenomenon has been investigated particularly thoroughly in the case of 2-phenylindanedione [39]. In strongly acidic solutions there are two waves on the polarograms for this compound; the height of the first of them ($E_{1/2}$ in the vicinity of -1.0 V) corresponds to diffusion-controlled current with transfer of two

electrons. The height of this wave decreases with increase in pH
value above 5.0 (Fig. 5) and acquires a kinetic character. At a pH
value of about 7.0 the wave disappears, and the new wave which
forms at a more negative potential attains maximum height. The
pH value at which the height of the first wave falls to half its initial
value (i. e., the "apparent" or "polarographic" dissociation constants
pK') under various experimental conditions, particularly in solutions
with different ethanol contents, is two to three units greater than the
true values of the protolysis constants pK_A. Consequently the height
of the kinetic wave is in this case limited by the rate of proton addition to the mesomeric enolate anion, which is the common anion of
the two tautomeric forms (diketo and enol) of 1,3-indanedione [40, 41].
According to data from [41], investigation of the change in limiting
current of the kinetic wave with drop time showed that this wave corresponds either to preceding protonation of the anion in the bulk of
the solution or to surface protonation with the establishment of adsorption equilibrium. The electrocapillary curves indicate considerable adsorption of the anions, which is nevertheless comparatively
small in the region of potentials corresponding to the appearance of
the kinetic wave. (The electrocapillary curves for the supporting
electrolyte and the solution of the anion combine shortly before the
potential corresponding to the appearance of the kinetic current.)
Nevertheless the adsorption is considered quite sufficient to ensure
the surface character of the process [43]. The surface character of
the protonation of 2-phenyl-1,3-indanedione anions is also demonstrated by the appreciable displacement of the $i - E_{1/2}$ plots into the
region of more acidic solutions and especially by the sharp drop in
the anion reduction current when the ethanol content of the solution is
increased (Fig. 5); like neutral organic molecules these anions are
evidently desorbed from the electrode under the influence of the
ethanol.

    During electrochemical reduction of aldehydes and ketones
other chemical reactions take place in addition to protonation in the
region in the vicinity of the electrode. Thus, with aromatic aldehydes and ketones in sufficiently acidic solutions at the potentials of
the first reduction wave the free radicals dimerize to pinacols immediately after the electron transfer, and under conditions where the
electrode process proper is reversible the dimerization affects the

position and form of the first wave [45]. A similar reduction mechanism also exists with other cyclic carbonyl-containing compounds (e. g., [33, 35, 46-50]). However, the reversible nature of the electron transfer in the reduction of aldehydes and ketones can only be observed under rigidly defined conditions where the retarding effect of the dimeric reaction products adsorbed on the electrode is reduced to a minimum [45], e. g., with increase in the concentration of organic solvent in aqueous – organic mixtures [51]. Thus, to obtain a reversible wave with benzophenone it is necessary to increase the ethanol concentration to 40% [52], but in the reduction of 1-benzoylnaphthalene, which gives a strongly adsorbed pinacol, the wave is irreversible even in 60% (by volume) ethanol [53].

Investigation of the reduction of unsaturated aliphatic aldehydes of general formula $CH_3(CH = CH)_n CHO$ has shown [54] that, under conditions where the reaction products have minimum hindering effect (at very low depolarizer concentrations, short drop times, and large

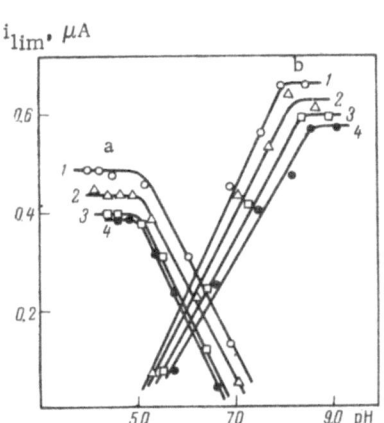

Fig. 5. Dependence of limiting currents for reduction of undissociated molecules (a) and anions (b) of 2-phenyl-1, 3-indanedione on pH of various contents of ethanol in solution [39]: 1) 10 vol. %; 2) 20 vol. %; 3) 30 vol. %; 4) 40 vol. %.

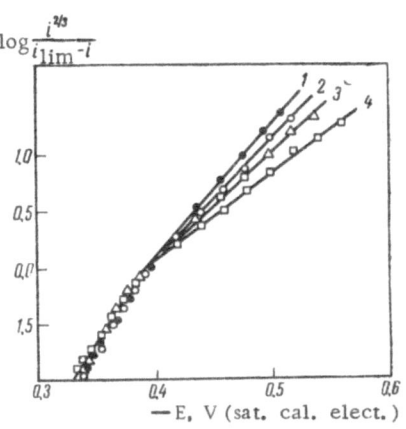

Fig. 6. Dependence of $\log [i^{2/3}/(i_{lim} - i)]$ on E for first wave in reduction of $CH_3(CH = CH)_4CHO$ in 50% dimethylformamide at pH 2.0 (HCl + KCl) and with various aldehyde concentrations: 1) $1.0 \cdot 10^{-4}$ mole /liter; 2) $1.5 \cdot 10^{-4}$ mole/liter; 3) $2.0 \cdot 10^{-4}$ mole/liter; 4) $2.6 \cdot 10^{-4}$ mole/liter.

organic solvent contents), it is also possible to observe in the first
wave of these compounds a reversible one-electron electrochemical
process, followed by bimolecular interaction of the products.   The
slope corresponding to such a process on the curve against the coordinates E and log $[i^{2/3}/(i_{lim} - i)]$ [45], i. e. , 60 mV, can only be
seen in part of the wave at its base (Fig. 6), where the current density
is small and the amount of surface-active products formed on the
electrode is consequently also relatively small.   When the current is
increased the coverage of the surface with reaction products causing
retardation of the electrode process increases, and this gives a bend
in the logarithmic analysis curve.   The higher the depolarizer concentration, the more the upper part of the curve deviates from the
theoretical straight line, but in its lower part the experimental points
obtained at various depolarizer concentrations lie on the same
straight line, in accordance with the theory [45] (Fig. 6).   Reduction
of the aldehydes is facilitated with increase in the length of the poly-
ene chain [54, 55], but at the same time the retarding effect of the
electrode products increases (Fig. 7) on account of their increased
adsorbability, due both to shift of the reduction potential into the region of maximum adsorption and to increase in the size of the particles formed.

In the case of systems where the carbonyl group is conjugated
with a series of double bonds, a shift of multiple bonds is possible in
the course of a multistep reduction process.   The rate of such isomerization of the intermediate products can limit the height of the
first of the succeeding waves on the polarograms.   In the reduction
of aromatic dicarbonyl compounds (terephthalic aldehyde, p-diben-
zoylbenzene, p-diacetylbenzene, p-dibenzoyldiphenyl, trans-diben-
zoylethylene) in buffered aqueous ethanolic solutions the first wave
corresponds under certain conditions to a reversible two-electron
and two-proton process to form a diol structure with a quinoid double
bond system [56].   The next wave which appears after this wave
arises [56] from reduction of the corresponding monocarbonyl compound, formed by rearrangement of the diol:

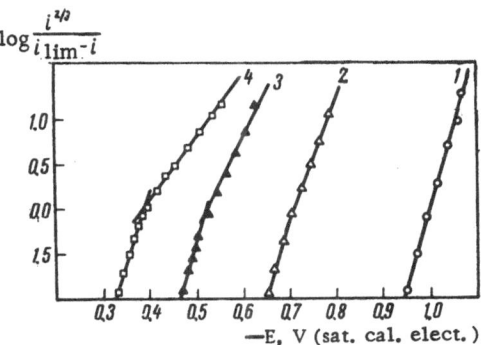

Fig. 7. Plots of log $[i^{2/3}/(i_{lim} - i)]$ against E for reduction waves of aldehydes $CH_3(CH = CH)_nCHO$ in KCl—HCl buffer solution in 50% water—dimethylformamide mixture. The figures on the curves represent the numbers of double bonds n.

Dibenzoylethane is formed as a result of a rearrangement (in the reduction of dibenzoylethylene). The height of the second wave falls with decrease in drop time and depends largely on the pH value of the solution, increasing to a two-electron diffusion-controlled process in the acidic and alkaline regions [56]. This indicates that H^+ and OH^- ions take part in the rearrangement (acid—base catalysis), and the process evidently has a surface character.

In a number of cases, e. g. , with the first wave of terephthalic aldehyde, the limiting current is restricted owing to sluggish dehydration of the carbonyl group.

The height of the reduction wave in α-hydroxyisobutyraldehyde [57] is limited by the dehydration rate. The wave is only observed in alkaline medium and increases with increase in pH value (owing to dehydration catalyzed mainly by OH^- ions) to a constant value, corresponding to a quasidiffusion current, at pH > 12. At pH < 12 the limiting current has clearly defined kinetic character (with a temperature coefficient of 5-7% per degree).

In the polarography [58] of buffered aqueous methanolic solutions of pyrimidone derivatives:

$$R'\!-\!\!\underset{N}{\overset{R}{\underset{\|}{\bigcirc}}}\!\!N\!-\!R'',$$

where $R'' = H$, $R' = R = H$ or CH_3, a single one-electron reduction
wave is observed in acidic and neutral solutions. With increase in
the pH value its half-wave potential is shifted towards negative potentials, and its form (at least in the lower part of the wave) corresponds
to the equation which holds for reversible electrode processes involving rapid dimerization of the electrode products. In alkaline
solutions the height of this wave decreases with increase in the pH
value and at the same time a new wave appears at more negative
potentials, so that the sum of the two waves remains constant. The
first wave corresponds to the reduction of a protonated form (evidently at a nitrogen atom) of the pyrimidone derivative, and the second (observed in alkaline solutions) corresponds to the unprotonated
form [58]. Pyrimidone derivatives which contain sulfur instead of
oxygen, and also those which contain large substituents R'' or two
pyrimidone rings linked by a $-CH_2-CH_2-$ bridge at the nitrogen (in
place of R''), exhibit on polarograms in acidic and neutral solutions
adsorption prewaves or subsequent waves which are absent in strongly alkaline solutions [58]. Following the reduction wave the thio
analog of pyrimidone and the compound with two pyrimidine groups
give catalytic hydrogen waves which have surface character.

From an analysis of the product obtained in preparative electrolysis of pyrimidone, in conjunction with other data, it has been
suggested that a dimer with the carbon – nitrogen double bond of the
pyrimidone rings hydrogenated is formed as a result of the electrode
reaction [58].

Kinetic Currents Limited by Protonation
Rate of Anions from Weak Acids

In acidic buffered solutions benzophenone-3,3',4,4'-tetracar-
boxylic acid is reduced in two steps at the dropping mercury electrode. This corresponds to two quasidiffusion waves [59], of which
the first represents reduction of the carbonyl group ($n = 2$) and the
second corresponds to the benzene nuclei ($n = 4$). Both waves

decrease in height with increase in the pH value and change to kinetic waves; the pH values at which the limiting kinetic currents reach half the diffusion currents are equal to 5.88 for the first wave and 3.22 for the second wave (with an ionic strength of 1.0 and addition of sodium perchlorate).

With increase in the ionic strength of the solution the first kinetic wave (pH 6.0) increases in height and the second wave (pH 3.5) decreases. Evidently both waves have surface character. The protonation rate constant of the benzophenonetetracarboxylate anions was evaluated formally on the assumption that the principal proton donor is the hydrogen ion and with allowance for the change in concentration of charged particles at the electrode surface under the influence of its field. In these calculations account was not taken of the variation in the acid dissociation constant during its adsorption on the electrode surface (see Section 1, Point f).

In the polarography of unbuffered solutions of dipotassium maleate the observed kinetic wave is determined by the formation rate of maleate monoanions from reaction of the polarographically inactive dianions with water. This reaction takes place in the volume reaction layer, the thickness of which is determined by the rate of the reverse reaction – the reaction between the maleate anions and OH^- ions. The effect of the indifferent electrolyte concentration (potassium chloride, between 0.05 and 2.0 mole/liter) and the structure of the double electric layer on this reaction and on the electrochemical reduction of maleate monoanions has been investigated [60]. In qualitative agreement with theory, increase in ionic strength leads to displacement of the observed kinetic wave to the region of positive potentials, and the shift of $E_{1/2}$ considerably exceeds the change in the effective potential jump between the electrode and the discharging particle. The greatest change in the ion concentration under the influence of the electrode field is observed at the electrode surface itself. Therefore the effect of the double layer structure must manifest itself the more strongly, the greater the part of the reaction layer (μ) which falls within the limits of the diffusion part of the double layer (δ). The value of μ decreases with increase in current along the wave (owing to the increased concentration of OH^- ions in the vicinity of the electrode), while δ does not depend on current and is only determined by the concentration of the indifferent electrolyte. At

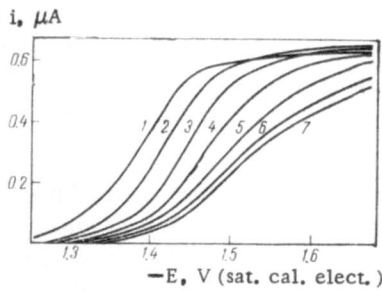

i, μA

—E, V (sat. cal. elect.)

Fig. 8. Polarograms for dipotassium maleate (c = 15 mmole/liter) in neutral solutions with various potassium chloride concentrations: 1) 2 mole/liter; 2) 1.0 mole/liter; 3) 0.5 mole/liter; 4) 0.25 mole/liter; 5) 0.1 mole/liter; 6) 0.075 mole/liter; 7) 0.05 mole/liter. The electrode had the following characteristics: m = 0.95 mg/sec; t = 0.34 sec (at 25°C).

constant potassium chloride concentration the μ/δ ratio decreases with increase in current, and the effect of the double layer becomes stronger. With increase in ionic strength therefore the upper part of the wave is shifted more towards positive potentials than the lower part, and the wave becomes steeper (Fig. 8). Thus, with a tenfold increase in the potassium chloride concentration, $E_{3/4}$ (i.e., the potential at which the current is equal to three quarters of the limiting value) is shifted by 165 mV into the region of positive potentials, while $E_{1/4}$ changes by only 75 mV.

The volume nature of the process, and also the fact that in sufficiently concentrated indifferent electrolyte solutions the thickness of the diffusion part of the double layer is small and does not affect the value of the limiting kinetic current, made it possible to calculate the true rate constants of the reaction between the maleate dianion and water (k_1) and the reverse reaction of the monoanion with hydroxyl ions (k_2). The value of k_1 varies little with increase in the ionic strength of the solution, whereas k_2 increases from $3.4 \cdot 10^9$ liter /mole·sec at $c_{KCl} = 0.1$ to $14.3 \cdot 10^9$ liter/mole·sec at $c_{KCl} = 2.0$ (Fig. 9), which is probably due to the primary salt effect. This effect is described quantitatively by the empirical equation of Perlmutter-Hayman [61], which takes account of the specific effect of cations on

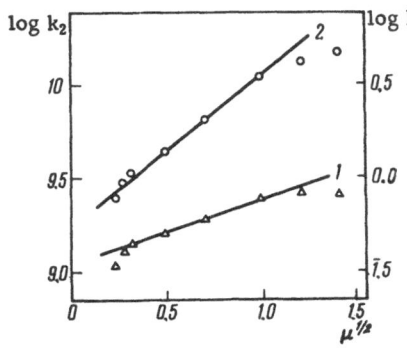

Fig. 9. Variation of k_1 (rate constant of reaction between maleate dianions and water) (1) and of k_2 (rate constant of reaction between maleate mono-anions and hydroxyl ions) (2) as functions of square root of potassium chloride con-. centration in solution.

reactions between anions. For an infinitely dilute solution this equation gave $k_2 = 9.7 \cdot 10^8$ liter/mole·sec, which is consistent with the value of $7.4 \cdot 10^8$ liter/mole·sec which Eigen and Kruse [62] obtained for this reaction using nuclear magnetic resonance.

So far only diffusion waves for the reduction of undissociated molecules or anions have been observed on polarograms for α-halogen-substituted carboxylic acids, and it has not been possible to observe a kinetic wave limited by the protonation rate of the anions [63, 64]. With α-bromopalmitic acid in buffered aqueous ethanolic solutions, however, there is a kinetic wave with a clearly defined surface character [44], and its height falls with increase in the pH value of the medium (Fig. 10).

When small amounts of substances which are adsorbed on the electrode are added to the solution the surface component of the kinetic current decreases [65]. For such a system the form of the i −t curves was calculated theoretically and was found to be in agreement with the limiting kinetic current observed for the i −t curves in buffered solutions of phenylglyoxalic acid with very small amounts of polyvinyl alcohol and triton X-102. The variation of half-wave potential (for the case where the electron transfer is retarded by adsorbed surface-active substances) with increase in the degree of surface coverage was calculated [65], with allowance for the change in ψ_1 potential. The relationships obtained were applied to the changes in $E_{1/2}$ for the reduction wave of maleic acid in the presence of amyl and hexyl alcohols.

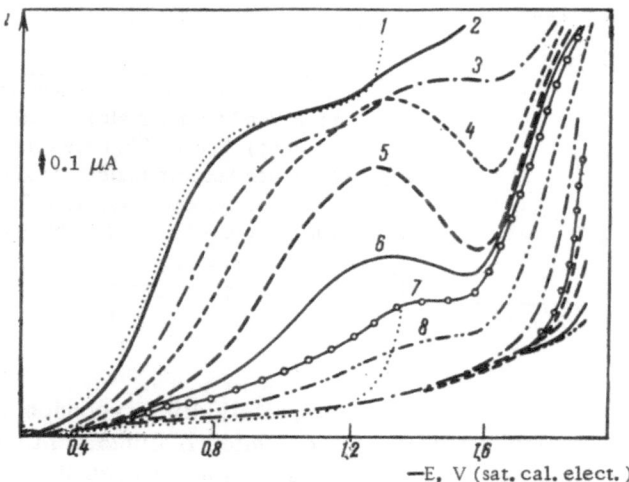

Fig. 10. Polarograms for α-bromopalmitic acid (1.5 mmole
/liter) in 50% ethanolic solutions with 0.1 N nitric acid (1),
citrate —phosphate buffer mixtures (2-6), and borate buffer
mixtures (7, 8) as supporting electrolytes at various pH
values: 2) 3.1; 3) 4.6; 4) 5.5; 5) 6.4; 6) 6.9; 7) 9.8; 8) 12.8.

Reduction Waves of Organic Substances
in Solutions with Low Proton-Donating
Activity

It is well-known that the reduction waves of many organic compounds in aprotic solvents differ considerably in form and character
from the waves for the same compounds in solutions with fairly high
proton-donor ability (e.g., [14-16, 66-68]). Addition of water or
other acids to the aprotic solvents changes the mechanism of the
electrode process in such a way that the corresponding waves approximate in their character the waves observed in solutions with
high proton-donating activity [69, 70].

In aprotic solvents, after transfer of the first electron to the
depolarizer, comparatively stable radical anions may be formed
which can only be subjected to further reduction with difficulty; if the
radical anion combines with a proton, the uncharged radical formed
is frequently more reactive than the original depolarizer.

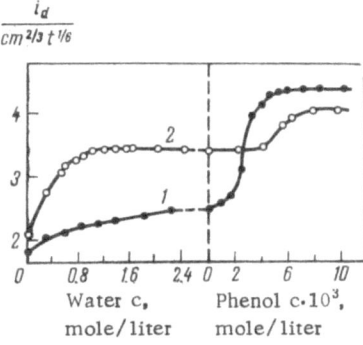

Fig. 11. Dependence of wave heights for acrylic (1) and methacrylic (2) esters in dimethylformamide on concentration of water and phenol in the solution [71].

In polarography in anhydrous dimethylformamide some of the aromatic dicarbonyl compounds which have already been mentioned above each give two reversible one-electron waves [56]. Addition of phenol (which is a proton donor) to the solution leads to increase of the first wave to a two-electron wave, and the electrode process becomes irreversible. In this process a monocarbonyl compound, which is reduced at more negative potentials, is formed. The reduction of dibenzoylethylene and dimethyl fumarate takes place differently and three waves are observed, the second of which is smaller than the first (one-electron) and decreases with shortening of the drop time. Proton donors added to the solution hardly react at all with the radical anion formed at the potentials of the first wave. On the basis of further experiments, in particular those obtained with the Kalousek commutator, it has been suggested that dianions, formed in the second electron transfer at the potentials of the second wave, react with the original compound to give a product, reduction of which gives rise to the third wave. Thus, the height of the two first waves corresponds to the transfer of $2/(m + 1)$ electrons, where m is the number of original depolarizer molecules which react with one dianion [56].

On polarograms for diallylaminoethyl esters of acrylic and methacrylic acids in anhydrous dimethylformamide a one-electron reduction wave appears, which gradually increases to the two-electron level when proton donors (water, phenol, propargyl alcohol) are added to the solution [71]. The stronger the proton donors, the lower the concentrations at which the wave increases; at the same time, the

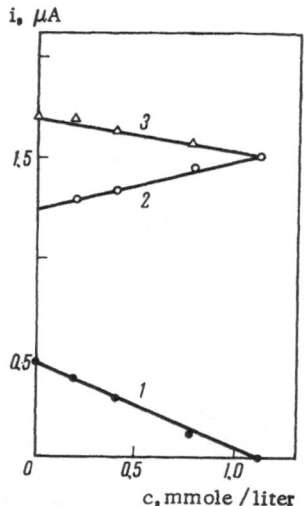

Fig. 12. Dependence of heights of first (1), second (2), and total (3) wave of maleic acid in dimethylformamide with 0.1 N lithium chloride solution as supporting electrolyte on concentration of lithium hydroxide added.

stronger the base formed as a result of electron transfer to the depolarizer, the more readily it reacts with the donors. Figure 11, taken from [71], shows the effect of successive addition of water and phenol on the wave heights of these esters. It is seen that the radical anion derived from the methacrylic ester is a stronger base than the radical anion from the acrylic ester, and phenol is a considerably stronger proton donor than water.

In the polarography of maleic acid in almost anhydrous (about 0.04% water) dimethylformamide, with 0.1 N lithium chloride as supporting electrolyte [72], two reduction waves are observed (the first is approximately half the second in height); their half-wave potentials are close to -1.1 V and -1.75 V, respectively (referred to the aqueous saturated calomel electrode). The heights of both waves are proportional to the maleic acid concentration. The limiting current of the total wave has diffusion character and corresponds to the transfer of two electrons. When sulfuric acid is added to the solution the height of the first wave increases at the expense of the second and limits towards the height of a two-electron diffusion wave; when lithium hydroxide solution is added, the first wave decreases until it disappears, while on neutralization of one equivalent of acid the second wave increases to the height of the overall diffusion current (Fig. 12). Further addition of lithium hydroxide leads to a decrease in this wave.

The first wave thus corresponds to reduction of undissociated acid molecules (in the case of low proton-donating activity, possibly with the transfer of one electron to the radical anion); the second corresponds to reduction of the monoanion. This is also shown by the nature of the shift in $E_{1/2}$ for both waves when the lithium chloride concentration is increased.

With neutral salts as supporting electrolyte, the limiting current of the first wave for maleic acid, to judge from its temperature coefficient and its dependence on the height of the mercury column, is nearly diffusion-controlled. Under these conditions the reduction process clearly involves "self-protonation" of the maleic acid molecule which enters into the electrode reaction at the expense of two other molecules which dissociate to manoanions, and the height of the first wave is limited by the delivery rate of these molecules, acting as proton donors, by diffusion. Two protons participate in the reduction of maleic acid, and this is why the height of the first wave is half that of the second. When weak acids (compared with the first step of maleic acid dissociation), such as formic and benzoic acids, are added to the solution the height of the first wave remains practically unchanged. Increase in the water content of dimethylformamide solutions causes some increase in the first wave at the expense of the second.

The reduction wave of monolithium maleate in dimethylformamide with 0.1 N lithium chloride as supporting electrolyte is unsymmetrical; the slope of the upper part is greater than that of the lower. When the concentration of monoanions increases a decrease (trough) appears on the limiting current of this wave. With increase in temperature the trough at first becomes deeper, but gradually levels out at temperatures above 45°C and almost disappears at 65°C. The trough (and at low monoanion concentrations also the asymmetry of the wave) also disappears (levels out) when proton donors, e.g., formic acid, are added to the solution (Fig. 13). These effects can be explained in the following manner. Owing to alkalinization in the vicinity of the electrode during the electrode process some of the monoanions change into electrochemically inactive dianions, reduction of which requires their previous protonation (at the expense of water in the dimethylformamide). As shown by electrocapillary curves obtained from variation of drop time with potential, the adsorbability of the dianions on the electrode is higher in dimethylformamide than in water, and their protonation therefore has a mixed volume – surface character.

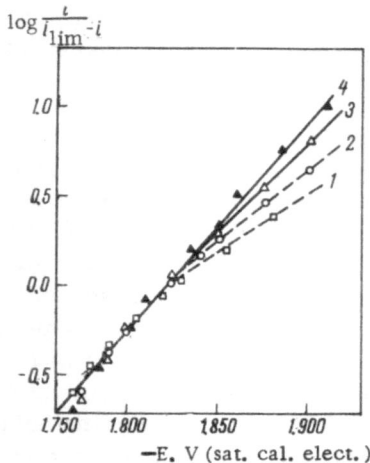

Fig. 13. Semilogarithmic curves for wave in dimethylformamide solution of mono-lithium maleate (0.7 mmole/liter) with 0.1 N lithium chloride as supporting electrolyte. 1) Without additions; 2)-4) with addition of 0.3 mmole/liter and 0.7 mmole/liter formic acid (2, 3) and 1.0 mmole/liter phenylacetic acid (4).

Many Schiff's bases are hydrolyzed in aqueous – organic mixtures, and it is therefore expedient to use dimethylformamide for their polarographic investigation [73-75]. In dimethylformamide, benzylidine-p-aminophenol and benzylidine-p-aminobenzenesulfonamide both give two one-electron diffusion waves [73], the first of which corresponds to the formation of a radical anion and the second to the formation of a dianion. When acids are added to the solution the first wave increases to a two-electron wave, and the second disappears. Preparative electrolysis at the potential of the first limiting current gave a dimer with a bond between the carbon atoms of the azomethine groups [73]. The corresponding substituted benzylanilines are formed at the potential of the second wave. A number of other benzylidineaniline derivatives with substituents in the aniline and benzaldehyde rings only give one-step two-electron diffusion

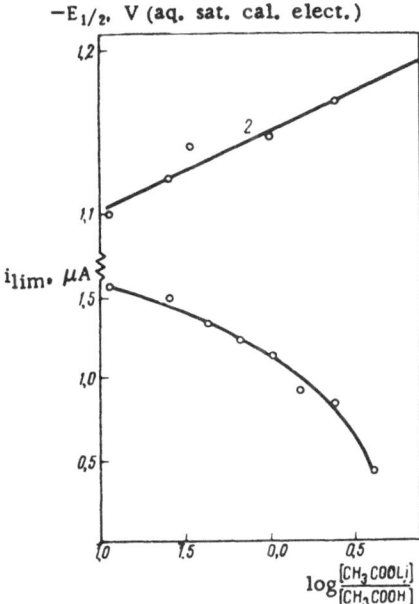

Fig. 14. Dependence of i_{lim} (1) and $E_{1/2}$ (2) of reduction wave of

$C_4H_9\!-\!\!\langle\!\!\begin{array}{c}\text{S}\end{array}\!\!\rangle\!\!-\!\!{}^{-CH=NC_6H_{11}}_{-SH}$ in 90% di-

methylformamide on logarithm of the [CH₃COOLi] / [CH₃COOH] ratio.

waves; this demonstrates the high reactivity of the radical anions, which facilitates their rapid reduction to dianions. Addition of acids to the solution of such compounds does not affect their waves [73].

In buffered aqueous ethanolic and dimethylformamide solutions Schiff's bases (prepared from aniline and salicylaldehyde or o-tosyl-aminobenzaldehyde derivatives) with an intramolecular hydrogen bond give two one-electron waves. In aqueous ethanolic solution both waves are shifted towards negative potentials with increase in the pH value, which indicates that electron transfer is preceded by protona-tion.

At pH > 10 the limiting current of the waves acquires a kinetic character [74].   The intramolecular hydrogen bond evidently stabilizes the free radical, and in aqueous ethanolic medium there are therefore two one-electron waves on the polarograms.

The possibility of intramolecular hydrogen bond formation with the nitrogen atom of the azomethine group considerably facilitates the reduction of the latter in an aprotic medium [75].   Thus, Schiff's bases obtained from 3-thiophenealdehyde with a mercapto group in the position 2 (I), on account of protonation of the azomethine nitrogen atom by the thiol hydrogen atom (II), are reduced in dimethylformamide considerably more readily than similar compounds where the hydrogen of the thiol has been substituted by an alkyl group (III) and intramolecular protonation is not therefore possible.

Polarography of compounds of type (III) in 90% dimethylformamide with $CH_3COOH + CH_3COOLi + LiCl$ as supporting electrolyte showed that the height of the reduction wave for the azomethine group increases with decrease in the $[CH_3COOLi]/[CH_3COOH]$ ratio, and its half-wave potential is shifted towards positive values (Fig. 14).

When hydrochloric acid is added to the solution of Compound II in 90% dimethylformamide three waves are observed on the polarograms: the first and second increase with increase in the hydrochloric acid concentration and the third decreases.   The first wave corresponds to reduction of the cationic protonated form (IV), which is formed comparatively slowly by reaction of Compound II with a hydrogen ion; this wave has a kinetic character.   The second wave is due to discharge of H^+ ion and the third to reduction of Compound II. With increase in temperature and drop time the contribution of the first wave to the diffusion current (the sum of the heights of the first and third waves) increases, and that of the third wave decreases [75].

Owing to hydrolysis of azomethines in acidic medium the true values for $E_{1/2}$ and i_{lim} were obtained [76] by recording their

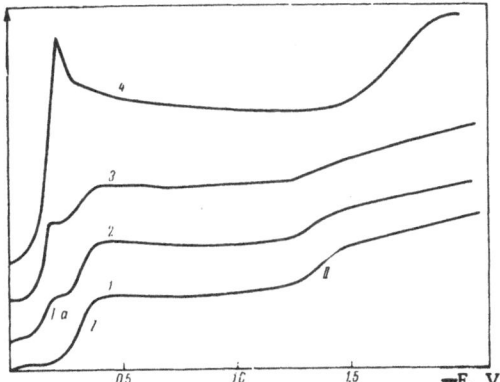

Fig. 15. Polarographic curves of oxygen in dimethylformamide with quaternary ammonium salts as supporting electrolyte. Phenyl isocyanate added in
concentration: 1) 0; 2) $0.36 \cdot 10^{-3}$ mole/liter; 3) 1.8
$\cdot 10^{-3}$ mole/liter; 4) $7.2 \cdot 10^{-3}$ mole/liter. I) First
diffusion wave for reduction of oxygen; II) second
wave; Ia) prewave.

variation with time and extrapolating the values obtained to zero
time, i.e., to the moment when the solutions were mixed.

The reaction forming azomethine and indoaniline dyes has been
investigated by generating quinonediimine at the dropping mercury
anode and recording its amount as a cathode-ray oscillogram before
and after addition of benzoylacetanilide and α-naphthol derivatives to
the solution [77].

Kinetic Quasidiffusion Waves for Reduction

of Oxygen in Nonaqueous Solvents and in

Presence of Proton Donors or Isocyanates

Polarograms of oxygen in aprotic solvents show two one-electron diffusion reduction waves [78,79], the first of which increases
to the level of a two-electron diffusion wave when an excess of strong
proton donors is added to the solution [79]. If a sufficiently strong
acid is added gradually to the aprotic solution a prewave can be observed on the polarogram before the first oxygen reduction wave [80].

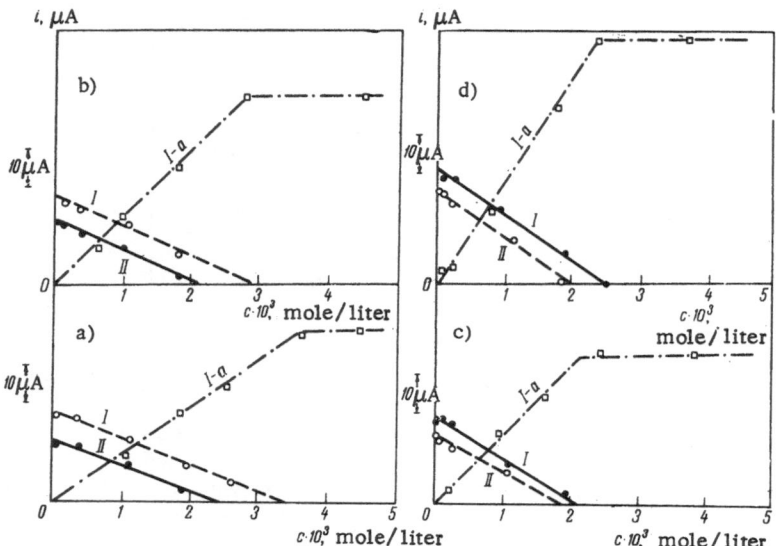

Fig. 16. Variation of limiting current for first (I) and second (II) waves and for prewave (Ia) on polarograms for oxygen as function of concentration of phenyl isocyanate (a, b, d) and toluylene diisocyanate (c) in various solvents. a) Dimethylformamide; b) and c) dioxane + dimethylformamide (3 : 1 by volume); d) acetone.

The height of the prewave increases with increase in the proton concentration, and the heights of both oxygen waves decrease.

A similar effect is also observed when isocyanates are added to an aprotic medium (Fig. 15). The height of the prewave (Ia, Fig. 15) corresponds to the quasidiffusion reduction current (with the transfer of two electrons) of a complex between the oxygen and the protons or isocyanates [80]; with a relative excess of oxygen the limiting current of the prewave is determined by the rate of transport of the protons or isocyanates, respectively, by diffusion to the electrode region (Fig. 16).

The heights of the first and second oxygen waves on the polarograms are governed by the amount of oxygen which has not entered into reaction with the protons or isocyanates, and the waves for the free oxygen therefore decrease with increase in the concentration of protons and isocyanates (Fig. 16, Curves I and II). When the solution contains a lower concentration of oxygen, a prewave and a wave for the reduction of excess protons or isocyanates are observed on the polarograms, but there are no waves for the reduction of free oxygen. In this case the height of the prewave is only determined by the transport of oxygen by diffusion and does not depend on the concentration of protons or isocyanates (horizontal section of lines in Fig. 16), and the waves for the reduction of excess protons or isocyanates vary linearly with their concentration under these conditions. It is clear that in these cases there is a "latent" limiting current of the reduction of protons or isocyanates [81, 82]. From the maximum height of the prewave and the concentration of protons or isocyanates at which it is reached, and allowing for the fact that two electrons are used in the reduction of monoisocyanates [83], it is found (by the method described [81, 82]) that two protons or one monoisocyanate molecule are required for one oxygen molecule in the formation of the complex. (In the case of toluylenediisocyanate one molecule requires two molecules of oxygen.)

The diffusion character of the prewave shows that the reaction rate of oxygen both with the protons and with the isocyanates is very high and does not control the rate of the process as a whole. Appreciable adsorbability of the isocyanates at the electrode in aprotic media at potentials of the prewave [80] indicates that the reaction between isocyanate and oxygen takes place mainly at the electrode surface. The requirement for protons or isocyanates in the electrode process shows that the prewave is not a catalytic wave but a kinetic quasidiffusion wave.

Some displacement of the half-wave potential of the prewave towards positive potentials with increase in the proton or isocyanate concentrations is explained by shift of the equilibrium towards the formation of the complex or protonated form, i. e., a decrease in the σ value, in agreement with, e. g., Eqs. (71) and (125) discussed earlier [9].

It should, however, be noted that (as follows from the slope of the logarithmic analysis for the prewave) the electrochemical reaction for reduction of the complex is close to reversible [80], and it is not therefore impossible that the prewave examined above is due to reaction with the protons or isocyanates not of molecular oxygen but of its radical anion, formed as a result of the reversible transfer of one electron to the oxygen molecule.

Variation of Dissociation Constants of Acids and Bases during Adsorption at Electrode Surface

The different adsorbabilities of acidic and basic components at the electrode surface lead to a change in the dissociation constants of the adsorbed acids and bases compared with their values in the bulk of the solution. This effect includes the change in the dissociation constants under the influence of the electrode field [84-86]. This change, together with the change in the concentration of charged substances near the electrode, must be taken into account when determining the rate constants of surface chemical reactions in the vicinity of the electrode.

To determine the effect of adsorption we shall consider equilibrium in the bulk of the solution (denoted by the subscript "o") and at the electrode surface with participation of adsorbed substances (ads):

$$K = \frac{[H^+]_o [A^-]_o}{[HA]_o} \quad \text{or} \quad K = \frac{[H^+]_o [B]_o}{[BH^+]_o} \tag{1}$$

and

$$K_{ads} = \frac{[H^+]_s [A^-]_{ads}}{[HA]_{ads}} \quad \text{or} \quad K_{ads} = \frac{[H^+]_s [B]_{ads}}{[BH^+]_{ads}}, \tag{2}$$

where A^- is the anion of the acid; BH^+ is the protonated (acid) form of the base; HA is the undissociated acid; B is the base.

We shall assume that hydrogen ions are not adsorbed at the electrode and shall denote their concentration directly at the electrode surface by $[H^+]_s$. For simplicity we shall only consider singly charged ions A^- and BH^+ and assume that adsorption of the components of the acid – base equilibrium obey the Henry isotherm, and we can therefore write for the uncharged particles:

$$\frac{[HA]_{ads}}{[HA]_0} = \beta_{HA} \Gamma_{HA(\infty)} \underline{and} \frac{[B]_{ads}}{[B]_0} = \beta_B \Gamma_{B(\infty)} , \tag{3}$$

where β is the adsorption constant; Γ_∞ is the amount of adsorbed substance on the completely covered electrode surface.

In the case of charged substances it is necessary to substitute the concentrations directly at the electrode surface $[BH^+]$ and $[A^-]_s$ in the equation for the adsorption equilibrium, and we can then write:

$$[A^-]_s = [A^-]_0 \exp\left(\frac{\psi_1 F}{RT}\right) and [BH^+]_s = [BH^+]_0 \exp\left(-\frac{\psi_1 F}{RT}\right) . \tag{4}$$

In exactly the same way,

$$[H^+]_s = [H^+]_0 \exp\left(-\frac{\psi_1 F}{RT}\right) . \tag{5}$$

Taking these equations into account, from Eqs. (1) and (2) we obtain respectively for the acid:

$$\frac{K_{ads}}{K} = \frac{\beta_{A^-} \Gamma_{A^-(\infty)}}{\beta_{HA} \Gamma_{HA(\infty)}} , \text{ where } \beta_{A^-} \Gamma_{A^-(\infty)} = \frac{[A^-]_{ads}}{[A^-]_s} , \tag{6}$$

and for the base:

$$\frac{K_{ads}}{K} = \frac{\beta_B \Gamma_{B(\infty)}}{\beta_{BH^+} \Gamma_{BH^+(\infty)}} , \text{ where } \beta_{BH^+} \Gamma_{BH^+(\infty)} = \frac{[BH]^+_{ads}}{[BH^+]_s} . \tag{7}$$

The area occupied by organic molecules at the electrode surface changes little (by not more than twice) when they are ionized, and therefore the change in adsorbability of the substance on ionization has the greatest effect on the value of K_{ads}/K. The β values are evidently much lower for ions than for the corresponding unionized molecules, and therefore the dissociation constants decrease in the case of adsorbed acids and increase in the case of bases. The maximum adsorption potentials for the ions and the corresponding undissociated molecules differ considerably and the effects of electrode potential on the β values for ions and molecules are also different; the K_{ads}/K value can therefore also change greatly with change in the electrode potential.

2. CATALYTIC ELECTROCHEMICAL PROCESSES

Catalytic Hydrogen Waves in Solutions of Noncomplex Catalysts

Volume catalytic hydrogen waves have been used [87] for determining the rate constants for transfer of a proton from the protonated amino group of amino acids to pyridine. For this purpose the heights of the catalytic waves for pyridine were determined in weakly alkaline buffer solutions, containing the investigated amino acid at constant ionic strength. Series of tests were performed with various pH values in the solution; within each series only the buffer capacity was varied and the pH value of the solution was kept strictly constant [88]. The apparent rate constant for transfer of a proton to pyridine was determined from the slope of the line for the relationship between the square of the limiting catalytic current and the concentration of the protonated amino acid [89]. The apparent constants obtained in this way at various pH values, and consequently with various thicknesses in the reaction layer, were extrapolated to a fairly thick reaction layer where the chemical reaction in the vicinity of the electrode was not affected by either the structure of the electric double layer or adosrption of the reaction components [90].

In aqueous solutions α-amino acids have the structure of a dipolar ion $NH_3^+CHRCOO^-$. In contrast to other series of carboxylic acids (and also other series of amines) the ionization constants of the carboxyl group (K_1) and the ammonium group (K_2) are almost independent of the substituent constant $\sigma*$ of R, i.e., the reaction constant $\rho*$ is practically zero. The rate constants obtained in [87] for transfer of a proton from the ammonium group in glycine (I), alanine (II), valine (III), methionine (IV), phenylalanine (V), and serine (VI) to pyridine were also found to have almost equal values (differing by not more than twice). The rate constants found for aliphatic amino acids (I-III) have rather smaller values than those for amino acids with electronegative side groups (IV-VI). No well-defined relationship was found between the logarithm of the rate constants and the $\sigma*$ values of the side groups. The sensitivity of the reaction rate constant to the inductive effect of the substituent is practically zero, as also with the ionization constants of amino acid carboxylic and amine groups. Aspartic acid gives an anomalously low rate constant for transfer of a proton to pyridine. These results are of great imporatance in the chemistry of α-amino acids.

While recording polarization curves (for the relationship between mercury electrode potential and current density) in acidic solutions of p-toluidine it was found [91] that the hydrogen overpotential decreased at very low current densities and increased at relatively high current densities (Fig.17). In this case the position of the curves is affected considerably not only by the concentrations of toluidine and acid but also by the nature and concentration of the anions of the indifferent electrolyte, and the greatest effect was given by anions I^- and ClO_4^- (compare curves in Figs.17 and 18). A similar effect was also observed for hydrogen evolution from acidic solutions containing small quantities of tribenzylamine [92]. Tedoradze and Asatiani [91] consider that in this case the mechanism of catalytic hydrogen evolution does not follow the scheme proposed [93], which supposes discharge of the protonated amine, and that the action of the amine merely amounts to a change in the structure of the double layer. Adsorption of the amine can either facilitate reduction of hydrogen ions (in the case of joint adsorption of the organic cation and the supporting electrolyte anions) or retard it (in the absence of anion adsorption).

If it is assumed that the hydrogen and catalytically active ions are reduced independently, and each according to its own rules [93], it follows from Fig. 17 that change in catalytic current density (lower part of Curves 2-4) leads to a considerably larger shift in the potential of the electrode than hydrogen ion reduction, and with increase in potential the catalytic wave is therefore simply overlapped by the normal hydrogen-ion reduction current.

Tedoradze and Asatiani [91] do not consider, moreover, that the chemical reaction of amine protonation can also be the governing step of the process; reduction of the protonated amine therefore takes place at current densities smaller than limiting current, and reduction of the free hydrogen ion (hydrated) at current densities above the limiting value.

−E, V (sat. cal. elect.)

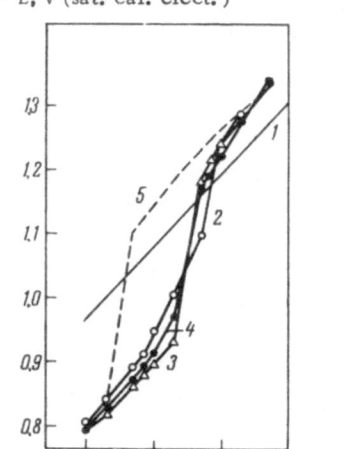

−E, V (sat. cal. elect.)

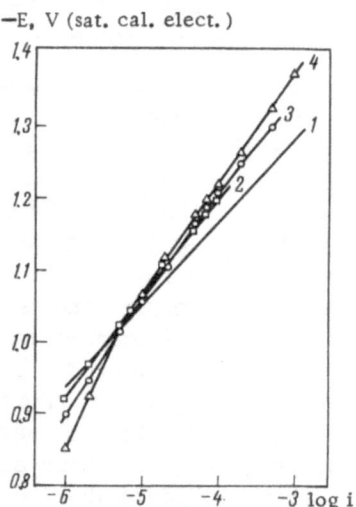

Fig. 17. Polarization curves on mercury electrode in 1 N hydrochloric acid with various concentrations of p-toluidine [91]: 1) 0; 2) 0.05 mole/liter; 3) 0.1 mole/liter; 4) 0.2 mole/liter.

Fig. 18. Polarization curves in 1 N perchloric acid solution with various concentrations of p-toluidine [91]: 1) 0; 2) 0.1 mole/liter; 3) 0.15 mole/liter; 4) 0.2 mole/liter; 5) 0.2 mole/liter, with variation of current density in reverse order (from large to small).

Inspection of the curves in Fig. 18 reveals areas which can be attributed to the limiting catalytic current (with log i about −4.5). The reason for the relatively low value of the limiting catalytic current observed here may be variation in the catalytic activity of the amines employed, resulting from formation of complexes with anions.

Under the conditions used [91] the ions of p-toluidine and hydrogen are reduced at almost the same potentials, and therefore in the general case it is possible to observe a certain degree of mixed current for the discharge of p-toluidine and hydrogen ions.

It can be supposed that adsorption of the amine and the resultant decrease in its basicity are equivalent to the introduction of an electronegative substituent into the amine molecule, which should shift the reduction potential of the particle to the positive side [94]. The amine protonated in the adsorbed state should then be discharged more readily than the amine protonated in solution, which is transported to the electrode and participates in the electrode process as it were in the unadsorbed condition [95] or is in another orientation less favorable for the electron transfer [96], at the electrode surface. Based on this assumption the reduction of the catalyst preceding hydrogen evolution should be attributed to its adsorbed particles, while reduction of the dissolved protonated amine, the concentration of which is very high, takes place at the same potential as or more negatively than reduction of hydrogen ions. The influence of indifferent electrolytes results in a change in the structure of the double layer and change in the adsorbability of the catalyst owing to salting-out [92]; it also supports the formation of associates between the anions and cations [97], which changes the ψ_1 potential and the reactivity of the amines.

The shift of the catalytic wave towards positive potentials, and of the following wave for discharge of "free" hydrogen ions towards negative potentials, with increase in the catalyst concentration is readily seen in the case of catalytic hydrogen waves, which are close in height to quasidiffusion waves (i.e., to waves with heights limited by the diffusion rate of hydrogen ions towards the electrode) as seen from Fig. 19 (taken from [98]). The phenomena shown in Figs. 17 and 18, on the one hand, and those in Fig. 19, on the other, are clearly very similar in nature.

As seen from Fig. 19, the height of the catalytic wave increases with increase in catalyst concentration. (In Fig. 18 this corresponds to a shift of the steeply rising lower part of the curves on the right.) At the same time the catalytic wave becomes more positive (shift of the lowest part of the curves in Figs. 17 and 18 downwards). The increase in the hydrogen overvoltage, following the catalytic process, should be explained by the fact that, although reduction of the protonated amine takes place quickly, the amine is continuously being protonated at the electrode, and this is one of the reaction steps participating in the catalytic cycle. Consequently, even at potentials beyond the limiting catalytic current a certain amount of the amine is in the protonated state at any moment of time, which reduces the

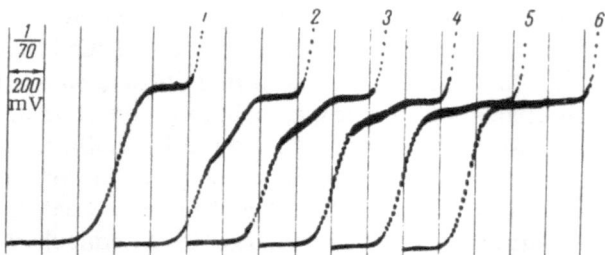

Fig. 19. Catalytic waves for hydrogen from 1.0 mmole/liter solution of hydrochloric acid and 0.1 N potassium chloride with various concentrations of riboflavin [98]: 1) 0; 2) $4.9 \cdot 10^{-8}$ mole/liter; 3) $9.6 \cdot 10^{-8}$ mole/liter; 4) $1.8 \cdot 10^{-7}$ mole/liter; 5) $3.4 \cdot 10^{-7}$ mole/liter; 6) $1.4 \cdot 10^{-6}$ mole/liter.

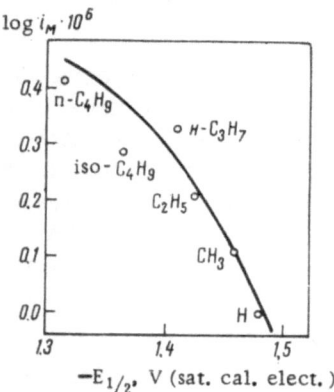

Fig. 20. Relationship between height (maximum) of catalytic curve and its "half-peak" potential $E_{1/2}$ in series of imidazole-2-thione derivatives with alkyl radicals of various lengths in the 4-position (plotted from data in [99]).

absolute value of the negative ψ_1 potential and impedes reduction of the hydrogen ions. Moreover, the retarding effect of the adsorbed amine film cannot be disregarded.

It is interesting that in acidic solutions of p-toluidine [91] and tribenzylamine [92] in the presence of iodide and perchlorate ions change in the direction of polarization gives a hysteresis loop on the polarization curves (Curve 5, Fig. 18). In [91, 92] this is explained by hindrance to establishment of the adsorption equilibrium of the anions in the presence of aromatic amines at the electrode surface.

It should be noted that only the inhibiting effect on hydrogen-ion reduction is observed in the case of relatively high tribenzylamine concentrations [92]. The absence of the catalytic wave is attributed to inhibition of the catalytic process by the film of adsorbed substance.

To conclude this section it is worth mentioning the interesting relationship between the height and position of the catalytic hydrogen wave in solutions of imidazole-2-thione derivatives

$$
\begin{array}{c}
HC\!=\!CR \\
|\quad | \\
HN\quad NH \\
\diagdown\diagup \\
C \\
\| \\
S
\end{array}
$$

on the length of the radical R in position 4 [99]; with increase in the size of this radical the wave becomes more positive and higher (Fig. 20, which we plotted from data in [99]). Although the basicity of the compound also increases to some extent with increase in the size of the radical, the increase is not so great as to lead to any observable increase in the wave. With such a change in basicity the wave should moreover become more negative [94]. The observed effect is clearly due to an appreciable increase in the adsorbability of the catalyst with increase in the length of the radical.

Catalytic Waves Due to Complexes of

Organic Bases with Metal Ions

Even from the time of the early work by Brdička [100] it has been known that addition of cysteine or proteins to a solution of a divalent cobalt salt leads to a shift of the reduction wave of Co (II) towards positive potentials and to the appearance, following the reduction of the cobalt, of a hump–shaped catalytic hydrogen wave; however, the nature and mechanism of the processes causing these effects have not yet been finally established. It is known that both the above-mentioned phenomena are connected with the formation of

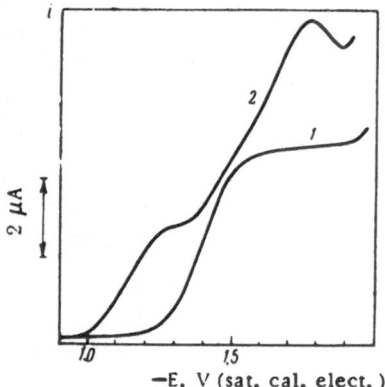

Fig. 21. Polarograms for solutions of
cobalt ions (1) and Co (II) — cysteine
system (2) in borate buffer solution
with ionic strength of 0.4.

complexes between divalent cobalt and cysteine [101]. In order to
obtain further information on the nature of these complexes the effect of change in the structure of the double electric layer on the
waves which they cause was investigated [102].

For this purpose the effect of the concentration of potassium
chloride added to a borate buffer solution (total concentration of
boric acid and potassium borate 0.10 mole/liter, pH 8.4) on polarograms for the Co (II) — cysteine system, with constant cobalt chloride (1.74 mmole/liter) and cysteine (0.08 mmole/liter) concentrations,
was investigated. The borate buffer solution was used in place of the
traditional ammoniacal buffer solution in order to avoid a number of
complications and, primarily, decrease in the buffer capacity of the
ammoniacal solution near the electrode when potassium chloride is
added, since ammonium ions (being proton donors) also participate
in the formation of the outer layer of the electric double layer. The
pH value of the solution was selected so that the solution would contain approximately equal amounts of anions and cysteine zwitterions
(pK_{am} = 8.33).

When the ionic strength of the solution μ is progressively increased from 0.1 to 1.6 mole/liter the cobalt reduction wave splits
into two, and its first step ("prewave") is shifted towards positive

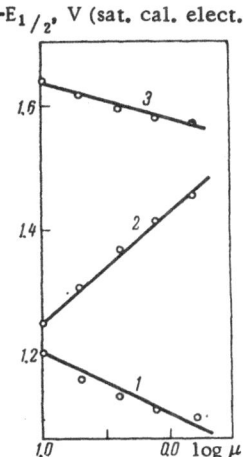

Fig. 22. Dependence of half-wave
potentials on logarithm of ionic
strength for kinetic reduction wave
of cobalt in Co (II) — cysteine system (1), wave of Co (II) aquo complexes (2), and catalytic hydrogen
wave of Co (II) — cysteine system
in borate buffer solution at pH 8.4 (3).

potentials and appreciably diminished in height, acquiring a kinetic
character; the second step is shifted to more negative potentials; the
total height of the wave remains practically equal to that of the original cobalt diffusion wave. (Curves for $\mu = 0.4$ are given as an example
in Fig. 21.) The catalytic hydrogen wave here is shifted, without substantial change in height, into the region of positive potentials and at
high ionic strengths combines with the second cobalt reduction step
to give a single common wave. Control tests with exactly the same
solution, but without the cysteine, showed that the second step corresponds to reduction of free divalent cobalt ions (in aquo complexes).
Logarithmic analyses for the waves of free cobalt ions show changes
of direction at currents of the order of $0.1 \, i_d$ (deflections of this
type were first observed by Tur'yan [103]); the upper (greater) part
of the wave is less steep than the lower part. (In Tur'yan's experiments the upper part was steeper.) With increase in ionic strength

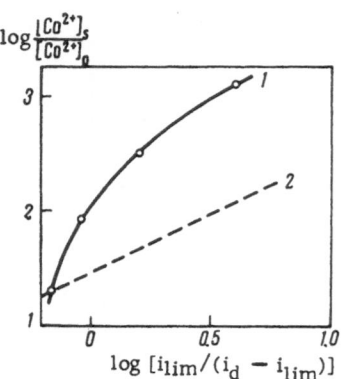

Fig. 23. Relationship between normalized height $[i_{lim}/(i_d - i_{lim})]$ of kinetic wave for cobalt and ratio of concentration of divalent cobalt ions near the electrode to their concentration in the bulk of the solution, determined by the ionic strength of the solution. 1) Observed relationship; 2) calculated without allowance for the salt effect or for variation in the adsorbability of cysteine.

the slope of the main part of the divalent cobalt wave gradually increases from 94 to 144 mV, and $\Delta E_{1/2}/\Delta \log \mu \approx -180$ mV (Line 2 in Fig. 22). The latter value (for a mean slope of 120 mV, i. e., $\alpha n_a = 0.5$) is equal to the theoretical value of the shift in the half-wave potential for doubly charged cations. The reasons for the inflection on the curves and the change in their slope have not been established.

The first step on the cobalt wave ("prewave") corresponds to discharge of Co (II) from its complex with cysteine. The height of this wave is limited by the formation rate of complexes between adsorbed cysteine anions and Co (II) ions in the layer adjacent to the electrode (henceforth written as $[Co]_s$). Under these conditions cysteine is clearly present in excess with respect to $[Co]_s$. Reduction of the complex Co (II) ions (compared with the free ions) evidently arises from increase in their concentration near the electrode owing to adsorption of the complexes. The decrease in the height of this surface kinetic wave with increase in ionic strength is due both to a reduction in the rate constant of the ionic reaction (salt effect) and to a decrease in $[Co]_s$ resulting from a decrease in the absolute value of the negative ψ_1 potential. However, the decrease in the corrected wave height $i_{lim}/(i_d - i_{lim})$ with increase in ionic strength (Curve 1, Fig. 23) is appreciably less than the reduction in $[Co]_5$ (Curve 2, Fig. 23); clearly the fall in the negative ψ_1 value promotes increased adsorbability in the cysteine anions. The value $\Delta E_{1/2}/\Delta \log \mu \approx 90$ mV for the kinetic wave (Line 1, Fig. 22) shows formally that the complex has a negative charge equal to one quarter of an electron; as a whole the complex is clearly electrically neutral [two cysteine anions are required for one Co (II) ion], but its anionic portion — one or two anions of the adsorbed cysteine — is oriented towards the electrode surface. As the ionic strength increases the slope of the wave [the value $bb'/(b' - b)$] decreases from 140 to 100 mV. The a value (the constant of the Frumkin equation, relating the adsorbability of a substance on the electrode to its potential), which was formally determined from the relationship between $bb'/(b' - b)$ and $E_{1/2}$, was found to be equal to 35 V^{-2}, which is considerably higher than the true value. Clearly in this case the change in the slope of the wave is affected not only by change in its position relative to the maximum adsorption potential but also by other factors.

The shift in the catalytic hydrogen wave to more positive potentials amounts to about 60 mV with tenfold increase in the ionic strength (Line 3, Fig. 22). Consequently the catalytically active complex which undergoes the electrochemical reaction and from which the cobalt is not reduced [9, 104], is a monoanion; it can be assumed that this complex, in addition to Co (II), contains three cysteine anions. It is not impossible, however, that the borate ion enters into the composition of the complexes. It was not possible to trace the effect of ionic strength on the slope of the catalytic wave owing to the proximity of the cobalt reduction wave.

In a number of cases [105, 106], e.g., in thioglycolic acid solutions [106], the catalytic hydrogen wave can be observed before discharge of the cobalt ions; this once again confirms that the waves arise from complexes of the organic compound with the cobalt ions and that the discharged metallic cobalt does not influence the catalytic effect. A catalytic wave due to Co (II) complexes with thioglycolic acid is observed right up to pH 2.0; the height of the wave increases with increase in pH value, reaches a maximum, and remains almost constant at pH 4-5.5; it then falls in a curve reminiscent of a dissociation curve (Fig. 24). Similar forms of relationship between wave

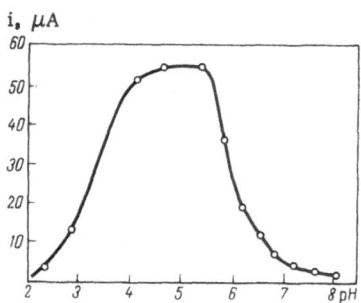

Fig. 24. Variation of limiting catalytic current as function of pH in solutions of thioglycolic acid in presence of cobalt chloride [106].

heights and pH value exist in the presence of 8-mercaptoquinoline [106] and sulfur-containing amino acids [107, 108]. The Co (II) complex with the thioglycolate anion is evidently catalytically active [106]; in acidic medium the amount of catalytically active complexes decreases owing to the appearance of the undissociated thioglycolic acid ($pK_{COOH} = 3.6$) and the wave decreases. With increase in pH value above 5.5 the catalytic current falls on account of decreased protonation rate. In the case of 8-mercaptoquinoline and its derivatives a catalytic hydrogen wave is also observed in the absence of cobalt salt, but the wave due to complexes with divalent cobalt is considerably higher and is situated at rather more negative potentials [106]; it can therefore be supposed that the formation of a complex with divalent cobalt increases the basicity (at the nitrogen) of the catalyst [94].

It has been shown by various methods [109] that $Co(NH_3)_6^{3+}$ only forms a complex with bovine serum albumin at very high concentrations of the latter, whereas Co^{2+} forms an extremely stable complex. Investigation of the variation with time of the wave heights (or more accurately the oscillopolarographic peaks) for reduction of the complex of Co (II) ion with the albumin and the free Co (II) ion (aquo complex) showed that the peak for the complex increases with time, whereas the peak for reduction of free Co (II) ions increases, reaches a maximum and then falls in proportion to $t^{-1/2}$. The area required by the albumin molecule was calculated on the assumption that the decrease in the height of the peak to zero corresponded to complete coverage of the electrode surface with albumin. This area was extremely large, which led to the conclusion [109] that the albumin was flattened out (with simultaneous denaturation) to a layer of 6-8 Å in thickness. The presence of two catalytic waves on the polarograms for the albumin in a solution of cobalt salt is explained by two forms of catalyst – one with flattened strongly adsorbed albumin (the first wave) and the other with weakly adsorbed globular albumin [109]. Investigation of the variation in height of the catalytic waves with time showed that the second wave decreases and disappears in tens of seconds, whereas the first wave at first increases due to the slow flattening of the abumin molecule and then decreases in the course of some tens of minutes. The decrease in the waves was related to formation of a multimolecular layer at the electrode surface.

Acetylation of the ε-amino group of lysozyme only slightly reduces the height of the catalytic hydrogen waves [109], and consequently this group does not participate in complex formation leading to formation of a catalyst. Some decrease in the catalytic activity of ε-acetylated lysozyme is explained by a small diminution in its adsorbability on the electrode [110].

Catalysis in Transfer of Protons from Donor (Acid) to Acceptor (Depolarizer)

The products formed in the reduction of certain aromatic nitro alcohols were found to be catalysts for the transfer of protons from acidic components of the buffer solution to the depolarizer [111]. When the current increases with increasing negative potential the amount of product (catalyst) which accelerates the process increases, and this gives the process an autocatalytic character. The polarographic wave corresponding to this process has a very steep slope, which depends on the pH value and the buffer capacity of the solution. The form of the i – t curves corresponding to such processes is also anomalous [112]. In Fig. 25 the i – t curves (recorded for the first drop) for reduction of 0.34 mmole/liter 1-(p-bromophenyl)-2-nitro-ethanol, using a citrate – phosphate buffer mixture with pH 5.4 as supporting electrolyte, in 16.7% (by volume) methanol [112] are given as an example. When the cathodic potential increases the growth rate

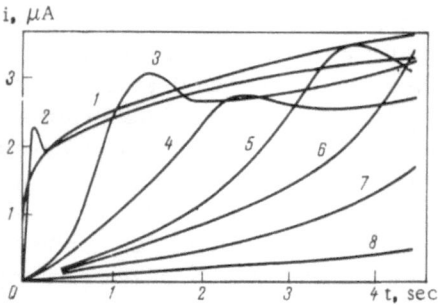

Fig. 25. Current – time curves for reduction
of 1-(p-bromophenyl)-2-nitroethanol at
various potentials (referred to saturated
calomel electrode): 1) 0.7 V (limiting
current); 2) 0.575 V; 3) 0.525 V; 4) 0.515
V; 5) 0.510 V; 6) 0.505 V; 7) 0.500 V;
8) 0.495 V.

of the current with time rapidly increases and at potentials corresponding to rise of the wave a peak is observed on the i – t curves,
where the current momentarily becomes even higher than the limiting
diffusion current (compare Curves 2 and 3 with Curve 1 in Fig. 25).
Curves of similar form were observed in the reduction of $S_2O_8^{2-}$ anions
catalyzed by adsorbed tetrabutylammonium cations [113]. The appearance of peaks arises from the following effects. In the initial
period of the life of the drop, when there is hardly any catalyst (reaction product) at its surface, the electrode process rate is comparatively small, and an excess of unreacted adsorbed depolarizer (which
approaches the electrode by diffusion) accumulates at the electrode
surface. The depolarizer is strongly adsorbed, and with small
degrees of surface coverage (in the initial period of the drop life) its
transport to the electrode surface corresponds to the limiting diffusion current. As the electrode product (catalyst) accumulates the
rate of the process increases autocatalytically and becomes so high
that the depolarizer which has accumulated at the electrode surface
is reduced in a comparatively short time. The current then falls to
a level determined by its transport rate by diffusion at the respective
concentration gradient. It is interesting that, owing to retention of

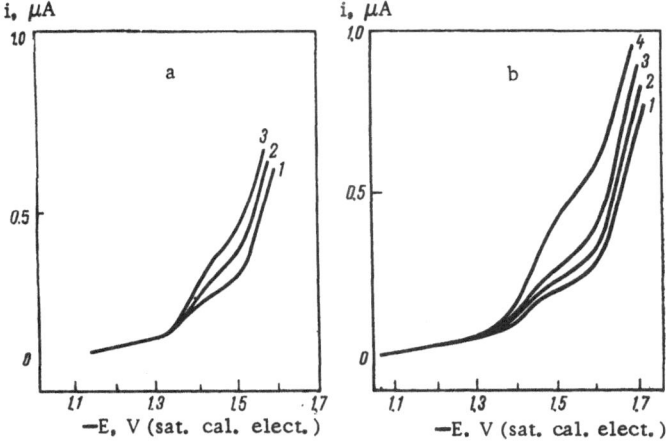

Fig. 26. Polarograms for 0.5 mmole/liter solution of acetophenone in acetate buffer solution at pH 5.5 with 20% ethanol and with addition of various amounts of diethylaniline (a) and pyridine (b): a) 1) 0; 2) 1.0 mmole/liter; 3) 2.0 mmole/liter; b) 1) 0; 2) 0.05 mmole/liter; 3) 0.1 mmole/liter; 4) 0.5 mmole/liter.

catalyst formed on preceding drops of the electrode tip, the current at the initial section of the i − t curves for subsequent drops is appreciably higher than the current for the first drop [112]. The rate of this electrode process also increases when certain amines, which act as catalysts, are added to the solution. Here the wave is shifted towards positive potentials and becomes less steep, approaching the usual form of reduction waves of nitro compounds; the autocatalytic process changes to a catalytic process, where the catalysis appears equally along the whole wave, irrespective of the current intensity.

The catalytic activity of various amines was found to be different; the strongest catalysts are quinine, ephedrine, pyridine and its homologs, and cysteine. Aniline, N-methylbenzylamine, and cyclohexylamine showed practically no catalytic activity [111]. The catalytic activity is not related to the pK values of the amines and is to some extent specific. It is interesting that those catalysts which facilitate transfer of protons to nitro alcohols also facilitate the electrochemical reduction of hydrogen ions, i.e., give rise to catalytic hydrogen waves. The parallelism between these effects goes even

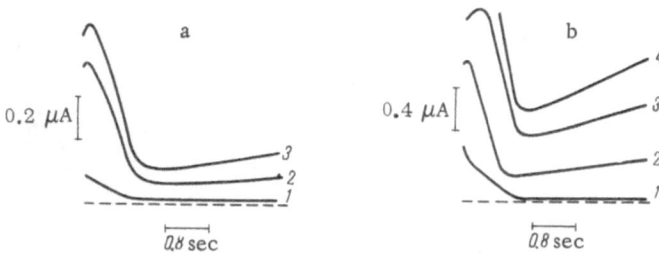

Fig. 27. Current — time curves for first drop at potentials of —1.35 V (a) and —1.40 V (b) near the limiting kinetic current, controlled by protonation of propiophenone in acetate buffer solution with 0.02% polyvinyl alcohol and various amounts of pyridine: 1) 0; 2) 2.0 mmole/liter; 3) 5.0 mmole/liter; 4) 10 mmole/liter.

further. Thus, in many cases, addition of cobalt salts to the solution greatly facilitates catalytic evolution of hydrogen in exactly the same way as the catalytic transfer of protons to nitro compounds under the influence of cysteine is considerably accelerated in the presence of cobalt salts [111].

The amine catalysts facilitate transfer of protons not only to complex nitro alcohols but also to simple aliphatic nitro compounds and to ketones [114]. Thus, for example, addition of diethylaniline and particularly of pyridine appreciably increases the height of the kinetic waves of acetophenone and propiophenone (Fig. 26) caused by their inhibited protonation. Pyridine has a greater effect than di-ethylaniline (Fig. 26).

Like catalytic hydrogen evolution, catalytic proton transfer may have surface or volume characteristics, i.e., the transfer may occur onto a compound adsorbed at the electrode or onto a depolarizer in the vicinity of the electrode. If small amounts of the strongly ad-sorbed polyvinyl alcohol are added to the solution, at the moment when the electrode surface is completely covered with this compound the surface kinetic reduction current of protonated propiophenone decreases almost to zero (Curves 1, Fig. 27). If pyridine catalyst is added to the solution, reduction current for the protonated propiophe-none is also observed on the completely covered electrode surface (Curves 2-4, Fig. 27), and even a considerable increase in the concentration of polyvinyl alcohol has practically no effect on the magnitude of the current observed in this case: this is evidently due to reduction of the propiophenone, which is mostly protonated in the volume of the solution near the electrode, by pyridine [114].

It should be mentioned that a similar catalytic effect was observed [115], independently of [113], in the reduction of certain nitro compounds in the presence of N,N-dimethyl-p-phenylenediamine. (The report [115] was published in September 1967, while [111]was received in April 1967.)

LITERATURE CITED

1. R. Brdička and K. Wiesner, Naturwiss., 31:247 (1943).
2. K. Wiesner, Z. Elektrochem., 49:164 (1943).
3. R. Brdička and K. Wiesner, Coll. Czechosl. Chem. Communs., 12:39 (1947).
4. R. Brdička and K. Wiesner, Coll. Czechosl. Chem. Communs., 12:138 (1947).
5. R. Brdička, Coll. Czechosl. Chem. Communs., 12:212 (1947).
6. K. Veselý and R. Brdička, Coll. Czechosl. Chem. Communs., 12:313 (1947).
7. J. Koryta and I. Kösler, Coll. Czechosl. Chem. Communs., 15:241 (1950).
8. S. G. Mairanovskii, J. Electroanalyt. Chem., 4:166 (1962).
9. S. G. Mairanovskii, Catalytic and Kinetic Waves in Polarography, Plenum Press, New York (1968).
10. L. Holleck and H. J. Exner, Sbórnić 1 Mezinar. Polarogr. Sjezdu., Prague, Vol. 1 (1951), p. 97.
11. B. Kastening and L. Holleck, Z. Elektrochem., 63:166 (1959).
12. L. Holleck and B. Kastening, Z. Elektrochem., 63:177 (1959); 64, 823 (1960).
13. L. Holleck, Z. Naturforsch., 186:439 (1963).
14. D. H. Geske and A. H. Maki, J. Am. Chem. Soc., 82:2671 (1960).
15. L. Holleck and D. Becher, Electroanalyt. Chem., 4:321 (1962).
16. É. S. Levin and Z. I. Fodiman, Papers of First Conference on Organic Analysis [in Russian], Izd. MGU, Moscow (1961), p. 108; W. Kemula and R. Sioda, Bull. Acad. Polon. Sci., Ser. Chim., 10:513 (1962); 11:395 (1963).
17. L. Holleck, B. Kastening and R. D. Williams, Z. Elektrochem., 66:396 (1962).
18. S. G. Mairanovskii, Dokl. Akad. Nauk SSSR, 132:1352 (1960).
19. Ya. P. Stradyn', G. O. Reikhmanis, and R. A. Gavar, Élektrokhimiya, 1:955 (1965).
20. Ya. P. Stradyn' and I. Ya. Kravis, Report presented at Sixth Conference on Electrochemistry of Organic Compounds [in Russian], Inst. Élektrokhimii Akad. Nauk SSSR, Moscow (1968).
21. J. E. Page, J. W. Smith, and J. G. Waller, J. Phys. and Colloid. Chem., 53:545 (1949).
22. S. F. Dennis, A. S. Powell, and M. J. Astle, J. Am. Chem. Soc., 71:1484 (1949).

23.   J. Tirouflet, Bull. Soc. Chim. France, 274 (1956).
24.   M. Person, Bull. Soc. Chim. France, 1832 (1966).
25.   Ya. P. Stradyn', Polarography of Organic Nitro Compounds [in Russian], Izd.
      Akad. Nauk LatvSSR, Riga (1961).
26.   S. G. Mairanovskii, V. M. Belikov, Ts. B. Korchemnaya, and S. S. Novikov,
      Izv. Akad. Nauk SSSR, Otd. Khim. Nauk, 523 (1962).
27.   S. G. Mairanovskii and Ya. P. Stradyn', Izv. Akad. Nauk SSSR, Otd. Khim.,
      Nauk, 2239 (1961).
28.   V. A. Petrosyan, S. G. Mairanovskii, V. I. Slovetskii, and A. A. Fainzil'berg,
      Izv. Akad. Nauk SSSR, Ser. Khim., 928 (1968).
29.   V. A. Petrosyan, S. G. Mairanovskii, V. I. Slovetskii, A. A. Fainzil'berg, and
      S. I. Kvachenok, Izv. Akad. Nauk SSSR, Ser. Khim., 1721 (1968).
30.   Ya. P. Stradyn' and G. O. Reikhmanis, Élektrokhimiya, 3:178 (1967).
31.   Ya. P. Stradyn' and I. Ya. Kravis, in: Electrochemical Processes Involving Or-
      ganic Substances [in Russian], Nauka, Moscow (1969).
32.   S. G. Mairanovskii and V. N. Pavlov, Zh. Fiz. Khim., 38:1804 (1964).
33.   T. S. Ivcher, E. N. Zilberman, and E. M. Perepletchikova, Zh. Fiz. Khim.,
      39:749 (1965).
34.   Ya. P. Stradyn' and V. V. Teraud., Izv. Akad. Nauk LatvSSR, Ser. Khim., 43
      (1965).
35.   E. Laviron and J. C. Ludy, Bull. Soc. Chim. France, 2202 (1966).
36.   Nguen van Tkhan', I. A. Avrutskaya, and M. Ya. Fioshin, in: Advances in
      Electrochemistry of Organic Compounds [in Russian], Nauka, Moscow (1968), p. 41.
37.   Ya. P. Stradyn', I. K. Tumane, and G. Ya. Vanag, Zh. Analiticheskoi Khim.,
      20:1239 (1965).
38.   Ya. P. Stradyn' and I. K. Tumane, Zh. Obshch. Khim., 37:1956 (1967).
39.   Ya. P. Stradyn' and V. P. Kadysh, Report Presented at Sixth Conference on
      Electrochemistry of Organic Compounds [in Russian], Inst. Élektrokhimii Akad.
      Nauk SSSR, Moscow (1968).
40.   M. I. Kabachnik, Dokl. Akad. Nauk SSSR, 83:407, 859 (1952).
41.   Ya. Linaberg, O. Neiland, A. Veis, and C. Ya. Vanag, Dokl. Akad. Nauk SSSR,
      154:1385 (1964).
42.   S. G. Mairanovskii, E. D. Belokolos, V. P. Gul'tai, and L. I. Lishcheta,
      Élektrokhimiya, 2:693 (1966).
43.   S. G. Mairanovskii, Dokl. Akad. Nauk SSSR, 142:1120 (1962).
44.   S. G. Mairanovskii and A. D. Krasnova, in: Electrochemical Processes Involving
      Organic Substances [in Russian], Nauka, Moscow (1969).
45.   S. G. Mairanovskii, Izv. Akad. Nauk SSSR, Otd. Khim. Nauk, 2140 (1961).
46.   V. F. Lavrushin, V. D. Bezuglyi, and G. G. Belous, Zh. Obshch. Khim.,
      33:1711 (1963).
47.   J. P. Stradins, Electrochim. Acta, 9:711 (1964).
48.   S. G. Mairanovskii, N. V. Barashkova, and Yu. B. Vol'kenshtein, Izv. Akad.
      Nauk SSSR, Ser. Khim., 1539 (1965).
49.   E. Laviron, Coll. Czechosl. Chem. Communs., 30:4219 (1965).
50.   A. A. Pozdeeva and S. I. Zhdanov, Polarography 1964, London, MacMillan
      (1966), p. 781.

51. S. G. Mairanovskii and V. N. Pavlov, in: Polarography and Kinetics of Chemical Reactions [in Russian], Information Bulletin No. 6, Goskomiteta po Khim. Prom. Pri Gosplane SSSR, IREA, Moscow (1964), p. 10.

52. K. Vettig, Élektrokhimiya, 3:269 (1967).

53. K. Wettig, Z. Phys. Chem. (Leipzig), 233:423 (1966).

54. S. G. Mairanovskii, L. A. Yanovskaya, N. V. Kondratova, and G. V. Kryshtal', Élektrokhimiya, 4:1127 (1968).

55. M. Fields and E. R. Blout, J. Am. Chem. Soc., 70:930 (1948).

56. Yu. M. Kargin, Ya. A. Levin, V. Z. Kondranina, and R. T. Safin, in: Advances in Electrochemistry of Organic Compounds [in Russian], Nauka, Moscow (1968), p. 38.

57. Ya. I. Tur'yan, M. S. Rusakova, B. F. Ustavshchikov, and V. A. Podgornova, in: Advances in Electrochemistry of Organic Compounds [in Russian], Nauka, Moscow (1968), p. 43.

58. G. K. Budnikov, in: Advances in Electrochemistry of Organic Compounds [in Russian], Nauka, Moscow (1968), p. 44.

59. Ya. I. Turyan, G. G. Kryukova, and A. V. Bondarenko, in: Advances in Electrochemistry of Organic Compounds [in Russian], Nauka, Moscow (1968), p. 42.

60. M. K. Polievktov and S. G. Mairanovskii, in: Advances in Electrochemistry of Organic Compounds [in Russian], Nauka, Moscow (1968), p. 36.

61. B. Perlmutter-Hayman and G. Stein, J. Phys. Chem., 63:734 (1959); J. Chem. Phys., 40:848 (1964).

62. M. Eigen and W. Kruse, Z. Naturforsch., 18b:857 (1963).

63. P. J. Elving, J. C. Komyathy, R. E. Van Atta, C. S. Tang, and I. Rosenthal, Analyt. Chem., 23:1218 (1951).

64. I. Rosenthal, C. S. Tang, and P. J. Elving, J. Am. Chem. Soc., 74:6112 (1952); P. J. Elving, I. Rosenthal, and L. M. Markovitz, J. Electrochem. Soc., 101:195 (1954).

65. B. N. Afanas'ev, in: Advances in Electrochemistry of Organic Compounds [in Russian], Nauka, Moscow (1968), p. 76.

66. S. Wawzonek, E. W. Blaha, R. Berkey, and M. E. Runner, J. Electrochem. Soc., 102:235 (1955).

67. P. H. Given and M. E. Poever, J. Chem. Soc., 385 (1960).

68. P. H. Given and M. E. Poever, Nature, 184:1064 (1959).

69. É. S. Levin, Dokl. Akad. Nauk SSSR, 151:1375 (1963).

70. S. Lasarov, A. Trifonov, and T. Vitanov, Z. Phys. Chem. (Leipzig), 226:221 (1964).

71. V. S. Bezuglyi and Yu. P. Ponomarev, Report Presented at Sixth Conference on Electrochemistry of Organic Compounds [in Russian], Inst. Élektrokhimii Akad. Nauk SSSR, Moscow (1968).

72. V. P. Gyl'tyai, S. G. Mairanovskii, and N. K. Lisitsina, in: Advances in Electrochemistry of Organic Compounds [in Russian], Nauka, Moscow (1968), p. 37.

73. L. V. Kononenko, V. D. Bezuglyi, and V. N. Dmitrieva, in: Advances in Electrochemistry of Organic Compounds [in Russian], Nauka, Moscow (1968), p. 57.

74. Yu. A. Davydovskaya and Yu. I. Vainshtein, in: Advances in Electrochemistry of Organic Compounds [in Russian], Nauka, Moscow (1968), p. 61.

75. S. G. Mairanovskii, I. A. D'yachenko, and Ya. L. Gol'dfarb, Élektrokhimiya, 5 (1969).

76. S. G. Mairanovskii, V. M. Belikov, I. B. Korchemnaya, V. A. Klimova, and S. S. Novikov, Izv. Akad. Nauk SSSR, Otd. Khim. Nauk, 1787 (1960).

77. G. P. Senikov and L. M. Karpova, Report Presented at Sixth Conference on Electrochemistry of Organic Compounds [in Russian], Inst. Élektrokhimii Akad. Nauk SSSR, Moscow (1968).

78. D. L. Maricle and W. G. Hodgson, Analyt. Chem., 37:1562 (1965).

79. E. L. Johnson, K. H. Pool, and R. E. Hamm, Analyt. Chem., 38:183 (1966).

80. G. S. Shapoval, S. G. Mairanovskii, N. P. Markova, and E. M. Skobets, in: Electrochemical Processes Involving Organic Substances [in Russian], Nauka, Moscow (1969).

81. W. M. Kemula and Z. R. Grabowski, Compt. Rend. Soc. Sci. et Lettre de Varsovic, Classe III, Sci. Math. et Phys., 44:78 (1951).

82. L. G. Feoktistov and S. I. Zhdanov, Izv. Akad. Nauk SSSR, Otd. Khim. Nauk, 45 (1963).

83. G. S. Shapoval, E. M. Skobets, and N. P. Markova, Dokl. Akad. Nauk SSSR, 173:392 (1967).

84. Z. Grabowski and E. Bartel, Roczn. Chem., 34:611 (1960).

85. A. Vincenz-Chodkowska and Z. R. Grabowski, Electrochim. Acta, 9:789 (1964).

86. V. D. Bezuglyi, Theory and Practice in Polarographic Analysis [in Russian], Izd. Shtiintsa, Kishinev (1962), p. 14.

87. R. G. Baisheva, U. I. Khurgin, and S. G. Mairanovskii, Report Presented at Sixth Conference on Electrochemistry of Organic Compounds [in Russian], Inst. Élektrokhimii Akad. Nauk SSSR, Moscow (1968).

88. S. G. Mairanovskii and L. I. Lishcheta, Izv. Akad. Nauk SSSR, Otd. Khim. Nauk, 227 (1962).

89. Ya. Koutetskii, V. Ganush, and S. G. Mairanovskii, Zh. Fiz. Khim., 34:651 (1960).

90. S. G. Mairanovskii, Ya. Koutetskii, and V. Ganush, Zh. Fiz. Khim., 36:2621 (1962).

91. G. A. Tedoradze and A. L. Asatiani, Report Presented at Sixth Conference on Electrochemistry of Organic Compounds [in Russian], Inst. Élektrokhimii Akad. Nauk SSSR, Moscow (1968).

92. A. L. Asatiani and G. A. Tedoradze, Report Presented at Sixth Conference on Electrochemistry of Organic Compounds [in Russian], Inst. Élektrokhimii Akad. Nauk SSSR, Moscow (1968).

93. S. G. Mairanovskii, Izv. Akad. Nauk SSSR, Otd. Khim. Nauk, 615 (1953).

94. S. G. Mairanovskii, Dokl. Akad. Nauk SSSR, 142:1327 (1962).

95. S. G. Mairanovskii and A. D. Filonova, Élektrokhimiya, 3:1397 (1967).

96. A. B. Érshler, G. A. Tedoradze, and S. G. Mairanovskii, Dokl. Akad. Nauk SSSR, 145:1324 (1962).

97. L. Hirst, J. Tonder, R. Cornelissen, and F. Lami, in: Fundamental Problems in Modern Theoretical Electrochemistry [Russian translation], Mir, Moscow (1965), p. 425.

98. E. Knobloch, Coll. Czechosl. Chem. Communs., 31:4503 (1966).

99. R. A, F. Bullerwell, J. Polarogr. Soc., 10:55 (1965).

100. R. Brdička, Coll. Czechosl. Chem. Communs., 5:112, 148 (1933).

101. R. Brdička, Research, 1:25 (1947).

102. S. G. Mairanovskii and É. F. Mairanovskaya, Report Presented at Sixth Conference on Electrochemistry of Organic Compounds [in Russian], Inst. Élektrokhimii Akad. Nauk SSSR, Moscow (1968).

103. Ya. I. Tur'yan, Dokl. Akad. Nauk SSSR, 113:631 (1957).

104. S. G. Mairanovskii, Usp. Khim., 33:75 (1964).

105. V. F. Toropova, T. A. Preobrazhenskaya, and V. E. Bashkirtseva, Élektrokhimiya, 3:1250 (1967).

106. V. F. Toropova and L. A. Anisimova, Report Presented at Sixth Conference on Electrochemistry of Organic Compounds [in Russian], Inst. Élektrokhimii Akad. Nauk SSSR, Moscow (1968).

107. M. Březina and V. Gultjaj, Coll. Czechosl. Chem. Communs., 28:181 (1963).

108. M. Kuik and A. Basinski, Roczn. Chem., 39:1327, 1701 (1965).

109. B. A. Kuznetsov, Report Presented at Sixth Conference on Electrochemistry of Organic Compounds [in Russian], Inst. Élektrokhimii Akad. Nauk SSSR, Moscow (1968).

110. B. A. Kuznetsov and M. I. Yakusheva, Report Presented at Sixth Conference on Electrochemistry of Organic Compounds [in Russian], Inst. Élektrokhimii Akad. Nauk SSSR, Moscow (1968).

111. V. N. Leibzon, V. M. Belikov, and S. G. Mairanovskii, Élektrokhimiya, 4:290 (1968).

112. V. N. Leibzon, V. M. Belikov, and S. G. Mairanovskii, Report Presented at Sixth Conference on Electrochemistry of Organic Compounds [in Russian], Inst. Élektrokhimii Akad. Nauk SSSR, Moscow (1968).

113. A. N. Frumkin, O. A. Petrii, and N. V. Nikolaeva-Fedorovich, Dokl. Akad. Nauk SSSR, 136:1158 (1961).

114. S. G. Mairanovskii and O. M. Dolgaya, Élektrokhimiya, 4:879 (1968).

115. D. Jannakoudakis, A. Wildenau, and L. Holleck, J. Electroanalyt. Chem., 15:83 (1967).

Stereochemistry of Electrode Processes

L. G. Feoktistov

The present review has the aim of comparing available published data on the effect of stereochemical factors on the kinetics and mechanism of electrode reactions involving organic compounds and on the stereochemistry of the electrode reaction products, but setting aside the numerous data on steric hindrance of coplanarity in molecules caused by substituents. A considerable number of papers have now been published either completely or partially devoted to these topics, but the last attempt to generalize information about the effect of stereochemical factors on the polarographic behavior of organic compounds was made a fairly long time ago [1]; information on the stereochemistry of electrochemical reaction products has not been treated systematically at all.

1. KINETICS AND MECHANISM OF REDUCTION IN STEREOISOMERS*

Geometrical (cis-trans) Isomers

The first observation of the effect of stereoisomerism on kinetics in the electrochemical reduction of organic compounds was made in 1928 by Herasymenko [2], who found a difference in the behavior of maleic and fumaric acids at the dropping mercury electrode. This pair of geometrical isomers, which are distinguished by fairly rigidly fixed orientation:

$$\begin{array}{cc} R_1 \quad\ \ R_1 & R_1 \quad\ \ R_2 \\ \diagdown\, C{=}C\,\diagup & \diagdown\, C{=}C\,\diagup \\ \diagup\ \ \ \ \ \ \diagdown & \diagup\ \ \ \ \ \ \diagdown \\ R_2 \quad\ \ R_2 & R_2 \quad\ \ R_1 \end{array}$$

---

*Data for anodic reactions are only available in papers by Zuman [86, 87] (phenyl-cysteine diastereoisomers) and Vincenc-Chodkowska and Grabowski [127] (cis- and trans-stilbenediols).

135

of the substituents in one plane, then became the topic for numerous electrochemical investigations [3-14].

It was found that electrochemical reduction of both isomers went as far as succinic acid [13] and that reduction took place at less negative potentials for the cis isomer than for the trans isomer. Since the state of the molecules in solution (their protonation and ionization) depends on the pH of the medium, the difference between the half-wave potentials in classical polarography or between the peak potentials in oscillographic polarography depends on the pH value and amounts to several tens of millivolts at the least negative reduction potentials (in acidic solutions) and several hundreds of millivolts at the most negative reduction potentials (in alkaline solutions) [7, 10].

Difference in the reduction rates of maleic and fumaric acids occurs not only at the mercury electrode but also at lead [12, 13], zinc [11, 12], and silver [12] electrodes. It is interesting to note that the differences in reduction rate are not the same on different faces of a lead single crystal. This effect is not noticeable with silver, although in both cases selectivity in reduction of the acids from their mixtures depends on the index of the crystal face.

Korchinskii [14] has measured the electrocapillary activity of maleic and fumaric acids and has shown that the cis isomer is adsorbed more strongly.

The polarographic behavior of other unsaturated acids and their derivatives [15-23] which exist in two isomeric forms has also been investigated. Frenkl [19], who investigated the polarographic behavior of cis,cis- and trans,trans-muconic acids ($COOHCH = CH - CH = CHCOOH$) in buffer solutions with pH values between 1.3 and 5.7, did not find any differences in the half-wave potentials ($E_{1/2} = -0.48-0.106$ pH). Since only one wave is observed and the wave heights for trans,trans-muconic and fumaric acids are identical, it must be assumed that 2-butene-1,4-dicarboxylic acid is evidently formed as a result of the reduction. Markman and Zinkova [20] investigated the polarographic behavior of four pairs of geometrically isomeric acids. In the case of citraconic (cis isomer) and mesaconic (trans isomer) acids, $COOHC(CH_3) = CHCOOH$, and cis- and trans-aconitic acids, $COOHCH_2C(COOH) = CHCOOH$, the difference in half-wave potentials amounted to 0.03-0.09 V depending on the pH value in aqueous solutions and increased to 0.11-0.21 V in aqueous ethanolic solutions. For example, $E_{1/2}$ is −0.68 V for citraconic acid and −0.72 V for mesaconic acid at pH 1.4 in aqueous solution and −0.75 and −0.96 V, respectively, in 75% ethanol, with 1 N hydrochloric acid as supporting electrolyte. Greater ease of reduction in cis isomers compared with trans isomers was also found with cinnamic ($C_6H_5CH = CHOCOH$) and crotonic ($CH_3CH = CHCOOH$) acids, the reduction of which takes place at more negative potentials than that of the dibasic acids. It should be noted that catalytic hydrogenation of the cis isomers of these acids on palladium and platinum also takes place more readily than hydrogenation of the trans isomers [21]. Citraconic and mesaconic acids were also investigated by Schwarz [15] and MacIntyre [18], and cis- and trans-aconitic acids by Semerano [16, 17]. In all cases considerable differences were also found in the limiting currents of the isomers. Elving and his co-workers [22], while investigating the polarographic behavior of diethyl esters of fumaric and maleic acids, found that in this case the cis and trans isomers were reduced in the opposite order, i.e., the cis isomer was reduced more difficultly than the trans isomer in acidic buffer solutions ($\Delta E_{1/2}$ ≈ 0.15 V); the difference in the half-wave potentials disappeared on passing to alkaline solutions. The conclusion about greater ease of reduction of the trans isomer in acidic solutions has been confirmed by our own latest measurements [41]. Meanwhile Takahashi and Elving [23] have obtained the following half-wave potentials in pyridine solution: −1.16 V (with reference to the pool and with 0.1 mole/liter lithium perchlorate as supporting electrolyte) for diethyl maleate; −1.25 V for diethyl fumarate. (The acids themselves, which exist in ionized form in pyridine, were not reduced before the discharge potential of the pyridine ion was reached.) Reduction of these esters in pyridine takes place in two steps; since diethyl succinate is polarographically inactive, the second wave must correspond to reduction of the radical anion. It is interesting to note that the order of reduction for the second waves is the same as for the first waves: $E_{1/2}$ = −1.35 V for diethyl maleate and −1.43 V for diethyl fumarate, i.e., the radical anions formed from the isomeric forms retain their configuration or two-electron transfer to the original molecules takes place at the second wave.

The isomers of stilbene ($C_6H_5CH = CHC_6H_5$) are also reduced at different potentials. Wessely and Wratil [24] found that in ethanolic solution the trans isomer was reduced at a potential 0.05 V less negative than the cis isomer, while in aqueous dioxane solution there was practically no difference. Data obtained by Wawzonek and co-workers [25], relating to anhydrous acetonitrile, show that cis-stilbene is reduced at a rather more negative potential than trans-stilbene, and that the reduction takes place in two steps ($E_{1/2}$ = −1.81 V, with reference to the pool, for the cis isomer and −1.73 V for the trans isomer; $E_{1/2}^{"}$ = −2.11 and −2.06 V respectively). In acetonitrile containing 2.5% water the reduction takes place in one step but in the same sequence ($E_{1/2}$ = −1.84 and −1.81 V), although the difference in the half-wave potentials is appreciably smaller. Data obtained by Sioda and co-workers [26] fall somewhat out of line with these results. In dimethylformamide, where two reduction waves are also observed with stilbene, $E_{1/2}^{'}$ = − 2.07 V (referred to the saturated calomel electrode) and $E_{1/2}^{"}$ = −2.36 V for the cis isomer and −2.08 V and −2.38 V, respectively, for the trans isomer. It is possible that no special significance should be attached to these results, since the $\Delta E_{1/2}$ values are very small. The following scheme has been adopted for reduction in anhydrous solvents [25]:

$$C_6H_5CH{=}CHC_6H_5 + e \rightarrow C_6H_5\dot{C}H\dot{C}HC_6H_5;$$

$$C_6H_5\dot{C}H\dot{C}HC_6H_5 + e \rightarrow C_6H_5\dot{C}H\dot{C}HC_6H_5;$$
$$C_6H_5\dot{C}H\dot{C}HC_6H_5 + \text{solvent} \rightarrow C_6H_5CH_2CH_2C_6H_5.$$

The second wave corresponds to the addition of an electron to the radical anion, and to explain the dependence of its half-wave potential on the structure of the original molecule within the framework of this scheme it is necessary to resort to the idea that the configuration of the original molecule is retained in the radical anion.

Evidence for the greater ease of reduction in the molecules of unsaturated hydrocarbons with phenyl substituents in the trans configuration is also provided by the data of Bezuglyi and his co-workers [27] for 1,4-distyrylbenzene $C_6H_5CH = CH - C_6H_4 - CH = CHC_6H_5$. The half-wave potentials of the first waves are −1.92 V for the cis,cis isomer, −1.88 V for the cis,trans isomer, and −1.86 V for the trans,trans isomer. At the same time the half-wave potentials of the second waves are practically the same (-2.12 to -2.14 V) for all three isomers.

Small differences in the half-wave potentials of the isomeric crotononitriles in dimethylformamide were obtained by Tomilov and co-workers (-1.90 V for the cis isomer and -1.89 V for the trans isomer) [28]. It was concluded from the infrared spectra of these compounds that in the cis isomer the π-electrons of the nitrile group interact with the carbon of the methyl group, leading to decreased polarity in the double bond.

Kemula and Kornacki [29] found agreement between the half-wave potentials of all three reduction waves for cis- and trans-1,4-diphenyl-1,4-ditert.-butylbutatrienes,

$$[(CH_3)_3C] [C_6H_5] C = C = C = C [C_6H_5] [C (CH_3)_3],$$

in dimethylformamide ($E'_{1/2} = -1.27$ V referred to the pool, $E''_{1/2} = -1.5$ V and $E_{1/2} = -2.16$ V). They proposed the following scheme for the reduction of butatriene:

$$>C=C=C=C< + e \rightleftarrows [>C=C=C=C<]^-$$

$$[>C=C=C=C<]^- + e \rightleftarrows [>C=C=C=C<]$$

$$[>C=C=C=C<] + \text{ solvent } \rightarrow >C=C=CH-CH<$$

$$>C=C=CH-CH< + 2e + \text{ solvent }$$

$$\rightarrow >CH-CH=CH-CH< + >C=CH-CH_2-CH<$$

$$>C=CH-CH_2-CH< + 2e + \text{ solvent } \rightarrow >CH-CH_2-CH_2-CH<$$

Baizer and co-workers [30] investigated the polarographic behavior of some geometrical isomers of two substituted diolefines, $C_2H_5OCOCH = CH - (CH_2)_n - CH = CH - COOC_2H_5$ (n = 3 and n = 4), and found that the trans,trans and cis,trans isomers were reduced at the same potential in dimethylformamide and formed ring compounds in accordance with the following proposed scheme:

$$
\begin{array}{ccccc}
CH{=}CH{-}COOC_2H_5 & & CH{-}\overset{\cdot\cdot}{C}H{-}COOC_2H_5 & & \overline{}CH{-}CH_2{-}COOC_2H_5 \\
\diagup & & \diagup & & | \\
(CH_2)_n & +e \rightarrow (CH_2)_n| & & +e+\text{solvent} \rightarrow (CH_2)_n & | \\
\diagdown & & \diagdown| & & \underline{}CH{-}CH_2{-}COOC_2H_5 \\
CH{=}CH{-}COOC_2H_5 & & CH{-}\overset{\cdot}{C}H{-}COOC_2H_5 & &
\end{array}
$$

Pasternak and co-workers [31, 32] investigated the polarographic behavior of geometrical isomers of dibenzoylethylene, $C_6H_5COCH=CHCOC_6H_5$. The half-wave potentials of the cis and trans isomers of this compound differ greatly in acidic solutions, where the cis isomer is reduced at more negative potentials, but the difference decreases on passing to alkaline solutions owing to a different relationship between the half-wave potentials and the pH value. According to more complete data obtained by Ryvolova [33], the half-wave potential at pH values between 1 and 9 is given by the relationships $E_{1/2} = -0.31-0.03$ pH (V, referred to saturated calomel electrode) for the cis isomer and $E_{1/2} = -0.07-0.06$ pH for the trans isomer. Ryvolova noticed that the wave of the trans isomer was very steep compared with that of the cis isomer, and the latter was moreover unsymmetrical. The differing relationships between half-wave potentials and pH value, the unequal slopes [33], and the unequal wave heights [32] of the stereoisomers give reason to suppose that in acidic solutions protonated molecules of the cis isomer exist and react at the electrode in the form of a seven-membered ring with a hydrogen bond:

$$
\begin{array}{c}
\text{H} \qquad\quad \text{H} \\
\text{C}=\text{C} \\
\text{C}_6\text{H}_5-\text{C} \qquad\quad \text{C}-\text{C}_6\text{H}_5 \\
\diagdown\text{O} \qquad \text{O}\diagup \\
\ddots\ \text{H}^+\ \ddots
\end{array}
$$

while molecules of the trans isomer exist and react in open form:

$$
\begin{array}{c}
\text{H}^+ \\
\text{O}\cdots \\
\text{C}_6\text{H}_5-\text{C} \\
\text{C}=\text{C}\ \ \text{H} \\
\text{H}\quad \text{C}-\text{C}_6\text{H}_5 \\
\text{O} \\
\text{H}^+
\end{array}
\qquad \text{or} \qquad
\begin{array}{c}
\text{C}_6\text{H}_5 \\
\text{O}=\text{C}\quad\ \text{H} \\
\text{H}^+\cdots\quad \text{C}=\text{C} \\
\text{H}\quad\ \text{C}=\text{O} \\
\text{C}_6\text{H}_5 \quad \text{H}^+
\end{array}
$$

Nevertheless the reduction evidently results in the same product (a saturated diketone). Ryvolova noticed that an appreciable quantity of the cis isomer appeared in a solution of the stable trans isomer after several minutes exposure to sunlight. Similar behavior is exhibited by esters of diketodicarboxylic acids, $ROCO(CH_2)_nCOCH$ $=CHCO(CH_2)_nCOOR$, [33] and by the methyl ester of 1-buten-3-one-1-carboxylic acid [34]; the isomers of the latter compound gave $E_{1/2}$ values of -0.68 and -0.82 V (saturated calomel electrode) at pH 5.53 and -0.71 and -0.92 V at pH 9.02, where the more negative potential belongs to the trans form. The behavior of the isomers of various unsaturated ketones was also investigated [35].

Kůta [36] has reported on the difference in polarographic behavior between the two isomers of retinol:

H_3C CH_3

H

H—CH=CH—C(CH_3)=CH—CH=CH—C(CH_3)=CH—CH$_2$OH,

H—

H CH_3

  H H

for which 16 stereoisomeric cis,trans forms are possible. The isomer with the trans disposition in all the substituents was found to be less polarographically active than the 13-cis isomer.

Mairanovskii and Bergel'son [37] investigated three pairs of isomeric compounds — the cis and trans isomers of 1-(1'-hydroxycyclohexyl)-1,2-dibromoethylene $C_6H_{10}(OH)CBr$ =CHBr, 3-hydroxy-1,2-dibromo-3-methyl-1-butene $(CH_3)_2C(OH)CBr$ =CBr(OH)C$(CH_3)_2$. The half-wave potentials in aqueous methanolic solution were -1.80 V (saturated calomel electrode) for the cis isomer and -1.15 V for trans isomer of the first compound, about -2 V (the wave merges with the discharge of the supporting electrolyte) and -1.42 V for the second compound, and -1.19 V and -0.72 V for the third compound. It was noticed that, while the slopes of the waves for the isomers differed appreciably in the case of the ethylene derivative (the wave of the cis isomer, reduced at more negative potentials, was steeper), the slopes were practically equal in the case of the hexene derivatives. The reduction involves two electrons and evidently forms the respective acetylene derivatives. According to data from Jura and Gaul [38], reduction of the isomeric 2,3-dichloroacrylonitriles $(CH_2Cl$ =CClCN) takes place at slightly different potentials and evidently leads to the formation of different products; the trans isomer is reduced in the usual way for vicinal dihalogen compounds to form a triple bond, while the cis isomer only eliminates one chlorine atom, which is replaced by hydrogen.

Elving and co-workers [39, 40] investigated the polarographic behavior of mono- and dibromosubstituted maleic and fumaric acids and their esters. Monobromomaleic acid is reduced to maleic, fumaric, and butadiene-1,2,3,4-tetracarboxylic acid, where the proportions of the products depend on the pH value. Monobromofumaric acid is reduced to fumaric acid. The esters of both acids are reduced to the esters of unsubstituted fumaric acid. Dibromomaleic and dibromofumaric acids and their esters are reduced to acetylene-dicarboxylic acid, and the half-wave potentials of the cis acid and its diethyl ester are less negative (by 0.04-0.18 and 0.10-0.12 V, respectively, depending on the pH value) than the half-wave potentials of the trans compounds.

Markova and Feoktistov [41] carried out a polarographic investigation of the three pairs of geometrically isomeric diesters of the simplest dihalogensubstituted dicarboxylic acids $C_2H_5OCOCX = CXCOOC_2H_5$ (where X = Cl, Br, or I) and the diethyl esters of chloroiodofumaric $(C_2H_5OCOCCl = CICOOC_2H_5)$ and chlorofumaric $(C_2H_5OCOCCl = CHCOOC_2H_5)$ acids. It was found that, in accordance with the earlier authors [39, 40], the diethyl esters of the dibromo-substituted acids were reduced to diesters of acetylenedicarboxylic acid regardless of the configuration, but the opposite order was found in ease of reduction; the trans isomers were reduced somewhat more readily than the cis isomers (by approximately 0.1 V) in both water–ethanol and water–dioxane solutions. The first wave, corresponding to the removal of two bromine atoms as a result of addition of two electrons without addition of protons (proved by means of microcoulometric measurements and measurement of "latent" currents of silver), does not depend on the pH value. Completely analogous behavior is found in another pair of isomers – diesters of diiodosubstituted acids – although in this case the difference in the half-wave potentials is very small (not exceeding 0.03 V). It was shown that the ability to suppress polarographic maxima of the second kind (on the reduction wave of Cu^{2+}) is greater with the cis isomers than with the trans isomers, which demonstrates the greater adsorbability of the cis isomers. The reduction mechanism of the chloroiodofumaric diester is similar to that of the dibromo and diiodo acids. The reduction mechanism for diesters of the dichloro-substituted acids was unexpectedly found to be completely different, although the order in which the isomers were reduced was the same. (The cis isomer is reduced more difficultly than the trans isomers, by approximately 0.1 V). In the case of the chloro acids in moderately acidic solutions (with pH values not less than 2) elimination of only one chlorine atom takes place on the first wave with the involvement of two electrons and one proton. In strongly acidic solutions (with pH values less than 2) diesters of the dichlorosubstituted acids are reduced with simultaneous addition of four electrons and three protons and elimination of one chlorine atom, which indicates formation of chlorosuccinic acid.

In accordance with this the half-wave potentials of the first waves for diesters of the dichlorosubstituted acids (corresponding to elimination of the halogens) reveal a dependence on the pH value. It is noteworthy that in this case it is the nature of the halogen and not the geometrical isomerism which has the deciding effect on the reduction mechanism in dihalogensubstituted acids and that the transitional states in the reduction of chloroiodofumaric and di-chlorofumaric acids are different. This is evidently explained by the fact that, in the molecules of the diesters of the dihalogen-substituted acids, there are several reaction centers (the halogen atoms and the oxygen atoms of the carbonyl groups which are capable of being protonated) which are favorably situated for conjugation:

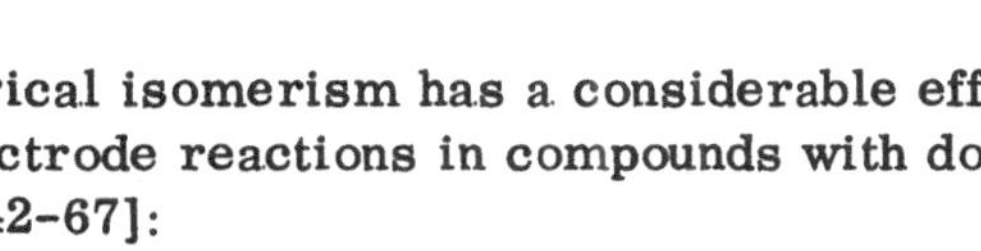

Geometrical isomerism has a considerable effect on the kinetics of electrode reactions in compounds with double-bonded nitrogen [24, 42-67]:

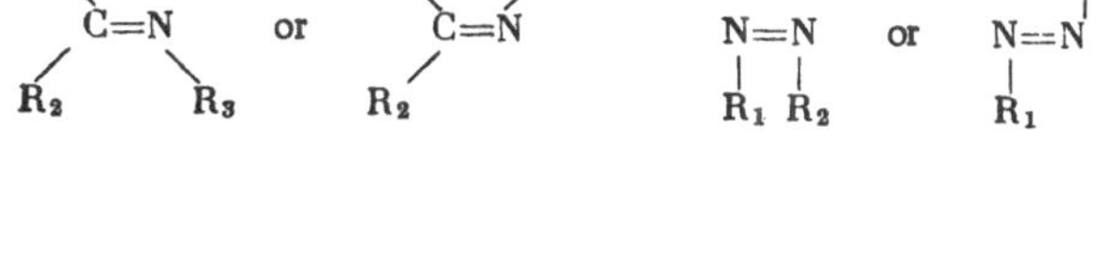

Winkel and Siebert [42] first investigated the cis and trans isomers of azobenzene, $C_6H_5N=NC_6H_5$. Azobenzene, and also azobenzenedisulfonic acid and helianthin, gave not one but two reduction waves. The half-wave potentials differed by approximately 0.2 V. While investigating the reduction of the usually more stable trans form they discovered that after irradiation of the solution with ultraviolet light a new wave appeared, and its half-wave potential coincided with the half-wave potential of the cis form. When a solution of cis-azobenzene was allowed to stand the wave of the trans isomer appeared. The considerable difference between the half-wave potentials of cis- and trans-azobenzenes [$E_{1/2} = -0.80$ V (saturated calomel electrode) for the cis isomer and -0.97 V for the trans isomer] was also preserved in 75% dioxane solution [24]. At the same time, on switching from alkaline to acidic solutions, the reduction was shifted to more positive potentials, but the difference between the half-wave potentials of the cis and trans isomers practically disappeared [48-50]. Elving and his co-workers [51] investigated the reduction of azobenzene at a pyrolytic graphite electrode, at which the azobenzene–hydrazobenzene system was found to be considerably less reversible than at the dropping mercury electrode. They also found no difference in the reduction potentials of the cis and trans isomers in acidic solution, but at pH 8-9 cis-azobenzene was reduced 0.2 V more positively than trans–azobenzene. In principle the behavior of the substituted azobenzenes does not differ from that of the unsubstituted compound [43-47, 52]. The decrease of the difference in reactivity of the isomers on switching from neutral to acidic solutions is evidently due both to an increase in the degree of reversibility of the azobenzene – hydrazobenzene system (which, however, cannot lead to complete coincidence of the equilibrium potentials for the cis-azobenzene–hydrazobenzene and trans-azobenzene–hydrazobenzene systems, since the free energy of the cis isomer is somewhat greater than that of the trans isomer) and to an increase in the interconversion rate of the isomers, as a result of which the most readily reduced isomer reacts at the electrode.

Freeman and Georgans [53] investigated the polarographic reduction of the isomeric diazosulfonates, $p-ClC_6H_4N=NSO_3K$. Here the trans (or anti) isomer is reduced at potentials 0.28 V more negative than the cis (or syn) isomer and is converted into the latter when exposed to ultraviolet light or sunlight.

The polarographic behavior of the geometrical isomers of aldoximes and ketoximes is much more complex [54-61]. Tyutyulkov [54] described the behavior of the α (syn or cis) and β (anti or trans) forms of benzaldoxime $C_6H_5CH=NOH$, p-tolyloxime $CH_3 \cdot C_6H_4 = NOH$, anisaldoxime p-$CH_3OC_6H_4CH = NOH$, piperonaldoxime

and o-chlorobenzaldoxime o-$ClC_6H_4CH = NOH$. The α forms give two waves, the first of which are kinetic. (Their heights depend on temperature and the nature of the solvent.) The β form gives one wave, the half-wave potential of which corresponds to that of the first wave of the α form. The difference in the half-wave potentials is very large (0.2-0.4 V), but it is related not so much to differing reactivities of the geometrical isomers in the electrode reaction as to nitrone – oxime tautomerism in the α form. Paspaleev [61] found that in aqueous ethanolic solutions of neutral salts both the α (syn or cis) and β (anti or trans) isomers of the ketoximes each give two waves, the first of which has both diffusion and kinetic components. The kinetic contributions are ascribed to tautomeric transformation of the $>C=N-OH$ group into the $>C=NH \rightarrow O$ grouping, which is reduced more readily. In addition to this reaction in solutions of the oximes another, more slowly established equilibrium occurs between the α and β forms, the reduction of which corresponds to the second waves. Changes thus take place according to the following scheme:

$$C_6H_5-\overset{\underset{\|}{O\leftarrow NH}}{C}-C_6H_4X \rightleftarrows C_6H_5-\overset{\underset{\|}{HO-N}}{C}-C_6H_4X \rightleftarrows C_6H_5-\overset{\underset{\|}{N-OH}}{C}-C_6H_4X \rightleftarrows C_6H_5-\overset{\underset{\|}{NH\rightarrow O}}{C}-C_6H_4X$$

The following half-wave potentials have been quoted for 4-chlorobenzophenone oxime and 4-methoxybenzophenone oxime: for the α isomers, $E'_{1/2} = -1.18$ V (sat. cal. elect.) and $E''_{1/2} = -1.51$ V for the chlorosubstituted compound in 50% ethanol, and $E_{1/2} = -1.08$ and -1.70 V for the methoxysubstituted compound in 30% ethanol; for the respective β isomers, $E'_{1/2} = -1.16$ V, $E''_{1/2} = -1.49$ V and $E''_{1/2} = -1.09$ V, $E''_{1/2} = -1.67$ V [16].

Lund [59] investigated the polarographic behavior of the syn and anti isomers of benzaldoxime and cinnamaldoxime C_6H_5CH $= CHCH = NOH$ and reduced them electrolytically at a mercury electrode with controlled potential. The half-wave potentials depend largely on pH at pH values between 1 and 13 and coincide for the isomers in the acidic region; small differences (0.01–0.09 V) in the half-wave potentials are only observed in the alkaline region, where kinetic contributions (which are more appreciable with the syn isomer) begin to take effect and where the syn isomer of cin-namaldoxime gives two waves.

The reduction products in acidic solutions differ from those in alkaline solutions. In the first case reduction of the anti isomer of cinnamaldoxime gives mainly the respective unsaturated amine with a small amount of the saturated aldehyde, while in the second case the saturated aldehyde is the main product.

The effect of stereoisomerism is clearly seen in the polaro-graphic behavior of arylhydrazones, and the relationships between the isomers are illustrated in the following scheme:

$$
\begin{array}{cccc}
\underset{\underset{R_1}{\overset{\|}{C}}\nearrow\underset{R_2}{}}{N-NH-R_3} &
\underset{\underset{R_2}{\overset{\|}{C}}\nearrow\underset{R_1}{}}{N-NH-R_3} &
\underset{\underset{R_1}{\overset{\|}{C}}\nearrow\underset{R_2}{}}{R_3-NH-N} &
\underset{\underset{R_2}{\overset{\|}{C}}\nearrow\underset{R_1}{}}{R_3-NH-N}
\end{array}
$$

An investigation of the polarographic behavior of the aryl-hydrazones of aldehydes and ketones was carried out by Arbuzov and Kitaev [62–64]. On the polarograms of benzaldehyde phenyl-hydrazone a new wave appeared at less negative potentials after illumination, and was attributed to the reduction of the more labile of the two stereoisomers. Brockman and Pearson [65] investigated the polarographic reduction of stereoisomeric pairs of substituted benzophenone semicarbazones. The difference in half-wave poten-tial between the α- and β-semicarbazones of 4-methoxybenzophenone, $NH_2CONH - N = C (C_6H_5) C_6H_4OCH_3$, and 4-bromobenzophenone, $NH_2CONH - N = C (C_6H_5) C_6H_4Br$, was found to be small ($\Delta E_{1/2} = 0.02$–0.03 V). Kitaev and Budnikov [66] observed a new wave at more negative potentials on the polarograms of semi- and thiosemi-carbazones of aromatic aldehydes and ketones after the solutions had been exposed to ultraviolet light.

Geometrical isomerism also occurs in azoxy compounds:

and their polarographic behavior has been investigated by Costa [44, 45] and Vainshtein[67]. In aqueous solution the half-wave potentials of the β isomers differ by 0.02-0.03 V from those of the α isomers, which are reduced earlier [44, 45]. In dimethylform-amide the reduction takes place in two steps, and the difference between the half-wave potentials of the benzeneazoxy-p-cresol and 2'-methoxy-6-hydroxy-3-methylazoxybenzene isomers amounts to 0.05-0.13 V for both the first and second waves.

The information given in this section about the effect of geo-metrical isomerism on the kinetics of electrolytic reduction in organic compounds shows that the less stable cis isomers, which have a large reserve of free energy, are by no means always more readily reduced than the trans isomers. Since reduction takes place irreversibly, it is frequently found that thermodynamic factors are less important than structural characteristics which facilitate trans-mission of the effect along the conjugation chain, ring formation, etc.

Cis,trans Isomers in Alicyclic Series

In the alicyclic series the inability of the substituents to shift from one side of the ring to the other gives rise to the existence of stereoisomers which, like the geometrical isomers in the ethylene series, are called cis,trans isomers. In addition, since the direc-tions of the bonds at the carbon atoms of the rings can be different (at each atom one bond, the equatorial or e-bond, is directed to-wards the periphery of the molecule and the other, the axial or a-bond, is directed along an axis perpendicular to the plane of the molecule), some isomers are capable of existing in different con-formations. The simplest example is provided by 1,2-disubstituted cyclohexane derivatives:

cis - 1a, 2e trans – 1a, 2a trans - 1e, 2e

Unlike the cis,trans change, the change from the 1a,2a conformation
to the 1e,2e conformation can be accomplished without breaking the
bonds. Thus, trans-1,4-dihalogenocyclohexanes exist exclusively
in the diequatorial conformation (1e,4e) in the solid state, while in
solution either this or the diaxial conformation (1a,4a) can predom-
inate depending on the solvent.

All examples in the literature [69-82] of the effect of stereoisomerism in this type of compound on the electrochemical behavior relate to halogen derivatives, and it is therefore appropriate
to examine briefly those problems in the mechanism of electrolytic
reduction of halogen derivatives which are of direct significance with
respect to the topic under discussion. From a critical examination
of the data it was earlier concluded [68] that reduction of the halogen
derivatives took place with the negative ends of the molecular dipoles
directed towards the electrode surface, i.e., the halogen atom was
directly adjacent to the electrode surface so that in some cases it
was even possible for surface compounds of the RHgX type to form
with the electrode metal. For some time another point of view on
the reduction of halogen derivatives, based on concepts developed
for homogeneous reactions, has been discussed in the literature.
In essence it assumes that during reduction electrons pass on the
molecule towards the carbon atom to which the halogen atom is
attached from the side opposite to the halogen. However, recent
investigations by Lambert and co-workers [69, 72], Sease and co-
workers [70], and Krupička and co-workers [71] have shown that
this point of view is untenable, as also is the idea of preceding dissociation of the carbon—halogen bond in the field of the double electric
layer to form a carbonium ion. Lambert finally concluded that
attack of the carbon—halogen bond by the electron took place in a
direction perpendicular to its axis and led to the formation of the
respective radical anion [72]. This conclusion is completely consistent with the idea put forward earlier by Cisak [76] and ourselves
[73] on the formation, in the reduction of halogen derivatives, of a
transitional complex involving the electrode metal atom.

Cisak [74-76] investigated the polarographic behavior of a
number of halogen-substituted cyclohexane derivatives. In 50%
ethanol cis-1,4-dibromocyclohexane gives a prewave, beginning at
−1.5 V, due to formation of some kind of surface compound (an

intermediate reduction product with mercury) and a main wave at
−2.0 V, while trans-1,4-dibromocyclohexane gives only one wave at
−2.2 V. It is interesting to note that reduction of both isomers takes
place in one step with the participation of four electrons and the
formation of cyclohexane, i.e., successive elimination of bromine
atoms to form monobromocyclohexane derivatives at the first stage
does not occur. Both isomers are distinguished by their ability to
be adsorbed on the mercury surface, which was established from
the suppression of polarographic maxima of the second kind and from
electrocapillary curves [76]. The cis isomer is adsorbed considerably more strongly than the trans isomer and approaches iodo-
cyclohexane in this respect. Different features are observed in the
stereoisomers of 1,2-dibromocyclohexane. The cis isomer is
reduced at −1.85 V, while the trans isomer is reduced much earlier
at −1.0 V and is more strongly adsorbed, although not so strongly
as the cis isomer of the 1,4-substituted cyclohexane. Reduction of
both compounds involves the participation of two electrons and
evidently leads to cyclohexene.

Compared with the dibromo derivatives the cis and trans
isomers of 1,4-diiodocyclohexane exhibit more complex but less
divergent behavior. In 50% ethanol both isomers give a two-step
prewave, which commences at potentials 0.1 V more negative with
the cis isomer than with the trans isomer. On switching to 90%
ethanol the commencement of reduction is shifted much more
strongly towards more negative potentials with the cis isomer than
with the trans isomer.

While dichlorocyclohexanes are not reduced even up to very
negative potentials, hexachlorocyclohexanes are reduced comparatively readily. Very great differences are found here in the half-
wave potentials of the various isomers, ranging from −1.1 V for
γ-1,2,3,4,5,6-hexachlorocyclohexane (aaaeee conformation) to −2.1
V for ε-1,2,3,4,5,6-hexachlorocyclohexane (aeeaee conformation).

Effects due to mutual repulsion of the substituents, which
weakens the carbon−halogen bond [76], and effects due to mutual
orientation of the carbon−halogen bonds [77] are not sufficient to
explain such diverse behavior in cyclohexane derivatives. A
considerably more definite relationship between the change in half-
wave potentials with change in the spatial distribution of the substituents was found by Zavada and co-workers [77], who investigated

the reduction of a number of vicinal dibromides in dimethylformamide. While the half-wave potentials of the monobromides (nine
compounds ranging from butyl bromide to endo-8-bromobicyclo-
[1,2,3]-oct-2-ene, were investigated) change at the most by 0.26 V,
which demonstrates the relatively small effect of steric hindrances
created by the hydrocarbon radical, the half-wave potentials of the
vicinal dibromides (21 compounds were investigated) vary between
−0.82 and −1.67 V. Particularly large differences are found with
such pairs of stereoisomers as cis- and trans-1,2-dibromocyclo-
hexane (0.6 V, which agrees well with data in [76]), exo- and endo-
cis-2,3-dibromobicyclo-[1,2,2]-heptane (0.32 V):

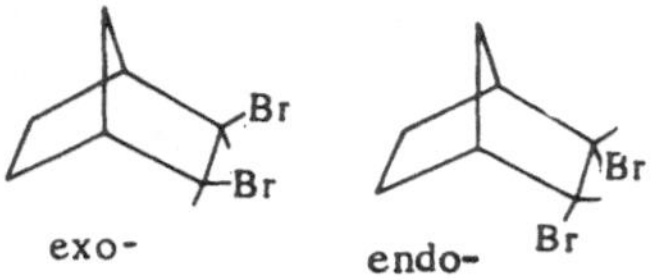

and 1-terbutyl-trans-3-cis-4-(two axial or aa-bonds) and 1-terbutyl-
cis-3-trans-4-dibromocyclohexane (two equatorial or ee-bonds)
(0.61 V):

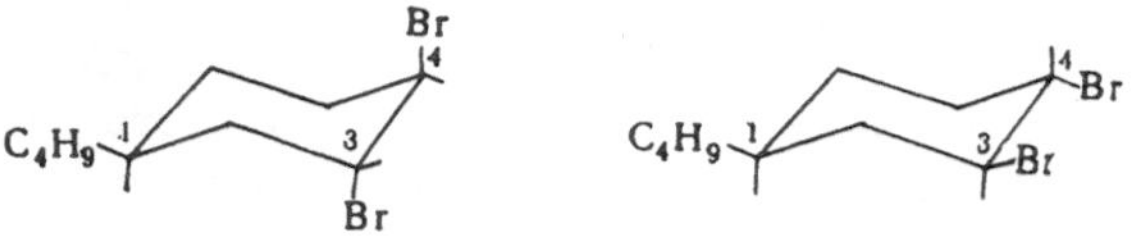

The principal reason for such a large difference in the half-wave
potentials of isomeric compounds is considered [77] to be the differing orientations of the carbon−bromine bonds with respect to each
other. In fact the graphical relationship between the half-wave
potential and the angle between the directions of the carbon−bro-
mine bonds (Fig. 1) shows that compounds in which the carbon−bro-
mine bonds are perpendicular to each other are more difficult to
reduce than compounds with parallel and particularly with anti-
parallel bonds. Similar relationships are also found with various
homogeneous bimolecular elimination reactions. It should be noted
that the slopes of the waves for the vicinal dibromides are different,
and the latter can be divided into two groups in respect of this feature; the slopes of one group (0.18 ± 0.04) are half those for the
other (0.36 ± 0.07), and individual members of each of the above-
mentioned stereoisomeric pairs sometimes fall into different groups.

Cisak [78] investigated three pairs of stereoisomers. With two of them [aldrin (endo-exo form) and isodrin (endo-endo form)]:

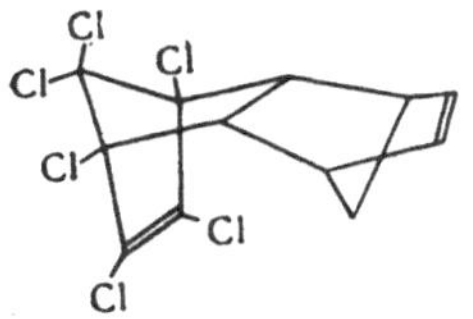

endo-exo 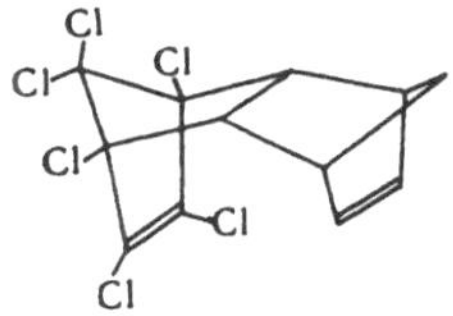   endo-endo

and their oxides (dieldrin and endrin) stereoisomerism does not lead to any substantial change in the state of the carbon—chlorine bonds and only affects the unsubstituted part of the molecules; the half-wave potentials of these isomers differ by 0.05 V (in 90% ethanol). In the third pair of isomers (α- and β-chlordanes, $C_{10}H_6Cl_8$):

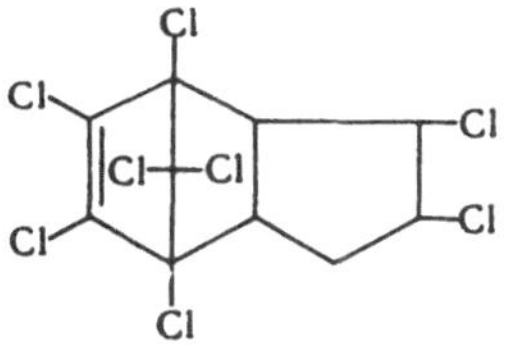

the isomerism arises from the mutual position of the two carbon —chlorine bonds in the least chlorinated ring, and this is largely reflected in the difference between the half-wave potentials, which amounts to 0.2 V.

Differences in half-wave potentials due to stereoisomerism are also observed in the case of halogenocyclohexanones [79-81]. In media with sufficient proton-donating activity their behavior is complicated by protonation of the keto group, while in aqueous solutions clearly defined effects arising from hydration of the keto group are observed. According to data obtained by Quan [79], hydration leading to a decrease in the limiting current is more strongly expressed with trans-2,5-dichlorocyclohexanone than with its cis isomer. In dimethylformamide solution the two isomers behave in the same way, and the half-wave potentials and diffusion current constants are identical.

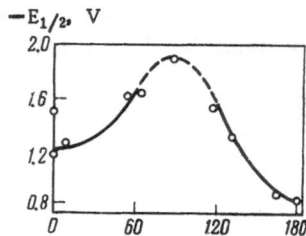

Fig. 1. Dependence of half-
wave potential for reduction
of vicinal dibromides on angle
(Θ) between directions of
carbon – bromine bonds [77].

Wilson and Allinger [80] observed different behavior in the
stereoisomers of 2-halogeno-4-
(tert.-butyl...)cyclohexanones in the
case of the bromine, chlorine, and
fluorine compounds, while investigating their polarographic reduction in
dimethylformamide. Owing to the
large size of the tertiary butyl group
it is situated almost exclusively in
the equatorial positions, and consequently the halogens are situated
at the equatorial positions in the cis
isomers and at the axial positions in
the trans isomers:

The half-wave potentials of the cis and trans isomers of the bromine
derivatives, which are readily reduced, almost coincide. The difference becomes more appreciable on passing to the chlorine and
particularly the fluorine compounds, where it amounts to 0.12 and
0.23 V, respectively. Here the trans isomers are reduced more
readily than the cis isomers. It should be noted that the reduction
evidently involves the keto group, which is the most electrophilic
part of the molecule:

$$O=C-C\overset{X}{\underset{}{<}} + 2e \rightarrow \bar{O}-C=C< \rightarrow O=C-CH$$

The relationships observed in the simple halogen-substituted
cyclohexanes are important for understanding the behavior of
steroids, where the cyclohexane ring forms one of the fragments in
the structure. The polarographic behavior of α-halogenoketo-
steroids was investigated by Delaroff and co-workers [81] and
Kabasakalian and McGlotten [82]. The latter came to the conclusion
that the thermodynamically more stable compounds were reduced

more difficultly, regardless of whether the halogen atom (in the five-membered or six-membered ring) was in the axial or the equatorial position.

In concluding this account of the published information on the effect of cis,trans isomerism in compounds of the alicyclic series on their reactivity at electrodes it can be mentioned that differences in the behavior of the isomers arise from both thermodynamic and nonthermodynamic reasons.

Optical Isomers (Diastereoisomers)

In the general case compounds containing two asymmetric carbon atoms each have two pairs of stereoisomers (diastereoisomers), where each member of either of the distereoisomeric pairs differs from the other as an object from its mirror image and forms with it an optically inactive racemic compound. If both asymmetric carbon atoms have the same set of substituents, one of the diastereoisomeric pairs degenerates and represents a single compound where one half of the molecule is the mirror image of the other. Such molecules are optically inactive (meso form) owing to internal compensation of the rotation of the plane of polarized light. Finally, if all the substituents at each of the asymmetric atoms are not identical a distinction can be made betweeen erythro and threo compounds, where the erythro series is analogous to the meso series.

The reactivity of these isomers in electrode reactions has been investigated in a number of papers [22, 27, 83-87]. Elving and co-workers [22, 77] investigated the polarographic behavior of racemic and meso-dibromosuccinic acids and their diethyl esters and found that the half-wave potentials of the diastereoisomers were somewhat different.

racemate   meso−

The racemic acid was reduced somewhat more readily than the meso acid in acidic solution and with greater difficulty in neutral and weakly alkaline solutions. The difference in the half-wave potentials amounted to 0.05-0.07 V and the half-wave potentials of the esters differed even less. It is interesting that the reduction products from the diastereoisomers differ in the case of the acids and that the same product is always obtained in the case of the esters (further details below). McKeon [84] found much greater differences in the half-wave potentials of the meso and racemic compounds, amounting to 0.2-0.3 V, in the case of 1,4-dihydroxy- and 1,4-dinitroxy-2,3-dibromobutanes. The latter give two waves, in the first of which the dihydroxy compound is formed; in the second, as in the single wave of a dihydroxy compound, the respective unsaturated compound is evidently formed:

$$NO_2OCH_2CHBrCHBrCH_2ONO_2 + 4e \rightarrow HOCH_2CHBrCHBrCH_2OH + 2e$$
$$\rightarrow HOCH_2CH = CHCH_2OH.$$

The reduction rate of the dihydroxy compounds does not depend on the pH value. The meso form ($E_{1/2} = -0.67$ V, saturated calomel electrode) is reduced more readily than the racemic compound ($E_{1/2} = -0.96$ V), which is characteristic of homogeneous elimination reactions. The meso form of 5,6-dibromodecane is also reduced more readily in dimethylformamide (at 0.10 V) than the diastereo-isomer [77].

Malspeis and Hung [85] investigated the isomeric N-nitro-soephedrine (erythro compound) and N-nitrosopseudoephedrine (threo compound):

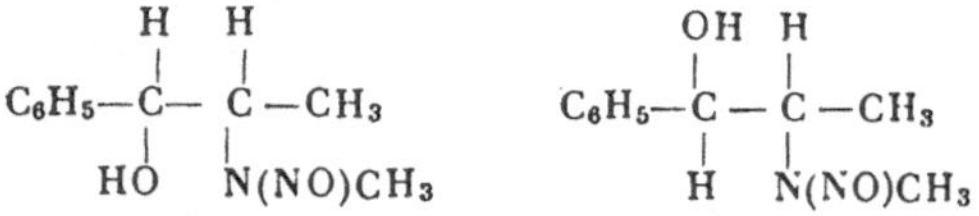

and found from the values of the kinetic currents at pH 3-8 that the protonation rate of the erythro compound, the configuration of which does not favor formation of an intramolecular hydrogen bond, was higher than that of the threo compound. Zuman [86, 87] investigated the catalytic hydrogen waves due to phenylcysteine in ammoniacal

solution of divalent cobalt salts. The diastereoisomers of phenyl-
cysteine were found to be unequal, and the erythro form gave a
higher wave than the threo form owing to differences in basicity and
in complex-forming ability with respect to the cobalt ion. Some
differences were also observed on the anodic waves of these diastereoisomers.

Dehydrotetramethylhollarhimine (α form) and dehydrotetra-
methylhollaridine (β form), which only differ in the position (axial
or equatorial) of one of the carbon—nitrogen bonds:

α-form   β-form (fragment)

were found to differ greatly in their adsorption characteristics.
Thus, the concentration needed to give the same (half) decrease in
the height of the polarographic oxygen maxima is six times greater
with the β form than with the α form. In alkaline ethanolic solution
the α form (unlike the β form) causes a large decrease in charging
current, which increases to the normal value fairly rapidly as a
result of desorption of the compound at −1.6 V [86].

The information in this section shows that structural differences in molecules with different composition have different
effects on their reactivity in electrode reactions. A substantial part
is played by difference in adsorption and acid—base characteristics
of the isomers, which are related both to mutual polarizability of
the bonds (which varies with their mutual orientation) and to steric
hindrances.

2. STEREOCHEMISTRY OF ELECTRODE REACTION PRODUCTS

Oxidation Reactions

In recent years several papers have appeared [88-98] in which
the stereochemistry of products from electrochemical oxidation of

organic compounds has been investigated. Inoue and co-workers [88] found that electrochemical oxidation of methanol on platinum electrodes in the presence of stilbene isomers gives a mixture of diastereoisomeric dimethyl ethers of hydrobenzoin: 20% meso and 40% racemic compound are obtained in the case of trans-stilbene and 26.3% meso and 17.5% racemic compound in the case of cis-stilbene. In the opinion of Inoue these results indicate that cis-addition of the primary methanol oxidation product to the unsaturated compounds predominates over trans-addition. The initial product is considered to be a methoxy radical, and the overall reaction is written in the following form:

$$CH_3O^- - e \rightarrow CH_3O^{\cdot} + C_6H_5CH = CHC_6H_5 \rightarrow C_6H_5CH(OCH_3)CH(OCH_3)C_6H_5.$$

The stereochemical relationships are illustrated in the following scheme:

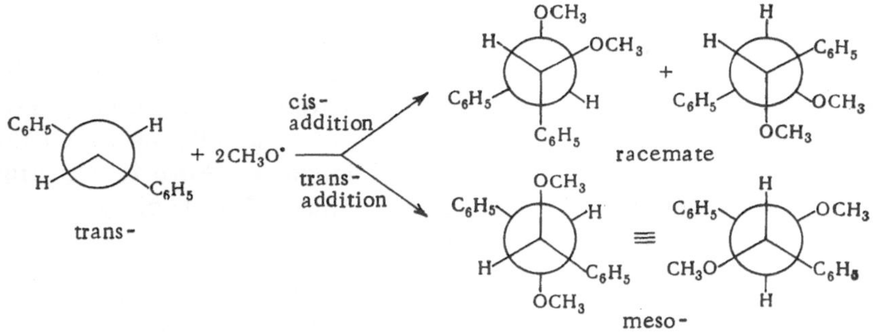

Further, it was shown [89] that electrolysis of benzoic acid in acetonitrile with triethylamine and cis- and trans-stilbene gives only meso-hydrobenzoin dibenzoate irrespective of the stilbene configuration. The authors attempted an explanation of this unexpected result on the assumption that stepwise oxidation and inversion of an intermediate unstable product occur. An important part is played here by adsorption and oxidation of stilbene itself at the electrode, although no evidence was obtained for its possible isomerization under the experimental conditions.

The stereochemistry of products from anodic oxidation of carboxylic acids has also been investigated [90-95]. Paudler and co-workers [90] found that β-truxinic acid (3,4-diphenylcyclobutane-1,2-dicarboxylic acid) gave a mixture of products, including 1,4-diphenyl-trans,trans-1,3-butadiene. The electrolytic oxidation products differ greatly from those obtained by oxidation of truxinic acid with lead tetraacetate. Dauben and Liang Khiao-Tian [91] investigated the stereochemistry of radicals formed during the anodic oxidation of methylethylacetic acid $CH_3CH_2(CH_3)CHCOOH$. In the oxidation of exo- and endo-norbornane-2-carboxylic acids at a platinum electrode in methanol Corey and co-workers [92] obtained only exo-norbornyl methyl ether:

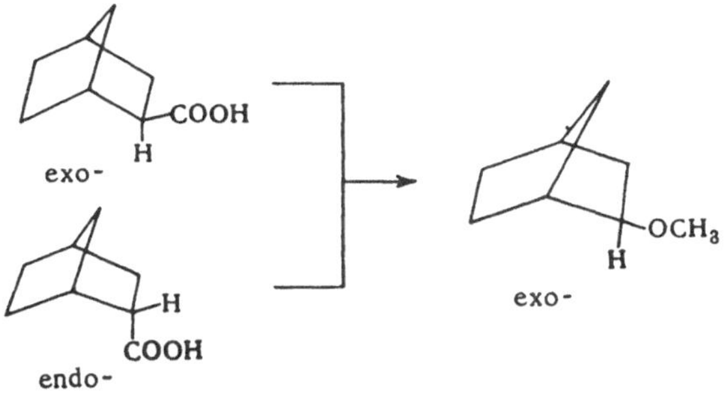

exo-

endo-

exo-

with a yield of 35-40%; oxidation of the optically active endo-norbornane-2-carboxylic acid gave a racemic mixture of ethers. Electrolysis of exo- and endo-5-norbornene-2-carboxylic acid leads to 3-methoxynortricyclene, and this evidently takes place through a carbonium ion as intermediate oxidation product:

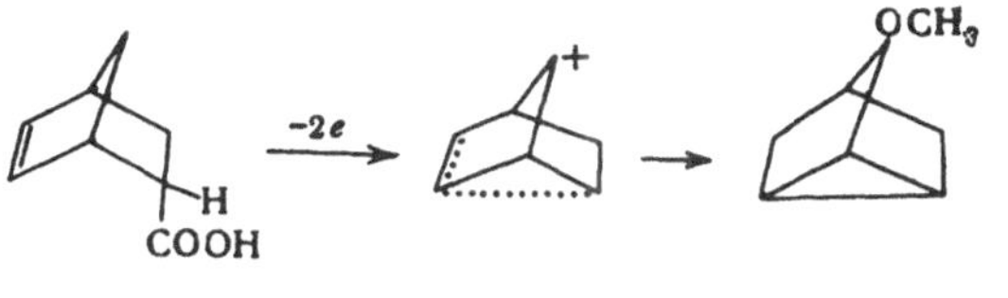

During oxidation of bicyclo-[0,1,3]-hexane-3-carboxylic acid in a mixture of pyridine and water (1 : 3) Gassmann and Zalar [95] obtained not the 3-substituted but the 2-substituted alcohol as major product (with a yield of 70-80%). The cis- and trans-isomeric acids give mainly the cis- and trans-alcohols respectively; thus, the cis acid:

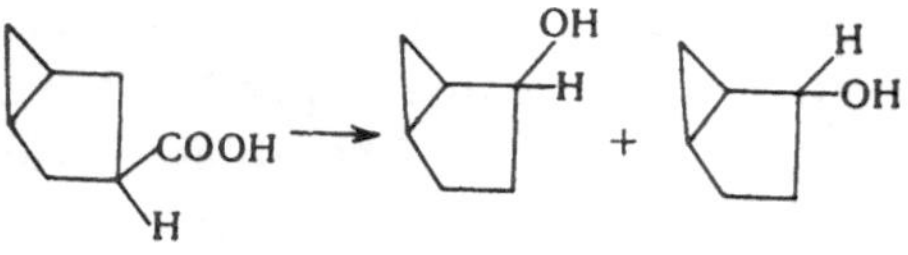

cis-   cis-   trans-

gives 37% cis and 30.8% trans alcohol. The authors conclude that
the process takes place through the formation not of a carbonium
ion but of a radical as intermediate product, and that the electrode
has a substantial effect on the stereochemistry of the process.
This also follows from the results obtained by Koehl [98]. In the
oxidation of monoesters of methylpropylmalonic and methyl-
isopropylmalonic acids Eberson [93] obtained a mixture of the
diastereoisomers of α,α'-dimethyl-α,α'-dipropylsuccinic and α,α'-
dimethyl-α,α'-diisopropylsuccinic acids. During electrolytic oxidation of meso- and racemic 2,3-diphenylsuccinic acids in a mixture
of pyridine and water (9 : 1) Corey and Casanova [94] obtained only
trans-stilbene, which is also formed in the oxidation of these acids
by lead tetraacetate. It is pointed out that these results exclude the
possibility of cis-elimination. In the case of anodic oxidation of
butylboric acid Humphrey and Williams [96] found that a mixture of
1-butene, cis-2-butene, and trans-2-butene in the proportions 3.3
: 1 : 1 was formed at a platinum electrode in aqueous alkaline solutions at low current densities.

Ponomarev and Markushina [97] showed that tricyclic spirans
are formed by intramolecular electrolytic alkoxylation of furandi-
alkanols. They can exist in the form of two geometrical isomers,
which differ in the mutual positions of the oxygen atoms in the two
terminal rings (arranged in planes perpendicular to the plane of the
central ring). Investigation of the product showed that the intramolecular alkoxylation process is stereooriented to form the trans
isomer:

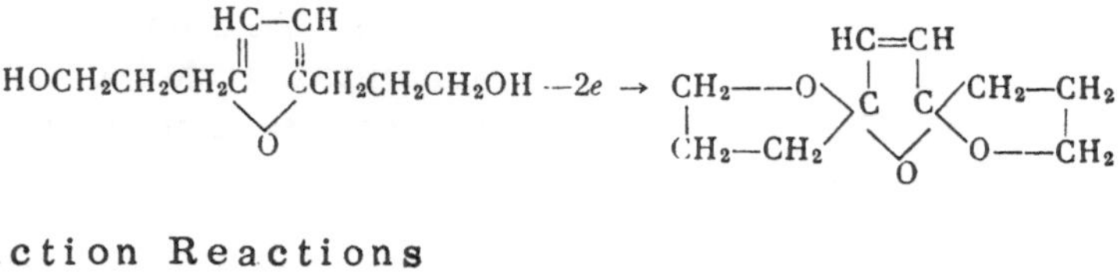

Reduction Reactions

Unsaturated Compounds. The first data on the stereochemistry of products from electrolytic reduction of unsaturated acids

apparently go back to 1906 [99]. Data on the stereochemistry of the reduction of a carbon—carbon double bond were then obtained by Campbell and Young [100]. Camilli [101] found that reduction of cyclohexene-1,2-dicarboxylic acid at various cathodes only gave the thermodynamically less stable cis-cyclohexane-1,2-dicarboxylic acid. Elving and co-workers [102] found that acetylenedicarboxylic acid and its monoethyl ester were reduced in acidic solution to racemic dimethylsuccinic acid and its diethyl ester:

$$2HOOCC \equiv CCOOH + 6e + 6H^+ \rightarrow HOOCCH(CH_3)CH(CH_3)COOH + 2CO_2.$$

Diethyl acetylenedicarboxylate is reduced to diethyl fumarate. Dimethylfumaric acid is reduced to meso-dimethylsuccinic acid, and dimethylmaleic acid is reduced to racemic dimethylsuccinic acid.

These results show that reduction of the multiple bonds takes place through trans-addition. By reduction of diphenylacetylene in dimethylformamide in the presence of carbon dioxide Wawzonek and Wearring [103] obtained diphenylfumaric acid, diphenylmaleic anhydride and meso-diphenylsuccinic acid in the proportions 2.5 : 1 : 7. If it is assumed that the meso acid is formed from fumaric acid, these results demonstrate the stereospecificity of the process, contrary to the assertions made by Wawzonek and Wearring [103]. From phenanthrene in the presence of carbon dioxide they obtained trans-9,10-dihydrophenanthrene-9,10-dicarboxylic acid,

but it is not impossible that the cis isomer is formed as well and is converted into the trans isomer during treatment of the reaction mixture to isolate the products.

Recently Peover [104] has again investigated the reduction of unsaturated compounds in aprotic solvents in the presence of carbon dioxide. The meso- and racemic-diphenylsuccinic acids which he obtained in the reduction of trans- and cis-stilbenes, respectively, provide direct evidence of the stereospecificity of the process.

Baizer and co-workers [105] obtained a series of ring compounds by means of the so-called intramolecular reductive coupling reaction:

$$
\begin{array}{c}
CH=C\begin{subarray}{c}H\\R\end{subarray}\\
Z\hspace{2em}R\\
CH=C\begin{subarray}{c}R\\H\end{subarray}
\end{array}
\;+2e+2H^+\rightarrow\;
\begin{array}{c}
CH-CH\begin{subarray}{c}H\\R\end{subarray}\\
Z\hspace{2em}|\\
CH-CH\begin{subarray}{c}R\\H\end{subarray}
\end{array}
$$

In the cases where $Z = C(C_2H_5)_2$ and $R = COOC_2H_5$, $Z = (CH_2)_2$ and $R = COOC_2H_5$, and $Z = (CH_2)_2$ and $R = CN$, mixtures in which trans, trans isomers predominated were submitted to electrolysis; the reduction products (the corresponding cyclopropane and cyclobutane derivatives) represented an approximately equimolar mixture of cis and trans isomers.

In the case where $Z = (CH_2)_3$ and $R = COOC_2H_5$ the trans isomer of the corresponding cyclopentane derivative predominated (with the cis and trans isomers in the ratio $1:3$). Electrolysis of diethyl cis-cyclopentane-1,3-diacrylate ($Z = C_5H_3$ and $R = COOC_2H_5$) gave two major products, one of which was the corresponding saturated compound (its cis isomer) and the other was the trans isomer of diethyl 2,3-norbornanediacetate:

$$
\begin{array}{c}
CH=CH-COOC_2H_5\\
CH=CH-COOC_2H_5
\end{array}
\;+\;2e+2H^+\longrightarrow\;
\begin{array}{c}
CH_2-COOC_2H_5\\
H\\
H\\
CH_2-COOC_2H_5
\end{array}
$$

Kato and co-workers [106] found that electrolytic reduction of 2-butyne-1,4-diol on silver deposited on a copper substrate gave only cis-butenediol.

Most results on the stereospecificity of electrolytic reduction of a double bond can be understood if it is assumed that at cathodes with high hydrogen overpotential, where electron transfer is the initial reaction, free charges in the intermediate product strive to occupy trans positions with respect to each other and that subsequent addition of hydrogen ions, carbon dioxide molecules, etc., does not change the molecular configuration. At cathodes with low hydrogen overpotential, where a considerable proportion of the surface is occupied by atomic hydrogen, hydrogenation of the double bond takes place by means of hydrogen from the cathode surface; in this case cis-addition products are obtained.

Halogen-Substituted Compounds. Elving and co-workers [22, 83] investigated the reduction of meso- and racemic dibromosuccinic acids at a mercury cathode. At all pH values the meso acid gave only fumaric acid, i.e., trans-elimination of bromine occurred. Racemic dibromosuccinic acid gives various products depending on the pH; fumaric acid alone is also formed in strongly acidic and alkaline solutions, i.e., cis-elimination of bromine occurs, but at pH values between 0.4 and 6.9 a mixture of fumaric acid and maleic acid is formed, and the latter is the main product at pH 4 (70% yield).

This behavior is explained by the fact that the monoanion of the racemic acid, which exists in weakly acidic solutions, is evidently capable of forming (through a hydrogen bond) a seven-membered ring with the bromine atoms in the trans position. In trans-elimination this favours the formation of the isomer, i.e., maleic acid. The analogous cyclic monoanion of the meso acid is energetically favorable, since the bromine atoms are in the cis position, i.e., in the position with maximum repulsion. It is difficult to find an explanation for the fact that the reduction is not stereospecific in very acidic and alkaline solutions, as in esters of the meso- and racemic acids which are always reduced to esters of fumaric acid). It should be noted that chemical reduction (iodide – iodine) of meso- and racemic 2,3-dibromobutanes leads, respectively, to trans- and cis-butenes.

It has already been mentioned that Jura and Gaul [38] came to the conclusion that various products are formed in the reduction of cis- and trans-2,3-dichloroacrylonitriles. These results indicate in favor of trans-elimination of halogen atoms to form a multiple bond. In the case where the halogen atoms are in the cis position they are successively replaced by hydrogen. Markova and Feoktistov [41] did not find any effect of geometric isomerism in dihalogen-substituted unsaturated acids on the composition of the reduction

products. Nevertheless Feoktistov and Gol'din [107], while investigating 1,2-dibromotetrafluoroethane and 1,2-dibromo-1-chloro-
1,2,2-trifluoroethane, found that different products were obtained
from the reduction of different conformations in these compounds.

While unsaturated compounds are formed from the anti (or
trans) conformations the bromine atoms are substituted successively
from the gauche (or syn) conformations:

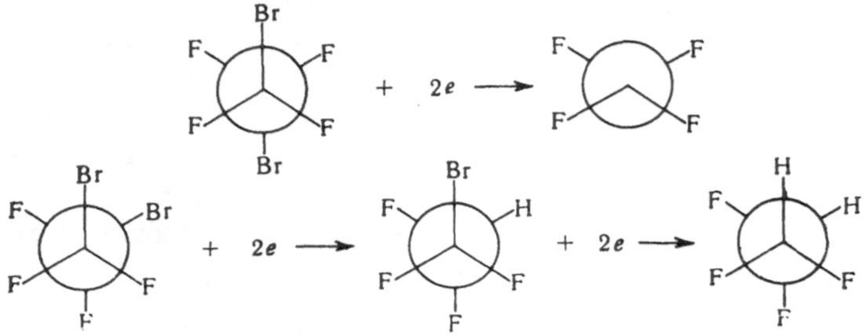

and the initially formed monobromide is reduced at a more negative
potential. From the ratio of the heights of the first wave (which for
any conformation corresponds to participation of two electrons in
the reduction) and the second wave (which also corresponds to participation of two electrons in the process, but for lower concentration
of the gauche conformation than the total concentration it is possible
to calculate the equilibrium concentrations of the conformations and,
consequently, the differences in the free energies of the two conformations.

Cisak [108-110] recently made a detailed investigation of the
reduction of the three stereoisomeric pairs of compounds aldrin
and isodrin, dieldrin and endrin, and α- and β-chlordanes (see p.
151). It was found that, in the case of the first four compounds,
three chlorine atoms are successively removed from the hexachloro-
cyclopentene ring and replaced by hydrogen; in the case of the
chlordanes, the same three atoms are removed and replaced by
hydrogen and two chlorine atoms in the least chlorinated cyclopentane ring are removed to form a new multiple bond. Here the
reduction product from each of the isomers evidently retains the
configuration of the original compound. Characteristic of these
examples is the fact that the usual course for reduction of vicinal

dihalogensubstituted compounds is not followed owing to the energetically unfavorable formation of a flat cyclopentadiene ring.

Recently, papers have appeared [111-113] in which the problem of retention or reversal of configuration has been investigated for the molecule of a hologensubstituted compound during electrolytic reduction leading to substitution of the halogen by hydrogen. Walborsky and co-workers [111] found that in the electrolytic reduction of (+)-S-1-bromo-1-methyl-1-2,2-diphenylcyclopropane on mercury in acetonitrile the configuration is retained to the extent of 63%. It is known that formation of the respective free cyclopropyl radical as an intermediate product leads to complete racemization of the final product, while formation of the cyclopropyl anion leads to almost complete retention of optical activity. Absence of the Walden inversion, which is usual in the case of homogeneous bimolecular nucleophilic substitution reactions, in this case may result from double inversion, since the formation of a mercury-containing intermediate product (probably R_2Hg_n, but not R_2Hg or $RHgBr$) was observed in the course of electrolysis.

Czochralska [113] investigated the reduction of the optically active 2-phenyl-2-chloropropionic acid at a mercury cathode in acidic ethanolic solution. It was found that inversion took place during electrolytic reduction, and the yield of optically active product amounted to 77.2-92.2%. Annino and co-workers [112] investigated the electrolytic reduction on mercury of the optically active 1-bromo-2,2-diphenylcyclopropanecarboxylic acid and its methyl ester, and also the reduction of 1-bromo-1-methyl-2,2-diphenyl-cyclopropane in neutral, acidic, and alkaline ethanolic solutions. It was found that reduction of the acid took place with 26-35% inversion of the configuration in acidic solutions and with 31-38% retention in alkaline solutions. Reduction of methyl 1-bromo-2,2-di-phenylcyclopropanecarboxylate took place with 30-56% inversion of the configuration irrespective of the composition of the solution, whereas reduction of 1-bromo-1-methyl-2,2-diphenylcyclopropane took place with 21% retention. These results can be explained on the assumption that the molecule is subject to attack by electrons from the side of the halogen atom and that the whole stereochemistry of the process is determined by stereoselective reaction with proton donors of the free carbanion or of the carbanion screened by the electrode.

Carbonyl Compounds. Anziani and co-workers [114] investigated the electrolytic reduction of α-methylcyclohexanone at various cathodes. It was found that the stereochemistry of the reduction product (the respective alcohol) depended substantially on the nature of the metal. Only the trans isomer was formed on mercury and lead, a mixture of isomers with preponderance of the trans compound was formed on nickel, and only the cis isomer was obtained with copper. It is interesting that reduction does not take place on platinum. Parker and Swann [115] also observed that the nature of the metal affected the yields of stereoisomers. Whereas glucose is mainly reduced to sorbitol at lead and mercury (and the products contain a small amount of mannitol), mannose is reduced to mannitol at all cathodes and only gives sorbitol with cadmium. Fructose is reduced to a mixture of the two stereoisomeric hexoses. Reduction of α-methyldesoxybenzoin, $C_6H_5COCH(CH_3)C_6H_5$, gives mainly the erythro alcohol (92%) [116]. Grimshaw and Ramsey [117] found that electrolytic reduction of hydroxybenzaldehyde in aqueous and aqueous ethanolic alkaline solutions at mercury takes place stereospecifically to form the corresponding meso-hydrobenzoin. Vanillin behaves in exactly the same way, whereas p-methylbenzaldehyde and veratric aldehyde give approximately equal amounts of the meso- and racemic hydrodimers. A high degree of stereoselectivity in reduction was also observed by other authors in the reduction of aldehydes containing hydroxy groups in alkaline solution [118-124]. This is due to repulsion of like charges during dimerization of the radicals.

Stereoisomeric pinacones were obtained by Lund [125] in the reduction of ketosteroids, and there was a preponderance of one or the other isomer depending on the pH value. Here the effect of pH was greater on the Δ-1,4-ketosteroids, which have an almost flat ring, than on the Δ-4-ketosteroids, where the ring can assume a "semichair" and a "semiboat" form. High stereospecificity in reduction of ketosteroids with the keto group in the nucleus was found by Kabasakalian and co-workers [126]. In reduction at mercury the ketosteroids and their α-hydroxy-substituted derivatives give the thermodynamically more stable alcohols with equatorial positioning of the hydroxyl groups; in these cases the elimination of the α-hydroxyl group precedes reduction of the keto group.

Vincenc-Chodkowska and Grabowski [127] have described interesting phenomena in the polarographic reduction of benzil and have attributed them to isomerization of the initial reduction product in the field of the double electric layer. In neutral and weakly acidic solutions stereoisomeric stilbenediols are formed from benzil, $C_6H_5COCOC_6H_5$, and their proportions vary according to the reduction potential. The trans isomer predominates at not very negative potentials, but the proportion of cis isomer (formation of which is promoted by increase in the intensity of the double layer field) increases with increase in the negative potential value. With further increase in potential the amount of cis isomer again decreases, owing to shortening of the time spent in the strong field. These results were recently confirmed by Bauer [128], who also found that benzil behaved similarly during reduction by a palladium sol.

Other Compounds. Horner and co-workers [129-131] have described the production of optically active phosphines and arsines in the cathodic cleavage of quaternary phosphonium and arsonium salts. This reaction, like its reverse quaternization reaction, takes place with retention of configuration. Shapoval, Skobets, and Markova [132, 133] have described the production of a stereoregular isocyanate polymer during reduction at a nickel cathode in dimethylformamide. The formation of a stereoregular product evidently demonstrates the orienting effect of the electrode on the molecules of the monomer in the polymerization process.

As seen from this review of the published data, stereospecificity in electrode reactions is still a little-investigated subject. In some cases the effect of purely electrochemical factors can be clearly traced, in other cases a much greater part is evidently played by factors which also arise in normal homogeneous reactions. Detailed investigation of the effect of electrode potential on the stereochemistry of oxidation and reduction products is of undoubted interest and may lead to methods for controlling the course of many of these reactions.

LITERATURE CITED

1. Yu. P. Kitaev and G. K. Budnikov, Usp. Khim., 31:670 (1962).
2. P. Herasymenko, Z. Elektrochem., 34:74 (1928).

3. E. Vopička, Coll. Czechosl. Chem. Comuns., 8:349 (1936).
4. G. Semerano and J. S. Rao, Mikrochem., 23:9 (1937).
5. G. Semerano, Mikrochem., 24:10 (1938).
6. J. S. Rao, J. Univ. Bombay, 10:56 (1941).
7. P. J. Elving and C. Teitelbaum, J. Am. Chem. Soc., 71:3916 (1946).
8. B. Warshowsky, P. J. Elving, and J. Mandel, Analyt. Chem., 19:161 (1947).
9. Ya. P. Stradyn', V. V. Teraud, and M. V. Shimanskaya, Izv. Akad. Nauk SSSR, Ser. Khim., 547 (1964).
10. S. D. Kamyshchenko and I. V. Kudryashov, Trudy MKhTI im. D. I. Mendeleeva, 54:194 (1967).
11. V. N. Nikulin and A. N. Shikin, Papers of Second Conference on Polarography [in Russian], Izd. Kazansk. Univ. (1962), p. 102.
12. V. N. Nikulin, Thesis [in Russian], Kazan University (1964).
13. P. Kanakam, M. S. V. Pathy, and H. V. K. Udupa, Electrochim. Acta, 12:329 (1967).
14. G. A. Korchinskii, Ukrainsk. Khim. Zh., 29:103 (1963).
15. L. Schwarz, Coll. Czechosl. Chem. Communs., 7:326 (1935).
16. G. Semerano and G. Sartori, Gazz. Chim. Ital., 68:167 (1938).
17. G. Semerano and G. Sartori, Mikrochem., 24:130 (1938).
18. MacIntyre, King. Fed. Proc., 1:160 (1942).
19. Z. Frenkl, Chem. Listy, 46:79 (1952).
20. A. L. Markman and É. V. Zinkova, Zh. Obshch. Khim., 27:1438 (1957).
21. A. L. Markman and É. V. Zinkova, Zh. Obshch. Khim., 32:353 (1962).
22. P. J. Elving, I. Rosenthal, and A. J. Martin, J. Am. Chem. Soc., 77:5218 (1955).
23. R. Takahashi and P. J. Elving, Electrochim. Acta, 12:213 (1967).
24. F. Wessely and J. Wratil, Mikrochem., 32:248 (1947).
25. S. Wawzonek, E. W. Blaha, R. Berkey, and M. E. Runner, J. Electrochem. Soc., 102:235 (1955).
26. R. E. Sioda, D. O. Cowan, and W. S. Koski, J. Am. Chem. Soc., 89:230 (1967).
27. L. V. Shubina, L. Ya. Malkes, V. N. Dmitrieva, and V. D. Bezuglyi, Zh. Obshch. Khim., 37:437 (1967).
28. A. P. Tomilov, S. K. Smirnov, I. G. Sevast'yanova, and O. G. Strukov, Zh. VKhO im. D. I. Mendeleeva, 12:357 (1967).
29. W. Kemula and J. Kornacki, Roczn. Chem., 36:1849 (1962).
30. J. P. Petrovich, J. D. Anderson, and M. M. Baizer, J. Org. Chem., 31:3897 (1966).
31. H. Keller, R. A. Pasternak, and H. Halban, Helv. Chim. Acta, 29:512 (1946).
32. R. A. Pasternak, Helv. Chim. Acta, 31:753 (1948).
33. A. Ryvolova, Coll. Czechosl. Chem. Communs., 22:1114 (1957).
34. E. D. Hartnell and C. E. Bricker, J. Am. Chem. Soc., 70:3385 (1948).
35. C. Prévost, P. Souchay, and J. Chauvelier, Bull. Soc. Chim. France, 5 (1951).
36. J. Kůta, Science, 144:1130 (1964).
37. S. G. Mairanovskii and L. D. Bergel'son, Zh. Fiz. Khim., 34:236 (1960).
38. W. H. Jura and R. J. Gaul, J. Am. Chem. Soc., 80:5402 (1958).
39. I. Rosenthal, J. R. Hayes, A. J. Martin, and P. J. Elving, J. Am. Chem. Soc., 80:3050 (1958).

40. P. J. Elving, I. Rosenthal, J. R. Hayes, and A. J. Martin, Analyt. Chem.,
     33:330 (1961).
41. I. G. Markova and L. G. Feoktistov, Zh. Obshch. Khim., 38:970 (1968).
42. A. Winkel and H. Siebert, Ber., 74B:670 (1941).
43. W. M. Lauer, H. P. Klug, and S. A. Harrison, J. Am. Chem. Soc., 61:2775
     (1953).
44. G. Costa, Gazz. Chim. Ital., 83:875 (1953).
45. C. Costa, Ricerca Sci., 27, Suppl. A. Polarographia, 3:94 (1957).
46. C. Prévost, P. Souchay, and C. Mallen, Bull. Soc. Chim. France, 78 (1953).
47. A. Volpi, Gazz. Chim. Ital., 77:473 (1947).
48. P. J. Hilson and P. B. Birnbaum, Trans. Faraday Soc., 48:478 (1952).
49. B. Nygard, Arkiv. Kemi, 20:163 (1962).
50. S. Wawzonek and J. D. Fredrickson, J. Am. Chem. Soc., 77:3985 (1955).
51. L. Chuang, I. Fried, and P. J. Elving, Analyt. Chem., 37:1528 (1965).
52. C. R. Castor and J. H. Saylor, J. Am. Chem. Soc., 75:1427 (1953).
53. H. C. Freeman and W. P. Georgans, Chem. a. Ind., 148 (1951).
54. N. Tyutyulkov, Zh. Fiz. Khim., 32:1389 (1958).
55. P. Souchay and S. Ser, J. Chim. Phys., 49:C172 (1952).
56. N. Tyutyulkov and S. Buduvoff, Compt. Rend. Acad. Bulgar. Sci., 6:5 (1953).
57. N. Tyutyulkov and W. Palikarov, Compt. Rend. Acad. Bulgar. Sci., 15:399
     (1962).
58. H. Gardner and W. Georgans, J. Chem. Soc., 4180 (1956).
59. H. Lund, Acta Chem. Scand., 13:249 (1959).
60. J. Nakaya, K. Mori, H. Kinoshita, and S. Ono, Nippon Kagaku Zasshi, 80:1212
     (1959).
61. E. Paspaleev, Monatsh. Chem., 97:230 (1966).
62. A. E. Arbuzov and Yu. P. Kitaev, Dokl. Akad. Nauk SSSR, 113:577 (1957).
63. Yu. P. Kitaev and A. E. Arbuzov, Izv. Akad. Nauk SSSR, Otd. Khim. Nauk,
     1037 (1957).
64. A. E. Arbuzov and Yu. P. Kitaev, Izv. Kazansk. Fil. Akad. Nauk SSSR, Ser.
     Khim., 3 (1957).
65. R. Brockman and D. Pearson, J. Am. Chem. Soc., 74:4128 (1952).
66. Yu. P. Kitaev and G. K. Budnikov, Theory and Practice of Polarographic
     Analysis [in Russian], Izd. STIINCA, Kishinev (1962), p. 395.
67. Yu. I. Vainshtein, V. M. Dziomko, K. A. Dunaevskaya, and M. D. Shirokova,
     Zh. Obshch. Khim., 32:2777 (1962).
68. L. G. Feoktistov, in: Advances in Electrochemistry of Organic Compounds [in
     Russian], Nauka, Moscow (1966), p. 135.
69. F. L. Lambert, A. H. Albert, and J. P. Hardy, J. Am. Chem. Soc., 86:3155
     (1964).
70. J. W. Sease, P. Chang, and J. L. Groth, J. Am. Chem. Soc., 86:3154 (1964).
71. J. Krupička, J. Závada, and J. Sicher, Coll. Czechosl. Chem. Communs.,
     30:3570 (1965).
72. F. L. Lambert, J. Org. Chem., 31:4184 (1966).
73. L. G. Feoktistov, A. P. Tomilov, Yu. D. Smirnov, and M. M. Gol'din, Élektro-
     khimiya, 1:887 (1965).

74. A. Cisak, Roczn. Chem., 31:337 (1957).

75. A. Cisak, Roczn. Chem., 36:1895 (1962).

76. A. Cisak, Roczn. Chem., 37:1025 (1963).

77. J. Závada, J. Krupička, and J. Sicher. Coll. Czechosl. Chem. Communs., 28:1664 (1963).

78. A. Cisak, Roczn. Chem., 40:1717 (1966).

79. D. Q. Quan, Compt. Rend., 251:2927 (1960).

80. A. M. Wilson and N. L. Allinger, J. Am. Chem. Soc., 83:1999 (1961).

81. V. Delaroff, M. Bolla, and M. Legrand, Bull. Soc. Chim. France, 1912 (1961).

82. P. Kabasakalian and J. McGlotten, Analyt. Chem., 34:1440 (1962).

83. I. Rosenthal and P. J. Elving, J. Am. Chem. Soc., 73:1880 (1951).

84. M. G. McKeon, J. Electroanalyt. Chem., 3:402 (1962).

85. L. Malspeis and N. G. Hung, J. Pharmac. Sci., 53:506 (1964).

86. P. Zuman, Acta Chim. Hungar., 18:141 (1959).

87. P. Zuman and K. Černý, Chem. Listy, 52:1468 (1958).

88. T. Inoue, K. Koyama, T. Matsuoka, and S. Tsutsumi, Tetrahedron Letters, 1409 (1963).

89. S. Tsutsumi, K. Koyama, T. Inoue, T. Ebara, T. Tani, and S. Fukuoka, 17th Meeting C. I. T. C. E., Extended Abstracts, Tokyo (1966), p. 109.

90. W. W. Paudler, R. E. Herbener, and A. G. Zeiler, Chem. a. Ind. (London), 47:1909 (1965).

91. H. J. Dauben Jr., Liang Xiao-Tian, Acta Chim. Sinica, 25:129 (1959).

92. E. J. Corey, N. L. Bauld, R. T. LaLonde, J. Casanova Jr., and E. T. Kaiser, J. Am. Chem. Soc., 82:2645 (1960).

93. L. Eberson, Acta. Chem. Scand., 14:641 (1960).

94. E. J. Corey and J. Casanova Jr., J. Am. Chem. Soc., 85:165 (1963).

95. P. G. Gassmann and F. V. Zalar, J. Am. Chem. Soc., 88:2252 (1966).

96. A. A. Humffray and L. F. G. Williams, Chem. Com., 24:616 (1965).

97. A. A. Ponomarev and I. A. Markushina, in: Advances in Electrochemistry of Organic Compounds [in Russian], Nauka, Moscow (1968), p. 80.

98. W. J. Koehl Jr., J. Am. Chem. Soc., 86:4686 (1964).

99. C. Mettler, Ber., 39:2933 (1906).

100. K. N. Campbell and E. E. Young, J. Am. Chem. Soc., 65:965 (1943).

101. C. T. Camilli, Diss. Abs., XVII, 826 (1958).

102. I. Rosenthal, J. R. Hayes, A. J. Martin, and P. J. Elving, J. Am. Chem. Soc., 80:3050 (1958).

103. S. Wawzonek and D. Wearring, J. Am. Chem. Soc., 81:2067 (1959).

104. M. E. Peover, The Chem. Engineer, No. 211, Trans. of the Institution of Chem. Engineers, 45:CE192 (1967).

105. J. D. Anderson, M. M. Baizer, and J. P. Petrovich, J. Org. Chem., 31:3890 (1966).

106. J. Kato, M. Sakuma, and T. Yamada, J. Electrochem. Soc. Japan, 25:331 (1957).

107. L. G. Feoktistov and M. M. Gol'din, Élektrokhimiya, 4:490 (1968).

108. A. Cisak, Roczn. Chem., 40:1727 (1966).

109. A. Cisak, Roczn. Chem., 41:809 (1967).

110. A. Cisak, Roczn. Chem., 41:963 (1967).

111. C. K. Mann, J. L. Webb, and H. H. Walborsky, Tetrahedron Letters, 2249 (1966).

112. R. Annino, R. E. Erickson, J. Michalovic, and B. McKay, J. Am. Chem. Soc., 88:4424 (1966).

113. B. Czochralska, Chem. Phys. Letters, 1:239 (1967).

114. P. Anziani, A. Anbry, and R. Cornubert, Compt. Rend., 225:878 (1947).

115. E. A. Parker and J. S. Swann, Trans. Electrochem. Soc., 92:343 (1947).

116. L. Mandell, R. M. Powers, and R. A. Day Jr., J. Am. Chem. Soc., 80:5284 (1958).

117. J. Grimshaw and J. S. Ramsey, J. Chem. Soc., 653 (1966).

118. H. Kaufmann, Z. Elektrochem., 4:461 (1898).

119. H. D. Law, J. Chem. Soc., 89:1512 (1906).

120. H. D. Law, J. Chem. Soc., 91:748 (1907).

121. H. D. Law, J. Chem. Soc., 99:1113 (1911).

122. J. Dewar and J. Read, J. Soc. Chem. Ind. (Trans.), 55:347 (1936).

123. M. J. Allen, J. Am. Chem. Soc., 72:3797 (1950).

124. I. A. Pearl, J. Am. Chem. Soc., 74:4260 (1952).

125. H. Lund, Acta Chem. Scand., 11:283 (1957).

126. P. Kabasakalian, J. McGlotten, A. Basch, and M. D. Yudis, J. Org. Chem., 26:1738 (1961).

127. A. Vincenc-Chodkowska and Z. Grabowski, Fundamental Problems in Modern Theoretical Electrochemistry [in Russian], Mir, Moscow (1965), p. 390.

128. E. Bauer, J. Electroanalyt. Chem., 14:351 (1967).

129. L. Horner and A. Mentrap, Ann., 646:65 (1961).

130. L. Horner, H. Fuchs, H. Winkler, and A. Rapp, Tetrahedron Lettters, 965 (1963).

131. L. Horner and H. Fuchs, Tetrahedron Letters, 1573 (1963).

132. G. S. Shapoval, E. M. Skobets, and N. P. Markova, Dokl. Akad. Nauk SSSR, 173:392 (1967).

133. G. S. Shapoval, N. P. Markova, and E. M. Skobets, in: Advances in Electrochemistry of Organic Compounds [in Russian] Nauka, Moscow (1968), p. 71.

Prewaves in Reversible Systems

G. A. Tedoradze, E. Yu. Khmel'nitskaya,
and Ya. M. Zolotovitskii

1. INTRODUCTION

Polarographic curves obtained for some typical reversible redox systems involving reduction at a dropping mercury electrode show certain differences from the expected form, but these differences are not observed at electrodes with a stationary surface. Curves of this type were first described by Brdička and Knobloch [1] in the reduction of riboflavin and then by Brdička [2] in the reduction of methylene blue. Müller [3] independently found a similar effect in the reduction of α-hydroxyphenazine in acidic solutions. These differences were manifested in the form of additional waves which preceded the normal reduction wave of the compound. The height of the prewave tended towards a maximum with increase in the concentration of the oxidized form. The relationship between its height and the outflow rate of the mercury also differed from the analogous relationship for a normal wave. An attempt by Müller [3, 4] to explain the formation of the prewave by the existence of reducible "tautomeric" forms of the substance was unsuccessful. Brdička proposed an explanation [2, 5, 6] which is now generally accepted.

He suggested that prewaves are formed when the oxidized form is practically unadsorbed on the electrode and the reduced form is strongly adsorbed. A smaller amount of energy is required for a reduction process where particles bound to the electrode surface by adsorption forces are formed than for a similar process leading to the formation of unadsorbed particles [6].

By means of the Brdička theory it is possible to calculate the adsorption equilibrium constant B in the equation of the Langmuir isotherm:

$$Bc = \frac{\Gamma}{\Gamma_\infty - \Gamma},$$

where c is the volume concentration, Γ is the surface concentration, and Γ_∞ is the maximum amount adsorbed.

The formulas proposed by Brdička are only applicable for reversible processes. In the case of an irreversible process the same equation for the quantity Γ_∞ is only obtained when the reaction product formed during electrolysis is not desorbed from the electrode surface. The Brdička theory is very logical and free from internal contradictions; the calculated values are consistent with the size of the adsorbed molecules.

The values obtained for the B constant, which is a function of the adsorption energy of the reduced form, range between 10^7 and 10^8 liter/mole. The enormous difference between the adsorption energies of the oxidized and reduced forms on mercury remains completely incomprehensible, since the molecules differ only slightly from each other. As yet no publication has appeared in which such a large difference between the adsorption energies of two forms in a redox system has been found experimentally. The work by Breyer and Biegler is an exception, but this will be analyzed in Section 3.

2. POLAROGRAPHIC INVESTIGATION
OF PREWAVES

Experimentally the prewaves formed in the reduction of the methylene blue cation have been better investigated than others [2]. In neutral and weakly alkaline solutions (the dye decomposes in strongly alkaline solutions) methylene blue gives a clearly defined prewave, which reaches a maximum at a concentration of the oxidized form (Ox) equal to $5 \cdot 10^{-5}$ mole/liter. The difference between the half-wave potentials of the main wave ($E_{1/2}$) and the prewave ($E_{1/2}^P$) is 125 mV at a concentration of [Ox] = $1.7 \cdot 10^{-4}$ mole/liter. The $E_{1/2}^P - E_{1/2}$ difference decreases with increase in temperature, and at 80-90°C the prewave combines with the main two-electron wave. This temperature effect is consistent with the Brdička theory. Brdička's experimental data on the leuco form of methylene blue give the value $\Gamma_\infty = 1.56 \cdot 10^{-10}$ mole/cm^2, which is extremely close to the values $1.7 \cdot 10^{-10}$ mole/cm^2 (obtained by Lorenz and Schmalz [7]) and $1.36 \cdot 10^{-10}$ mole/cm^2 (obtained by Los and Tompkins [8, 9]) (see Section 3).

According to the Brdička theory, a 10-fold increase in the concentration of the oxidized form should lead to an increase of 29 mV in the $E_{1/2}^P - E_{1/2}$ difference. At low concentrations of methylene blue the half-wave potential of the prewave in fact shifts towards the positive side with increase in the concentration of the oxidized form. At high concentrations, however, this difference, which amounts to 125 mV, undergoes no further change. The value $E_{1/2}^P - E_{1/2} = 125$ mV gives a B value of 10^8 liter/mole.

A very unusual pattern was observed during investigation of the reduction of methylene blue on mercury by the oscillographic method [10]. The prewave gives an oscillopolarogram of triangular shape. The relationship between the peak height and the concentration of oxidized form is given not by a smooth curve of the Langmuir isotherm type but a curve with a maximum. In other respects the oscillopolarographic data are not at variance with the results from ordinary polarographic investigations. In particular, as in the case of the ordinary polarographic curves, at methylene blue concentrations equal to or less than $5 \cdot 10^{-5}$ mole/liter there is only the prewave observed, and a peak corresponding to the main wave on polarographic curves appears only at higher concentrations. However, the peak characteristics are not given [10] in detail and it is not possible to estimate their relative heights.

The dependence of the instantaneous current on time (i, t curve) for the methylene blue prewave differs little in form from that predicted by the Brdička theory except in one detail: at short times the current is equal not to infinity but to zero, and therefore a maximum is formed on the curve. However, this deviation is not very substantial and may be due to the fact that the Brdička theory does not take account of the effect of diffusion of the oxidized form on the variation of the instantaneous current with time at the prewave potentials [12].

Prewaves formed in the reduction of riboflavin (lactoflavin, vitamin B$_2$) have been studied by Brdička and Knoblock [1], Brdička

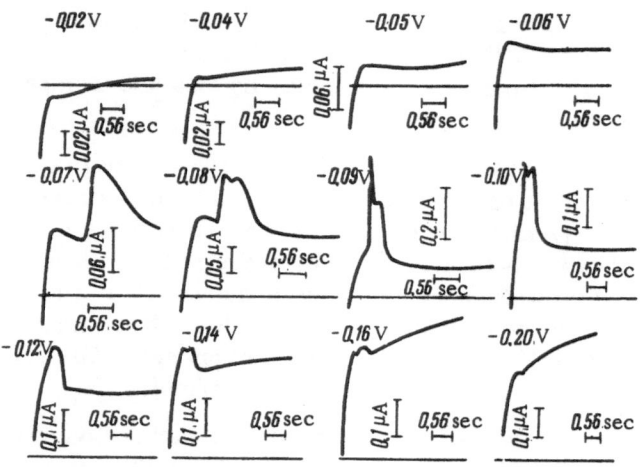

Fig. 1.   Dependence of instantaneous current on time at various
potentials for riboflavin in 0. 1 N perchloric acid solution with the
reduced compound at a concentration of $1.15 \cdot 10^{-4}$ mole/liter.

[5], Cohen and Kolthoff [11], and Muller [4]. The prewave of riboflavin differs in form from that of methylene blue: the slope of the
former is steeper. The i, t curves for riboflavin at the prewave
potentials present an extremely complex picture (Fig. 1). A semiquinone is mainly formed under these conditions. Brdička [2] explains the delay in the increase of the current at these potentials by
the fact that the semiquinone and the leuco form of riboflavin are
surface-inactive in statu nascendi but are converted at a definite
rate into surface-active isomers, which then behave autocatalytically towards the conversion of the surface-inactive isomer.

Oscillopolarograms for riboflavin in 0.1 N perchloric acid,
which we obtained at a dropping mercury electrode, gave two peaks
corresponding to the half-wave potentials of the prewave and the
main wave (Fig. 2c). The height of the prewave peak depends linearly on the rate of potential scanning (V), and the height of the main
peak depends on the square root of V.

Phenazine and its derivatives are reduced in acidic solutions
to give two one-electron waves. Here the first wave at more positive potentials is always higher than the second, although theoretically they should be equal. The difference in the limiting currents of

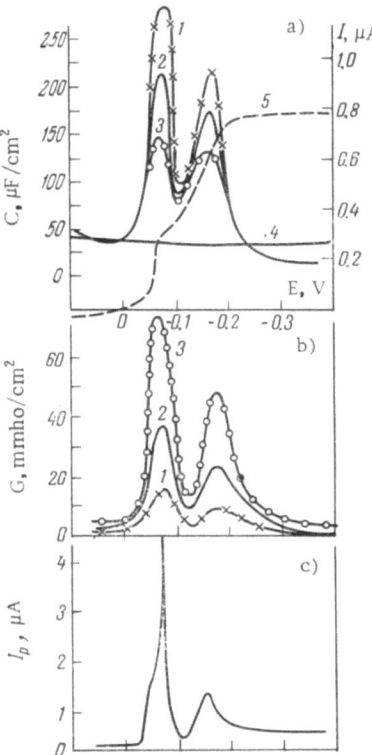

Fig. 2. Curves for dependence of differential capacity (a) and conductivity (b) of mercury electrode on potential in $1.2 \cdot 10^{-4}$ mole/liter solution of riboflavin in 0.1 N perchloric acid. Alternating-current frequency: 1) 200 rad/sec; 2) 500 rad/sec; 3) 1000 rad/sec; 4) curve for supporting electrolyte; 5) polarogram; c) oscillopolarogram of riboflavin in same solution.

the first and second processes does not depend on the concentration of the reactant but increases in proportion to the pressure under which the mercury flows from the capillary and tends to zero with increase in temperature. These waves merge in weakly acidic and neutral solutions, since the semiquinone formed at the first wave is only stable in strongly acidic solution. As a result a single two-electron wave, normally with a prewave, is formed. The similar prewave, formed in the reduction of rosinduline GG, was investigated in detail by Müller [4].

The reduction of pyocyanine (a phenazine derivative) in acidic solution has been investigated in [13]. In this case two one-electron waves are obtained, and the first (as in the case of phenazine) is higher than the second. In 0.01 mole/liter sulfuric acid, under the conditions used in [13], this difference amounted to 0.5 μA. With fairly high concentrations of pyocyanine (more than $3 \cdot 10^{-4}$ mole/liter) a prewave appears before the second wave, but the difference between the second wave and the first wave (together with the prewave) remains equal to 0.5 μA. The height of the prewave of the second wave does not depend on the concentration of the reactant, but its half-wave potential moves to the positive values with increase in the pyocyanine concentration. This prewave is rather better defined in hydrochloric acid solution than in sulfuric acid.

In [13] the disproportion between the first and second waves was explained in the following way: if the semiquinone (S) is stable, equilibrium is established between the reduced form R and the reacting compound Ox: Ox + R $\rightleftharpoons$ 2S, and the R form is preferentially adsorbed. In this case a prewave corresponding to the second wave is formed. This prewave, which is superimposed on the first wave, increases its height at the expense of the second wave. However, in addition to the prewave which merges with the first wave, another prewave appears at more negative potentials, and this (as already mentioned above) does not merge with either the first or the second wave. To explain the formation of this prewave Voriškova [13] assumes that a second adsorbed layer of compound R is formed.

A similar pattern was obtained in the reduction of α-hydroxyphenazine [3], but a second clearly defined prewave was not observed.

Among other reversibly reduced compounds which give prewaves we mention 8-hydroxyquinoline [14] and quinoline-8-carboxylic acid [15]. These compounds give the greatest difference between $E^{p}_{1/2}$ and $E_{1/2}$ of all the reversible systems described in the literature (260 mV for 8-hydroxyquinoline in two-electron reduction and 320 mV for quinoline-8-carboxylic acid in one-electron reduction). It was suggested in [16] that the prewave obtained in the

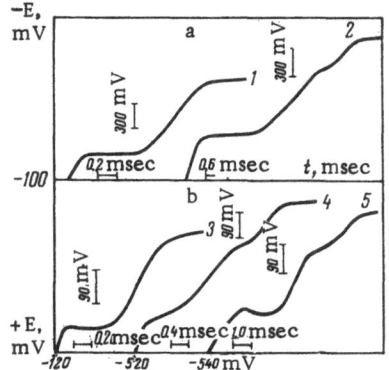

Fig. 3. Chronopotentiograms for cathodic (a) and anodic (b) processes. 1) and 3) for $1.28 \cdot 10^{-4}$ mole/liter solution of riboflavin in 0.1 N perchloric acid, $i = 16$ mA/cm^2; 2), 4), and 5) for $1.2 \cdot 10^{-4}$ mole/liter solution of riboflavin in phosphate buffer solution at pH 7.8, $i = 16$ mA/cm^2 (2, 4) and 4 mA/cm^2 (5).

reduction of quinine also arose for the reasons discussed by Brdička. However, our own calculations, based on impedance measurements, demonstrate conclusively the irreversible nature of this reaction.

Adsorption prewaves formed during anodic polarization of mercury in a solution of the reversible bis-(diethylcarbamyldisulfide)-diethylcarbonate system have been investigated in detail [17]:

$$(C_2H_5)_2N\!-\!\underset{\underset{S}{\|}}{C}\!-\!S\!-\!S\!-\!\underset{\underset{S}{\|}}{C}N(C_2H_5)_2$$

$$+2e \rightleftarrows 2\,[(C_2H_5)_2N\!-\!\underset{\underset{S}{\|}}{C}\!-\!S]^-.$$

The difference between $E^P_{1/2}$ and $E_{1/2}$ was 110-120 mV, which corresponds to a B value of 10^7 to 10^8 liter/mole.

The simplest compound to give a prewave is evidently juglone (5-hydroxy-1,4-naphthaquinone). The prewave is observed over a wide range of pH values between 2 and 10 [18]. However, the isomeric lawsone (2-hydroxy-1,4-naphthaquinone) does not give a prewave.

3. INVESTIGATION OF ADSORPTION OF REDOX-SYSTEM COMPONENTS

Adsorption of the methylene blue cation on a positively charged mercury surface has been investigated by Los and Tompkins [8, 9]. The measurements were performed in the absence of faradaic current by two methods: 1) direct determination of the adsorption from the change in concentration of methylene blue during desorption from the surface of mercury, transferred from the methylene blue solution into water; 2) from the dependence of the weight of the drop on the methylene blue concentration.

With the first method Los and Tompkins found that in a phosphate buffer solution at pH 7.0 the adsorption of the methylene blue cations satisfied the Langmuir equation. The following B values were obtained: $4 \cdot 10^5$ to $74 \cdot 10^5$ liter/mole (at 0.024 V, s.c.e.) and $12 \cdot 10^5$ to $30 \cdot 10^5$ liter/mole (at -0.076 V).

The results obtained by the second method showed that the adsorption rate of methylene blue was limited not by diffusion of the adsorbate to the electrode surface but by some kinetic stage, where the rate constant k_a in the equation:

$$\left(\frac{\partial \Gamma}{\partial c} \right)_{E, \, c} = k_a c \, (1 - \theta) - k_d \theta$$

(where k_d is the desorption rate constant) is equal to $(7.5 \pm 2.5) \cdot 10^{-5}$ cm/sec (at -0.024 V) and $(4.2 \pm 1.8) \cdot 10^{-5}$ cm/sec (at -0.076 V).

Adsorption of the leuco form of methylene blue on mercury was investigated by Laitinen and Mosier [19], who measured the differential capacity of the double electric layer in the absence of faradaic current. They found that at the maximum adsorption potential (-0.6 V, s.c.e.) the adsorption of the leuco form satisfied the Langmuir equation with a B value of $1.67 \cdot 10^5$ liter/mole. This is almost three orders of magnitude smaller than the value obtained from the Brdička theory.

Gupta [20] studied the relationship between the capacity of the mercury electrode (or more accurately the value of the alternating current passing through the cell) and the electrode potential in a solution of methylene blue. He found that the presence of methylene blue in the solution led to the appearance of two capacity peaks. The height of the peak situated at the more positive potentials increased with increase in the pH value. However, it decreased more rapidly with increase in frequency than would correspond to inverse proportionality to the square root of the frequency. The depression of the capacity caused by adsorption of the reaction product decreased with increase in the hydrogen ion concentration of the solution. According to the data of Gupta a second peak situated at more negative potentials is also due to electrochemical reaction.

TABLE 1. Dependence of Height of First Step and Combined Height of Two Steps on Current Density

Current density, mA/cm^2	$i\tau_1$, $\mu C/cm^2$ (first step)		$i(\tau_1 + \tau_2)$, $\mu C/cm^2$ (sum of two steps)	
	Experiment I	Experiment II	Experiment I	Experiment II
31.6	10.8 *	11.8 *	18.6	17.1
16.0	12.8 *	12.8 *	17.6	17.6
8.03	16.9	16.1	—	—
4.05	17.8	16.1	—	—
1.98	19.8	15.9	—	—
0.99	19.8	—	—	—

* At these current densities there are two steps on the chronopotentiograms, and in other cases there is one step. The riboflavin concentration amounts to $1.28 \cdot 10^{-4}$ mole/liter in Experiment I and $1.15 \cdot 10^{-4}$ mole/liter in Experiment II. In each case before the anodic pulse was applied the electrode was held at 0.09 V (the limiting current of the prewave).

An argument in favour of this suggestion is the existence of this peak in acidic solutions, where the adsorption of the leuco form of methylene blue is insignificant. The height of this peak rapidly reduces to zero with increase in the alternating-current frequency.

Somewhat later Lorenz and Schmalz [7] investigated the dependence of differential capacity of a mercury electrode on potential by the more modern bridge method in solutions of methylene blue at pH 10.8 and at frequencies between 0.8 and 20 kHz. Like Gupta, they consider that the capacity peak in the region of positively charged mercury corresponds to a pseudocapacity peak of the reaction, but the second peak they relate to desorption of the leuco form of methylene blue.

In addition Lorenz and Schmalz [7] investigated the adsorption of methylene blue and its leuco form by the method of capacity curves. As mentioned in Section 2, the Γ_∞ value for the reaction product, calculated in this way, is closely consistent with the polarographic data. It follows from the results obtained that the surface concentrations of methylene blue and its leuco form are similar at the equilibrium potential. This signifies similarity in the values of

the B parameter for both components, since the concentrations of
the oxidized and reduced forms near the electrode are equal at the
equilibrium potential.

Adsorption of riboflavin was investigated by Breyer and
Biegler [21] using the so-called tensometric method. There were
two peaks found on the curve for the relationship between the total
conductivity (admittance) and potential. One peak (at potentials
close to the half-wave potential) was considerably lower than the
second peak, which was situated at rather more positive potentials.
This pattern was observed not only in strongly acidic solutions,
where the prewave is clearly defined, but also at pH 4.7, where the
prewave is hardly noticeable.

In the weakly acidic, alkaline and neutral regions solutions of
riboflavin a decrease in the capacity of the double electric layer is
shown at potentials both more positive and more negative than the
potentials of the pseudocapacity peaks. This means that both the
oxidized and reduced forms in the riboflavin—dihydroriboflavin
system are strongly adsorbed under these conditions. In acidic
solutions with the same concentrations of riboflavin the diminishing
of the double-electric-layer capacity caused by adsorption of the
leuco form is considerably less. Breyer and Biegler consider that
this result provides evidence for the validity of the Brdička theory;
in conditions where a prewave is observed the oxidized form hardly
diminishes the double-electric-layer capacity, whereas there is
clearly defined adsorption of the reduced form.

One of us [22, 23] has already pointed out that determination
of the degree of surface coverage from data obtained by the capacity
method involves the use of a nonthermodynamic formula. As shown
by experiment, this formula is applicable to the adsorption of a number of aliphatic compounds both with positive charges and with negative charges on the mercury surface.

With the adsorption of aromatic compounds containing π-
electron systems the formula can only be used in the region of negative charges. It is not possible to assess the adsorption of these
compounds on the positive branch of the electrocapillary curve from
change in capacity of the double layer. For example, it follows from
electrocapillary curves that aniline is strongly adsorbed in the region of positive mercury charges, whereas no depression is observed

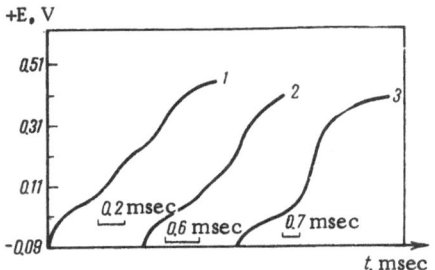

Fig. 4.   Chronopotentiograms of anodic process for dihydroriboflavin in 0.1 N perchloric acid solution, recorded from solution of $1.28 \cdot 10^{-4}$ mole/liter riboflavin at various current densities: 1) 31.6 mA/cm^2; 2) 16 mA/cm^2; 3) 8.03 mA/cm^2.

on the differential capacity curves in this region [24]. Under the conditions where riboflavin gives a prewave the mercury is positively charged, and it is not therefore possible to assess the adsorption of the oxidized form from the change in capacity of the double layer.

In fact investigations by Tatwawadi and Bard [25] have shown that riboflavin is adsorbed even from millimolar solutions. The method of capacity curves with low current densities was used, and calculation of the surface concentration of riboflavin therefore involved considerable error, which is unavoidable in the determination of Γ values by extrapolation.

Similar results were obtained by Hartley and Wilson [28] during investigation of the reduction of flavinomononucleotide (ester of phosphoric acid and riboflavin). In 0.1 N perchloric acid this compound gives a polarographic curve with a prewave. The electrocapillary curves of flavinomononucleotide show that both the oxidized and the reduced forms of the redox system are adsorbed under the conditions mentioned above. The surface concentration of the oxidized form varies between 0.6 and $1.53 \cdot 10^{-10}$ mole/cm^2 depending on the extrapolation method and the volume concentration.

By various electrochemical methods we investigated adsorption phenomena in the riboflavin—dihydroriboflavin system. The chronopotentiometric method gave relationships between the surface concentration of the redox-system components and the potential of the electrode, and these were then compared with the polarographic curves [27]. The capacity curves (chronopotentiograms) were recorded on a suspended mercury drop at such current densities that the contribution of diffusion was negligibly small. In order to determine differences in the adsorption behavior of the riboflavin—dihydroriboflavin systems under conditions with and without a

TABLE 2. Dependence of Length of Step on Potential*

E, V	Cathodic process		E, V	Anodic process	
	τ_{drop}, msec	τ_{susp}, msec		τ_{drop}, msec	τ_{susp}, msec
+0.10 (before rise of prewave)	1.0 1.0 1.0	1.2 1.2 —	−0.3 (limiting current of main wave)	0.8 0.8 1.0	1.2 1.1 —
−0.05 (~$E_{1/2}$ of prewave)	0.8 0.8 0.7	1.0 1.1 —	−0.05	0.2 0.2 0.2	0.3 0.3 —
			−0.10 (limiting current of prewave)	0.8 0.7 0.8	1.0 0.8 0.9

*The measurements were made in a solution of riboflavin at a concentration of $1.5 \cdot 10^{-4}$ mole/liter in 0.1 N perchloric acid, i = 16 mA/cm^2.

prewave on the polarograms, all the measurements in this system were carried out in both 0.1 N perchloric acid solution and phosphate buffer solution (pH 7.6-7.8).

The chronopotentiograms of riboflavin for the cathodic process in 0.1 N perchloric acid at current densities greater than 0.5 mA/cm^2 contains a single step corresponding to the conversion of riboflavin into dihydroriboflavin (Fig. 3, curve 1). The chronopotentiograms for the anodic process, corresponding to the conversion of dihydroriboflavin into riboflavin, also have a single step at low current densities (1 < i < 15 mA/cm^2). With increase in the current density, beginning at 16 mA/cm^2, two steps appear on the chronopotentiograms in place of one; with further increase in current density the second step is rapidly displaced towards positive potentials (Fig. 4) and at i > 32 mA/cm^2 merges into the step due to discharge of the supporting electrolyte. A single step is thus also observed on the chronopotentiogram at high current densities. The height of the step at low current densities (1 < i < 15 mA/cm^2), expressed in coulombs, hardly differs from the total height of the two steps in the region of current densities where splitting is observed on the chronopotentiogram (Table 1).

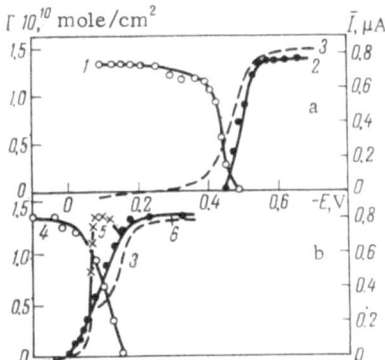

Fig. 5. Dependence of Γ on potential of electrode. a) In phosphate buffer solution at pH 7.8 and with riboflavin concentration of $1.2 \cdot 10^{-4}$ mole/liter: 1) cathodic; 2) anodic process; b) in 0.1 N perchloric acid with riboflavin conclentration of $1.2 \cdot 10^{-4}$ mole/liter: 4) cathodic; 5) and 6) anodic process; 5) variation of Γ for dihydroriboflavin, calculated from combined length of two steps on chronopotentiogram, i = 16 mA/cm^2; 3) polarographic curves.

The Γ_{DRF} values (surface concentration of dihydroriboflavin) at potentials corresponding to the limiting current of the prewave, calculated from the length of the step at low current densities ($1 < i < 15$ mA/cm^2) or from the sum of the steps at $16 \le i \le 32$ mA/cm^2, are close in value to Γ_∞ determined from the potential corresponding to the limiting current of the main wave. However, as seen from Fig. 5b, at these potentials the electrode is still covered with a considerable amount of the original oxidized form. The total degree of coverage θ of the electrode surface with the two components of the system can be expressed as the following ratio:

$$\theta = (\Gamma_{RF} + \Gamma_{DRF})/\Gamma_\infty,$$

where Γ_{RF} (surface concentration of riboflavin) and Γ_{DRF} are taken at the same potential. At potentials corresponding to the limiting current of the prewave θ amounts to 1.6. Such an increase cannot be due to the effect of diffusion on Γ_{DRF}, since dihydroriboflavin is generated at the electrode surface, and its concentration is considerably lower in the bulk of the solution than near the electrode. This could only lead to a decrease in the value of Γ_{DRF}.

Another reason for the increased value of θ could be the longer time spent by the suspended mercury drop in the riboflavin solution, compared with the dropping electrode, due to the time taken to transfer the drop to the stationary hook; this could lead to error in the determination of Γ_{RF} from chronopotentiometric data.

To check this assumption we compared the step heights obtained on the suspended drop and on the dropping electrode before removal of the drop in both cathodic and anodic processes. The results of this comparison are given in Table 2, from which it is seen that within the limits of experimental error the heights of the steps coincide in both cases. The somewhat lower values of τ for the dropping electrode arise from the fact that the current pulse in this case is applied to a drop which has not reached its maximum size.

Consequently the high value of θ cannot be due to experimental error. If the total degree of coverage is calculated from the Γ_{DRF} value obtained from the first step on the chronopotentiogram, the θ value obtained is very close to unity. This fact can only be explained on the supposition that, in the region of their joint existence, riboflavin and dihydroriboflavin cover the mercury surface by a layer more than one molecule thick.

A second step also appears on chronopotentiograms for the anodic process in highly dilute solutions of riboflavin, where one wave is obtained on the polarogram, but this step occurs in an even narrower range of current densities (from 32 to 27 mA/cm^2). The second step can be observed with i < 27 mA/cm^2 (27-19 mA/cm^2) by increasing the life of the suspended drop at the starting potential from ~4 sec (in all the chronopotentiometric measurements) to 30-40 sec.

In phosphate buffer solution (pH 7.8) there are two steps each on the chronopotentiograms for the cathodic and anodic processes (Fig. 3, curves 2, 4, and 5). From the potentials of the steps it can be deduced that the first steps result from electrochemical reaction of the compound at the electrode surface and that the second step on the cathodic chronopotentiogram results from desorption of the product of this reaction, since it coincides in potential with the desorption peak of the riboflavin reduction product on the curve for the dependence of the differential capacity of the double electric layer on the potential of the electrode. The absence of a second (desorption) step on the cathodic chronopotentiograms for riboflavin in perchloric acid clearly arises from the masking effect of hydrogen evolution. The second step on the anodic chronopotentiogram is greatly shifted from the first in potential and, unlike the second

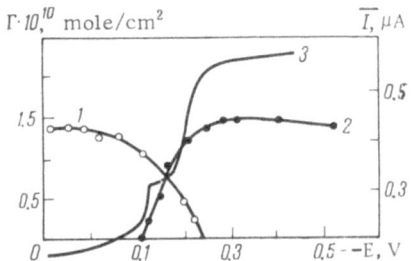

Fig. 6. Dependence of Γ for juglone (1) and hydrojuglone (2) on potential of electrode. Concentration $2 \cdot 10^{-4}$ mole/liter, pH 6.8. 3) Polarogram.

step on the anodic chrono-potentiograms for riboflavin in 0.1 N perchloric acid, is retained at all current densities.

To determine the nature of the steps the relationship between $i\tau$ and $1/i$, where i is current density (mA/cm^2) and τ is the transition time (msec), was plotted. Both for the steps on the cathodic and anodic chrono-potentiograms of riboflavin in perchloric acid and for both steps in the phosphate buffer solutions the amount of electricity corresponding to the step ($i\tau$) is practically independent of the current density employed. This shows that diffusion has little effect under the selected conditions.

Sharp potential drops are observed on the chronopotentiograms for the anodic processes in both media (Fig. 3, Curves 3 and 5), and these are retained at various current densities, including fairly high current densities where τ is small and the drop on the curve cannot be due to a polarographic maximum. As shown by Zolotovitskii, Tedoradze and Érshler [28], the presence of such drops on the E, t curves indicates the existence of an effect from large coverages. Consequently accumulation of dihydroriboflavin on the electrode surface retards the electrochemical oxidation of this compound both under conditions where a prewave is obtained on the polarographic curve and under conditions where there is no prewave.

By recording E, t curves using various potentials for preliminary polarization and preconditioning of the electrode relationships were obtained between the surface concentration of the electrochemically active substance and the potential. Figure 5 gives the relationship between Γ of each component and potential in the riboflavin—dihydroriboflavin system in perchloric acid and in phosphate buffer solution (pH 7.8). It follows from these curves that, in both cases, the two components are adsorbed at the electrode. At potentials where faradaic current is absent the Γ_{RF} values are very

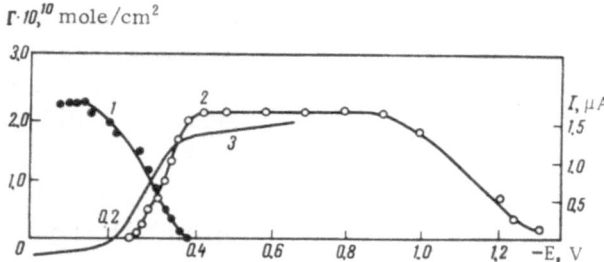

Fig. 7.   Dependence of Γ for lawsone (1) and hydrolawsone (2) on potential of electrode.   Concentration $3.1 \cdot 10^{-4}$ mole/litre, pH 6.8, phosphate buffer solution; 3) polarogram.

similar in both media (with and without a prewave). A similar conclusion follows from a comparison of the Γ_{DRF} values recorded at potentials corresponding to the limiting diffusion current. Great difference is only observed in the comparison of dihydroriboflavin concentrations at potentials more positive than $E_{1/2}$; under conditions where there is a prewave Γ_{DRF} considerably exceeds the analogous value in alkaline solution (if the $1 \leq i \leq 32$ mA/cm² region is considered).

If in plotting the Γ, E curves for the anodic process in perchloric acid account is taken of the second step (in addition to the main step), as mentioned above, the curve will have a form similar to that of curve 5 in Fig. 5b. In examining the variation of the surface coverage by adsorbed dihydroriboflavin on the combined curve a very sharp rise is noticed in the Γ, E curve towards the Γ^{∞}_{DRF} value. The second step disappears when practically no oxidized form remains on the surface.

The dependence of the surface concentration of riboflavin and dihydroriboflavin on their volume concentrations was also investigated by the chronopotentiometric method. It was found that the experimental curves readily satisfied the equation of the Langmuir isotherm with $B = 3.4 \cdot 10^5$ liter/mole for riboflavin (at $+0.10$ V, vs. s.c.e.) and $B = 2.4 \cdot 10^5$ liter/mole for dihydroriboflavin (at -0.5 V). The values are $1.25 \cdot 10^{-10}$ and $1.82 \cdot 10^{-10}$ mole/cm², respectively, for the oxidized and reduced forms [27].

Investigation of the adsorption of juglone and hydrojuglone also showed that the Langmuir equation could be applied to this

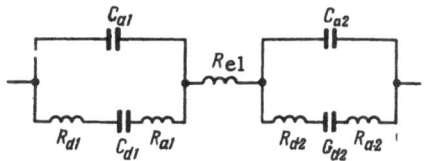

Fig. 8. Equivalent circuit for reaction Ox + ne ⇌ R with adsorption of both reaction components.

system with $B = 1.5 \cdot 10^5$ liter /mole for juglone (at -0.05 V) and $B = 1.1 \cdot 10^5$ liter/mole for hydrojuglone (at -0.45 V). The Γ_∞ values are $1.67 \cdot 10^{-10}$ and $2.2 \cdot 10^{-10}$ mole/cm^2 respectively for the oxidized and reduced forms [27].

Comparison of the Γ, E curves obtained at the hanging drop with the polarogram of juglone shows (Fig. 6) that in the region of the prewave limiting current the electrode is to a considerable degree covered by the oxidized form.

In the investigation of adsorption prewaves comparison of the two reversible redox systems juglone—hydrojuglone and lawsone (2-hydroxy-1,4-naphthaquinone)—hydrolawsone is of interest. It is interesting that two compounds so similar in structure as juglone and lawsone have quite different polarographic behavior. A polarogram with a prewave is obtained in the first system, but no prewave is obtained with the second. Here the Γ, E curves for lawsone are similar to those obtained for juglone, i.e., the electrode is covered with both components of the redox system (Fig. 7).

Investigation of the adsorption of lawsone and hydrolawsone also showed that the Langmuir equation could be applied to this system with $B = 1.09 \cdot 10^5$ liter/mole (at -0.10 V for lawsone and -0.50 V for hydrolawsone). The Γ_∞ values were found to be identical for both systems ($2.2 \cdot 10^{-10}$ mole/cm^2).

Thus, the adsorption characteristics of these systems are very similar and cannot explain the reasons for the formation of the prewave. Evidently the reason for the difference in polarographic behavior of these systems should be sought in a difference in the chemical properties of the compounds. Work is continuing in this direction.

4. INVESTIGATIONS OF REDOX SYSTEMS BY MEANS OF ALTERNATING CURRENT

As already mentioned above (Section 2), the Brdička theory only applies to reversible systems. The degree of reversibility in

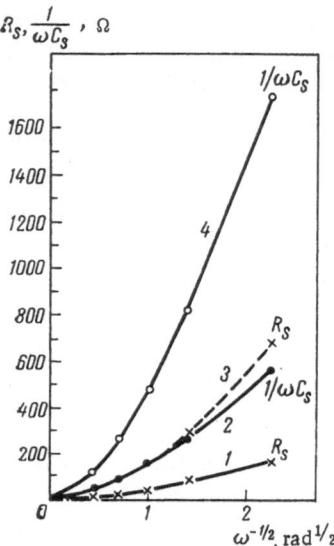

Fig. 9. Frequency dependence of extremal values of impedance components for mercury electrode in solution of $2 \cdot 10^{-4}$ mole/liter riboflavin in 1 N perchloric acid. 1) and 2) for $E_{1/2}$ of the prewave; 3) and 4) for $E_{1/2}$ of main wave of polarogram.

a given system is usually determined by polarography. Considerably more reliable is the faradaic impedance method. Where the reaction components are adsorbed, however, the relationships are somewhat different from those where adsorption does not occur.

In the faradaic impedance method the criterion of specific adsorption of an electrochemically active substance is the angle (φ) of the phase shift between the sinusoidal current and the potential. A reversible electrode process without adsorption of the reaction components is characterized by a φ value of 45°. Various slow stages (charge transfer, accompanying chemical reactions) can lead to a decrease in the phase angle $(\varphi < 45°)$, and only in the case of specific adsorption leading to increased surface concentration of electrochemically active substances does the phase angle become greater than 45°.

The increase in phase angle is the more considerable, the more rapid the adsorption process [29]. These qualitative considerations are at present illustrated in a number of theoretical and experimental papers. Faradaic impedance with allowance for adsorption of electrochemically active substances at the equilibrium potential was frequently [30-35] investigated theoretically. Originally [30, 35] it was assumed that the adsorption of the reaction components can be described by the Langmuir isotherm and that the adsorption and desorption rate constants did not depend on the potential. A more general analysis avoiding these assumptions was made by Senda and Delahay [31] on the supposition that only adsorbed particles were electrolyzed and by Lorenz [32], who took account of the electrolysis of both adsorbed and unadsorbed particles. An analysis was also given [31] for the case where the adsorption processes for various substances are mutually independent and the effect of change in potential on the adsorption process is insignificant compared with the effect on the charge-transfer process. On these assumptions the adsorption of compound i can be represented by a supplementary resistance R_{ai}, connection in series with a Warburg impedance W_i and a supplementary frequency-independent capacitance C_{ai} in parallel with the series $W_{ai}-R_{ai}$ combination. An equivalent circuit for the reaction $Ox + ne \rightleftharpoons R$, where both components are specifically adsorbed at the electrode, is given in Fig. 8. The following notation is used in the equivalent circuit: R_{di} and C_{di} are elements of the Warburg impedance,

$$R_{di} = \frac{1}{\omega C_{di}} = \frac{RT}{n^2 F^2} \frac{1}{(2\omega D_i)^{1/2}} \frac{1}{c_i} = \sigma_i \omega^{-1/2}. \tag{1}$$

R_{el} is the discharge resistance,

$$R_{el} = \frac{RT}{nF} \cdot \frac{1}{i_a^0}, \tag{2}$$

where i_a^0 is the exchange current; R_{ai} and C_{ai} are elements of the adsorption impedance

$$R_{ai} = \frac{RT}{n^2 F^2} \left(1 \Big/ \frac{\partial \Phi_{a_i}}{\partial c_i} \right) \frac{1}{c_i}, \tag{3}$$

$$C_{ai} = \frac{n^2 F^2}{RT} \left(\frac{\partial \Gamma_i}{\partial c_i} \right) c_i, \tag{4}$$

where Φ_{ai} is the flow of adsorbed substance; c_i is the concentration of the reacting substance in the solution; Γ_i is the surface concentration of adsorbed particles.

The resistance R_{ai} is determined by the adsorption kinetics, and the capacitance C_{ai} is determined by the adsorption equilibrium. If adsorption takes place infinitely quickly ($\partial \Phi_{ai}/\partial c_i \rightarrow \infty$), then $R_{ai} \rightarrow 0$. If the substance i is not specfically adsorbed ($\partial \Gamma_i/\partial c_i \rightarrow 0$), then $c_{ai} \rightarrow 0$. Thus, specific adsorption of an electrochemically active substance should lead to an increase in the pseudocapacity peak of the reaction. The maximum peak, equal to the sum of

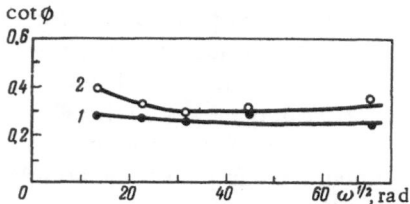

Fig. 10. Frequency dependence of phase shift angle. 1) At $E_{1/2}$ of prewave; 2) at $E_{1/2}$ of main wave. The composition of the solution is the same as in Fig. 9.

$C_{di} + C_{ai}$, is obtained with infinitely rapid charge transfer and infinitely rapid adsorption.

A phase angle greater than 45° was first observed experimentally by Randles and Somerton [35] in the reduction of tetramethyl-p-phenylenediamine on platinum and by Laitinen and Randles [30] in the reduction of the trivalent cobalt complex with ethylenediamine on mercury. Phase angles greater than 45° were also observed in the Tl^+−thallium amalgam [34, 36, 37] and Cd^{2+}−cadmium amalgam [38] systems. Although all these authors explain the anomalously large phase angle by specific adsorption of the electrochemically active substance, there is no common view regarding the appearance of specific adsorption in measurements with alternating current. Thus, according to Sluyters and co-workers [34], who studied the Tl^+−thallium amalgam system, unadsorbed thallium ions are discharged, and the $\varphi < 45°$ values are due to the effect of specifically adsorbed thallium ions on the capacity of the double electric layer. An explanation of the mechanism for the appearance of specific adsorption requires further development of the faradaic impedance theory to include discharge of both adsorbed and unadsorbed particles.

It was shown in [39, 40, 42], by use of the Delahay and Senda scheme, that the frequency dependence of the impedance and phase angle components make it possible to establish the nature of the electrode processes and to calculate their kinetic and adsorption parameters. In the absence of adsorption of the reaction components $\cot \varphi$ increases linearly with $\omega^{1/2}$ [29]:

$$\cot \varnothing = 1 \left(\frac{\omega D}{2} \right)^{1/2} \frac{1}{k_s} , \tag{5}$$

where k_s is the standard rate constant of the electrode reaction. In the case of specific adsorption the linear relationship between $\cot \varphi$ and $\omega^{1/2}$ is not obeyed. For an irreversible process with adsorption of one or both electrochemically active substances and a

TABLE 3. Γ and Γ_∞ Values for Riboflavin and Dihydroriboflavin

	$\Gamma_{RF} \cdot 10^{10}$, mole/cm^2	$\Gamma_{DRF} \cdot 10^{10}$, mole/cm^2
$E^{II}_{1/2}$	1	1
$E_{1/2}$	0	1.2
Γ_∞	1.3	1.3

reversible process with adsorption of one substance it passes
through a minimum in a low-frequency region. For a reversible
process with adsorption of both components the relationship between
$\cot \varphi$ and $\omega^{1/2}$ is a diminishing function. The criterion for reversibility of the process in the presence of specific adsorption is the
fulfilment of the inequality $\cot \varphi < 1$ at all (even high) frequencies.

The adsorption parameters can be calculated from the extrapolated values of the faradaic impedance components at $\omega \to 0$ and
$\omega \to \infty$. The relationship between the adsorption parameters and
the impedance components in the series R_s and C_s circuit were
given in [39]:

$$\left(R_s - \frac{1}{\omega C_s}\right) R_{el} + \sum R_{ai} - 2 \sum C_{ai} \sigma_i^2, \tag{6}$$

$$\left(1/C_s\right)_{\omega \to \infty} = \sum 1/C_{ai} - \omega^{-3/2} \sum \left(\sigma_i/R_{ai}^2 C_{ai}^2\right). \tag{7}$$

No systematic impedance measurements have been made for
systems with adsorption prewaves. Even by a.c. polarography only
riboflavin has been investigated [21]. We recorded the dependence
of the impedance components for riboflavin and juglone on frequency
and potential. The impedance measurements confirmed the data on
the reversibility of these systems and the adsorption of both components of the electrode reaction.

The C, E and G, E curves (G is conductivity) were recorded
in a parallel circuit on a hanging mercury drop (electrode area
$\simeq 0.03$ cm^2) with the oscillographic vector polarograph at the L'vov

TABLE 4. Some Results from Impedance
and Chronopotentiometric Measurements

Method	$\sum \frac{\partial \Gamma_i}{\partial c_i} c_i \cdot 10^{10}$, mole/cm^2	
	$E^{\text{II}}_{1/2}$	$E_{1/2}$
Chronopotentiometric	0.48	0.1
Impedance		
At $\omega \to 0$	0.32	0.8
At $\omega \to \infty$	0.16	0.046

Polytechnical Institute, and by this means it was possible to separately and simultaneously record the dependence of the impedance
components of the electrode on potential at frequencies between 32
and 159,000 Hz.

Figures 2a and b show the dependence of the impedance components of the mercury electrode on potential in a solution of $1.2 \cdot 10^{-4}$
mole/liter riboflavin + 0.1 N perchloric acid. The C, E and G, E
curves each contain two peaks, corresponding to the half-wave potentials of the prewave and the main wave of the polarogram. Figure 9
shows the frequency dependence of extreme values for the impedance components, converted to the series circuit, and Fig. 10
shows the frequency dependence of the corresponding phase shift
angles. The double layer capacity ($C_{dl} = 0.8 \mu F$) and electrolyte
resistance ($R_{\Omega} = 15 \Omega$) required for the conversion were measured
in a solution of the supporting electrolyte at frequencies of 200 and
64,000 rad/sec, respectively. In order to reduce the resistance of
the electrolyte the measurements were carried out in a solution of
1 N perchloric acid.

The frequency dependences of capacitance, conductivity, and
$\cot \varphi$ are the same for the prewave and the main wave. The pseudocapacity of the reaction corresponding to the prewave is three times
greater than the pseudocapacity corresponding to the main wave
(270 and 90 $\mu F/cm^2$), and both pseudocapacities considerably exceed
the pseudocapacity of diffusion, which must correspond to 72 μF
/cm^2 for $\omega = 200$ rad/sec and c = $2 \cdot 10^{-4}$ mole/liter. Our results
show that $\cot \varphi < 1$, the capacity resistance is greater than the ohmic

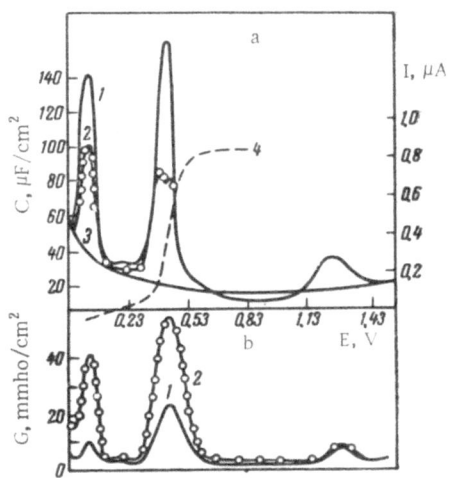

Fig. 11. Curves for dependence of differential capacity (a) and conductivity (b) of mercury electrode on potential for $1.2 \cdot 10^{-4}$ mole/liter solution of riboflavin, with phosphate buffer solution at pH 6.8 as supporting electrolyte. Alternating-current frequency: 1) 200 rad /sec; 2) 500 rad/sec; 3) capacity curve for supporting electrolyte; 4) polarogram.

resistance over the whole range of frequencies, and the dependences of both components on $\omega^{-1/2}$ commence at the orgin of the co-ordinates (Fig. 9). This means that the reduction of riboflavin is a reversible process involving adsorption of both reaction components. The small change in the pseudocapacity of the reaction with frequency (the pseudocapacity decreased by a factor of 1.5 when the frequency was increased by 25 times) clearly results from shunting of the diffusion capacity by the larger adsorption capacity. This allowed us with sufficient justification to use extrapolation formulas for determining the adsorption parameters, although the experimentally accessible frequency region for the given reaction is small (ω = 200–5000 rad/sec). While understanding that our calculations were to a certain degree approximate, we used Eqs. (6) and (7) to determine the parameter $\Sigma(\partial \Gamma_i / \partial c_i)c_i$, which characterizes the adsorption capacity. Here we assumed that $R_{el} = 0$ and that $\Sigma R a_i \to 0$.

(Justification for this assumption is found in the fairly large phase shift angle $-\cot\varphi = 0.3$.)

The calculations were made for the half-wave potential of the prewave and of the main wave. The values obtained for $\Sigma(\partial\Gamma_i/\partial c_i)c_i$ were compared with the results from adsorption measurements made by the chronopotentiometric method, by which means it was possible to determine the validity of the approximation. Since the adsorption of riboflavin and dihydroriboflavin is described by the Langmuir isotherm, by differentiating the isotherm equation we obtain:

$$\sum\frac{\partial\Gamma}{\partial c}c = \sum\frac{\Gamma(\Gamma_\infty-\Gamma)}{\Gamma_\infty} = \sum\theta(1-\theta)\Gamma_\infty .$$

The Γ (for $E_{1/2}^P$ and $E_{1/2}$) and Γ_∞ values of both components are given in Table 3.

The results from the calculation of the $\Sigma(\partial\Gamma/\partial c_i)c_i$ parameter from impedance and chronopotentiometric measurements are given in Table 4.

The comparatively good agreement between the results obtained by the various methods once again confirms our ideas regarding the fact that the reduction of riboflavin is a reversible process involving adsorption of both components. The fact that $\Sigma(\partial\Gamma_i/\partial c_i)c_i$ is greater for the prewave than for the main wave explains why the capacity peak is higher at $E_{1/2}^P$ than at $E_{1/2}$.

The decreased values of the $\Sigma(\partial\Gamma_i/\partial c_i)c_i$ parameter obtained from impedance measurements at high compared with low frequencies may be due to the fact that the reduction of riboflavin occurs through an intermediate semiquinone stage [6]. If the frequency of the alternating current is commensurate with the life of the intermediate compound [41, 42], the pseudocapacity peak of the reaction decreases (by a factor of not more than four) and even splits into two, and for a reversible process the minmum between the two peaks occurs at $E_{1/2}$. This type of splitting is in fact observed, beginning at $\omega = 2000$ rad/sec.

In phosphate buffer solution (pH 7.6) there are three peaks on the C, E curve for riboflavin; the first is at potentials far removed

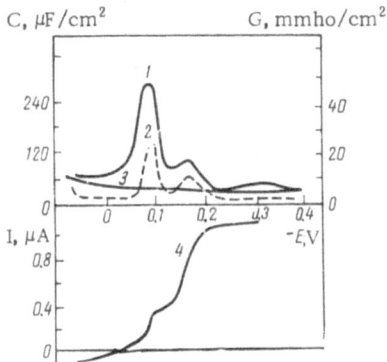

Fig. 12. Curves for dependence of differential capacity (1) and conductivity (2) of mercury electrode on potential in $2.2 \cdot 10^{-4}$ mole/liter solution of juglone, with phosphate buffer solution at pH 6.8 as supporting electrolyte. Alternating-current frequency 200 rad/sec; 3) capacity curve for supporting electrolyte; 4) polarographic curve.

from the reaction region (0.35 V more positive than $E_{1/2}$). In the literature this peak has been called the peak for interaction of riboflavin with mercury [21]. The second peak coincides with $E_{1/2}$ for the main wave on the polarogram, and the third peak is approximately 1 V more negative than $E_{1/2}$ (Fig. 11).

Unlike the first two peaks, the third peak is practically independent of frequency and corresponds to desorption of dihydroriboflavin from the electrode surface. Similar results are also obtained from chronopotentiometric measurements. The pseudocapacity peak of the reaction splits even at $\omega = 500$ rad/sec, which indicates that the protonated semiquinone is very stable in neutral solution. This complication makes it more difficult to determine the adsorption parameters, but the fact that $\cot \varphi < 1$ and the frequency dependence of the impedance components is similar to that at pH $\simeq 0$ means that the reduction of riboflavin in neutral solution, as in acidic solution, is a reversible process involving adsorption of both reaction components.

Figure 12 shows the potential dependence of capacity and conductivity of a mercury electrode in a $2.2 \cdot 10^{-4}$ mole/liter solution of juglone in phosphate buffer solution at pH 6.8. The capacity curve has three peaks: two peaks correspond to the half-wave potentials of the prewave and of the main wave on the polarogram, and the third peak is at E $\simeq -0.8$ V. The frequency dependence of the pseudocapacity peaks of the reaction and their corresponding conductivity peaks is similar to the frequency dependence of the impedance components of the electrode in a solution of riboflavin. The difference between juglone and riboflavin lies in the size of the phase shift angle, which is considerably less for juglone ($\cot \varphi = 0.8$). Since $\cot \varphi < 1$ over all the experimentally accessible frequency range (200–5000 rad/sec), decrease in the phase shift angle signifies slowness in the adsorption of the reaction components or difference in adsorbability between juglone and hydrojuglone. Slowness in the adsorption of juglone also follows from the chronopotentiometric measurements. The capacity peak at $E \simeq -0.8$ V is practically independent of frequency and represents a peak for desorption of hydrojuglone, which is consistent with the chronopotentiometric data.

Figure 13 show the potential dependence of capacity and conductivity for a mercury electrode in a $4.6 \cdot 10^{-4}$ mole/liter solution of lawsone in phosphate buffer solution at pH 6.8. A distinguishing feature of lawsone is the fact that the pseudocapacity peak splits into two at a frequency of only 200 rad/sec, while with juglone the splitting commences at 2000 rad/sec. This signifies that the semiquinone of lawsone is considerably more stable than the juglone radical. The peak at $E \simeq -1.1$ V represents a desorption peak for hydrolawsone, which is also confirmed by the chronopotentiometric data.

The results from impedance measurements once again demonstrate the difficulties encountered in the interpretation of the Brdička adsorption prewaves in the systems we have investigated. According to Brdička, only the reaction product is adsorbed and the coverage amounts to 0.5 at $E_{1/2}^{p}$ and to 1 at $E_{1/2}$. This corresponds to maximum adsorption pseudocapacity at $E_{1/2}^{p}$ (520 $\mu F/cm^2$ for $\Gamma_{\infty} = 1.3 \cdot 10^{-10}$ mole/cm^2) and $C_{ai} = 0$ at $E_{1/2}$. This means formally that at the potentials of the main wave charge transfer would take place onto unadsorbed particles, and the corresponding pseudocapacity peak could not exceed the value of the diffusion capacity. Consequently the pseudocapacity peaks corresponding to the prewave and the main wave would depend differently on frequency.

The experimental results, on the other hand, indicate that both reaction components are adsorbed both at the potential of the prewave and at the potential of the main wave.

5. POSSIBLE EXPLANATIONS OF FORMATION
OF PREWAVES

The Brdička theory, which has been used by electrochemists to explain prewaves in reversible systems, has not found experimental verification in the compounds for which it was developed, namely methylene blue and riboflavin. Since we have investigated adsorption processes in the riboflavin–dihydroriboflavin system in greatest detail, we are giving here our own explanation for the prewaves in this system.

It is well known that riboflavin, partially reduced, gives three types of free radical depending on the pH value; these are a radical anion at pH $\geq$ 11, a neutral radical between pH $\simeq$ 3 and 6, and a radical cation in the strongly acidic region [43]. When riboflavin is reduced to the extent of 50% the concentration of radicals depends on the pH value. It has been shown by electron paramagnetic resonance that the yield of radicals falls sharply with decrease in acidity from strongly acidic solutions, amounting to ~30% of the total quantity of riboflavin at pH 0, remains practically constant (3-5%) between pH $\simeq$ 3 and pH $\simeq$ 6, and again increases on passing into the alkaline region. Similar data on the stability of riboflavin semiquinones follow from calculations of their formation constants given by Brdička on the basis of polarographic data [44, 45]. It has been shown in a number of papers that dimerization of the semiquinones takes place in the pH region between $\simeq$ 3 and 6, i.e., in the region where the radicals have relatively low stability [46-50]. There are no indications that a dimer is formed in regions where radical ions exist, since the negative charge of the isoalloxazine ring (formed by alkaline dissociation of the $>C=O$ group) and the positive charge of the free radical in strongly acidic solutions prevent dimerization. Only the neutral electrically uncharged form of the free radical is capable of showing an appreciable tendency towards dimerization.

The dimer may be a complex of two identical radicals or a complex of the oxidized and reduced molecules oriented in parallel planes and stabilized by overlapping of the π-orbitals, or so-called

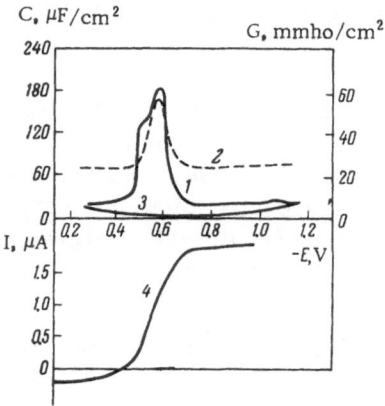

Fig. 13. Curves for dependence of differential capacity (1) and conductivity (2) of mercury electrode on potential in $4 \cdot 10^{-4}$ mole/liter solution of lawsone, with phosphate buffer solution at pH 6.8 as supporting electrolyte. Alternating-current frequency 200 rad/sec; 3) capacity curve of supporting electrolyte; 4) polarographic curve.

charge transfer [51-54]*. Here the molecules of the oxidized and reduced forms are also uncharged in the given pH region. In addition to such a biradical dimer further complexes exist between the radical and the initial or reduced forms of riboflavin [50, 55]. In particular in the region where neutral or uncharged semiquinones are formed riboflavin gives a prewave on the polarographic curves (pH values between 0 and $\simeq 6$).

The ability of riboflavin and dihydroriboflavin to form charge transfer complexes has also been explained by quantum-mechanical calculations [56]. Riboflavin is both a good donor and a satisfactory acceptor, while dihydroriboflavin possesses unusual properties; its highest filled molecular orbital is antibonding. The sign of its energy coefficient corresponds to orbitals which usually can be filled in the excited state of the molecules. This means that filling of the above-mentioned orbital in the ground state of dihydroriboflavin corresponds in principle to an unstable electronic configuration. Consequently dihydroriboflavin must exhibit a tendency to remove electrons from this orbital, i.e., possess unusually strong electron-donor characteristics. Similar characteristics are found in methylene blue and its leuco form and in thionine and its leuco form, which (like riboflavin) give polarographic curves with prewaves during reversible reduction [2, 4]. As with riboflavin, formation of a prewave with methylene blue and thionine is observed

---

*An example of a complex formed in this way is the molecular compound between benzoquinone and hydroquinone (so-called quinhydrone), in which the mutual disposition of the molecules was determined by x-ray analysis [54, 57].

in a region corresponding to relatively low stability in the semi-quinones, where their yield amounts to 5-10% [58] of the total amount of dye when it is reduced to the extent of 50%. This corresponds to the neutral or alkaline pH region, where the semi-quinones of the dyes are uncharged.

If it is assumed that formation of the above-described charge transfer complexes is also possible in the electrochemical reduction of riboflavin, this process will be considerably easier at an electrode surface covered with the oxidized form, since in this case the adsorbing surface will secure the close approach of the two molecules (riboflavin and dihydroriboflavin), which is a necessary condition for the overlapping of their π-orbitals. Formation of such a complex will lead to the formation of a second layer; in the cours of the electrochemical reaction the riboflavin adsorbed at the surface will become covered with a layer of dihydroriboflavin. In the case of the anodic process, dihydroriboflavin lying on the electrode surface and dihydroriboflavin in the second layer will both be subject to oxidation. In the latter case the electron transfer should be preceded by exchange of electrons between the second layer of dihydroriboflavin and the layer of the oxidized form lying directly on the surface. If the rate of this exchange is considerably greater than the polarizing current, a single step relating to reduction of both the layer of dihydroriboflavin situated directly on the mercury and the layer adsorbed on the riboflavin will be obtained on the chronopotentiogram. We observed such an effect with low current densities $(0.5 < i < 15 \text{ mA/cm}^2)$. With the rates in the inverse ratio only the layer directly adsorbed on the electrode will be reduced, and the $i\tau$ value will decrease sharply with increase in current density; this was obtained with large current densities $(i > 32 \text{ mA/cm}^2)$. In the intermediate region both layers may be successively reduced, and this would be expressed by splitting of the step on the chronopotentiogram. The absence of analogous splitting in the steps of the cathodic chronopotentiogram can be explained by absence of the oxidized form in the second layer.

We thus propose that the prewave results from adsorption of dihydroriboflavin at a surface occupied by riboflavin molecules, and that the latter serve as active centers, as it were, for adsorption of dihydroriboflavin. The adsorption energy of dihydroriboflavin, calculated from polarographic data, represents the sum of the specific adsorption energy of riboflavin on mercury and the subsequent adsorption energy of dihydroriboflavin on the riboflavin layer, i.e., the formation energy of an adsorption charge transfer complex. As we have shown in an earlier paper [27], adsorption of riboflavin on a mercury surface corresponds to a constant of approximately 10^5 liter/mole in the adsorption isotherm equation, whereas polarographic data give a value close to 10^8 liter/mole. The Γ_{DRF} values which we obtained in [27] at the potentials corresponding to the limiting current of the prewave were calculated from the values of the first step, where only molecules directly adsorbed on the mercury surface are oxidized. The Γ_{DRF} values obtained in this case are much smaller than those calculated from polarographic data. However, if the calculation is based on values of steps at low current densities or on the sum of the two steps, the results obtained are very close to the polarographic results.

The interpretation of the formation mechanism of prewaves would be also possible under assumption that riboflavin−semiquinone and dihydroriboflavin−semiquinone complexes are formed and are the cause of the appearance of prewaves which meet the requirements of Brdička theory. In this case, however, splitting of the steps on the chronopotentiogram should be observed both with the application of anodic (the semiquinone is oxidized) and with the application of cathodic (the semiquinone is reduced) pulses, and this was not confirmed in our experiments. If only the semiquinone is adsorbed on the electrode, the steps for the cathodic and anodic processes should have the same length at each potential, which is also inconsistent with the experimental results.

LITERATURE CITED

1. R. Brdička and E. Knobloch, Z. Elektrochem., 47:721 (1941).
2. R. Brdička, Z. Electrochem., 48:278 (1942).
3. O. H. Müller, J. Biol. Chem., 145:425 (1942).
4. O. H. Müller, Trans. Electrochem. Soc., 87:441 (1945).
5. R. Brdička, Z. Electrochem., 48:686 (1942).
6. R. Brdička, Coll., 12:552 (1947).
7. W. Lorenz and E. O. Schmalz, Z. Electrochem., 62:301 (1958).
8. J. M. Los and C. K. Tompkins, J. Chem. Phys., 24:630 (1956).
9. J. M. Los and C. K. Tompkins, Canad. J. Chem., 37:315 (1959).

10. A. M. Mirri and P. Favero, La Ric. Sci., 28:2307 (1958).
11. B. Cohen and I. M. Kolthoff, J. Biol. Chem., 148:711 (1943).
12. E. D. Belokolos, Thesis [in Russian], Inst. Elektrokhimii Akad. Nauk SSSR, Moscow (1965).
13. M. Voříškova, Coll., 12:607 (1947).
14. J. T. Stock, J. Chem. Soc., 586 (1949).
15. J. T. Stock, J. Chem. Soc., 763 (1949).
16. S. G. Mairanovskii, Zh. Fiz. Khim., 36:165 (1962).
17. E. C. Gregg and W. P. Tyler, J. Am. Chem. Soc., 72:4561 (1950).
18. P. Zuman, Coll., 19:1140 (1954).
19. H. A. Laitinen and B. Mosier, J. Am. Chem. Soc., 80:2363 (1958).
20. S. L. Gupta, Kolloid-Zeitschrift, 160:30 (1958).
21. B. Breyer and T. Biegler, Coll., 25:3348 (1960).
22. G. A. Tedoradze, in: Advances in Electrochemistry of Organic Compounds [in Russian], Nauka, Moscow (1966), p. 23.
23. G. A. Tedoradze, Doctoral Thesis [in Russian], Inst. Élektrokhimii Akad. Nauk SSSR, Moscow (1965).
24. B. B. Damaskin, I. P. Mishutushkina, V. M. Gerovich, and R. I. Kaganovich, Zh. Fiz. Khim., 38:1797 (1964).
25. S. V. Tatwawdi and A. J. Bard, Analyt. Chem., 36:2 (1964).
26. A. M. Hartley and G. S. Wilson, Analyt. Chem., 38:681 (1966).
27. G. A. Tedoradze, E. Yu. Khmel'nitskaya, and Ya. M. Zolotovitskii, Élektrokhimiya, 3:200 (1967).
28. Ya. M. Zolotovitskii, G. A. Tedoradze, and A. B. Érshler, Élektrokhimiya, 1: 828 (1965).
29. D. E. Smith, in: Electroanalytical Chemistry, Vol. 1, ed. A. J. Bard, Dekker, New York (1966), p. 1.
30. H. A. Laitinen and J. E. B. Randles, Trans. Faraday Soc., 51:54 (1955).
31. M. Senda and P. Delahay, J. Phys. Chem., 65:1580 (1961).
32. W. Lorenz, Z. Phys. Chem., 218:259 (1961).
33. J. Llopis, I. Fernandez-Biarge, and M. Perez-Fernandez, Electrochim. Acta, 1:130 (1959).
34. M. Sluyters-Rehbach, B. Timmer, and J. H. Sluyters, Rec. Trav. Chim., 82:553 (1963); M. Sluyters-Rehbach, J. H. Sluyters, Rec. Trav. Chim., 82: 525, 536 (1963).
35. J. E. B. Randles and K. W. Somerton, Trans. Faraday Soc., 48:937 (1952).
36. J. E. B. Randles, in: Transactions of the Symposium on Electrode Processes Philadelphia, May 1959, Wiley, New York (1961).
37. R. Tamamushi and N. Tanaka, Z. Phys. Chem. (N. F.), 28:158 (1961).
38. H. H. Bauer, D. L. Smith, and P. J. Elving, J. Am. Chem. Soc., 82:2094 (1960).
39. A. M. Baticle and F. Perdu, J. Electroanalyt. Chem., 12:15 (1966).
40. B. Kastening, H. Gartmann, and L. Holleck, Electrochim. Acta, 9:741 (1964).
41. H. L. Hung and D. E. Smith, J. Electroanalyt. Chem., 11:237, 425 (1966).

42.   M. Senda, M. Senda, and J. Tachi, Rev. of Polarography (Japan), 10:4, 142 (1962).

43.   A. Ehrenberg, Arkiv. for Kemi, 19:97 (1962).

44.   R. Brdicka, Z. Elektrochem., 47:314 (1941).

45.   J. Heyrovsky and J. Kůta, Principles of Polarography [Russian translation], Mir, Moscow (1965).

46.   H. J. Lowe and W. M. Clark, J. Biol. Chem., 221: 983 (1956).

47.   B. Ke, Arch. Biochem. Biophys., 68:330 (1957).

48.   L. Michaelis and G. Schwarzenbach, J. Biol. Chem., 123:527 (1938).

49.   H. Beinert, J. Am. Chem. Soc., 78:5323 (1956).

50.   Q. H. Gibson, V. Massey, and N. M. Atherton, Biochem. J., 85:369 (1962).

51.   R. S. Mulliken, J. Am. Chem. Soc., 74:811 (1952).

52.   K. H. Hausser, Z. Naturforsch., 11a:20 (1956).

53.   K. H. Hausser and J. N. Murrel, J. Chem. Phys., 27:500 (1957).

54.   A. Szent-Gyorgyi, Introduction to Submolecular Biology [in Russian], Nauka, Moscow (1964).

55.   A. Ehrenberg, Electronic Aspects of Biochemistry, Academic Press, New York– London (1964), p. 379.

56.   B. Pulman and A. Pulman, Quantum Biochemistry [Russian translation], Mir, Moscow (1965), p. 382.

57.   K. Nakamoto, J. Am. Chem. Soc., 74:1739 (1952).

58.   L. Michaelis, M. R. Schubert, and S. Granik, J. Am. Chem. Soc., 62:204 (1940).

Reduction of Organic Molecules in Region of Low Electrode Surface Coverages

A. B. Érshler

We will consider some relationships which prove useful for investigating the kinetics of electrode processes involving organic molecules. Work on such reactions has been based to a considerable degree on theoretical ideas formulated as a result of the investigation of the electrochemical characteristics of the simplest inorganic substances. There is, however, no doubt that organic compounds possess a number of special features which are important from the point of view of the electrochemist, namely:

a) considerable surface activity;

b) large size of the molecule, frequently exceeding the thickness of the dense part of the double layer;

c) possibility of the reaction nearly always proceeding by several parallel routes.

Equations will be obtained for the polarization curve which gives the course of the reduction of an organic compound by the following reaction:

$$A + e \rightarrow A^-, \tag{1}$$

together with experimental data suitable for checking these equations. Such equations find fairly wide use; they can be applied to irreversible processes, as steps following Eq. (1) do not in fact have any effect on the form of the polarization curve. These equations can therefore be used for the examination of fairly complex electrode reactions, provided that the slow step is represented by electron transfers according to Reaction (1).

1. EQUATION OF POLARIZATION CURVE.
REDUCTION OF UNADSORBED PARTICLES

In electrical units a first-order reaction rate is described as follows [1]:

$$i = \varkappa n F \left(\frac{kT}{h}\right) \frac{a_O}{f^{\neq}} \exp\left(-\frac{\Delta G_0^{\neq}}{RT}\right),$$ (2)

where a_O is the activity of the reacting substance in the initial state; f is the activity coefficient of the transitional state; $\Delta G_0^{\neq}$ is the standard free energy of activation.

In the case of electrode reactions [2]:

$$i = \varkappa n F \left(\frac{kT}{h}\right) \frac{a_O}{f^{\neq}} \exp\left[-\frac{(\Delta G_0^{\neq})_n}{RT} - \frac{\alpha n_a F}{RT}(E - E_e) - \frac{z - \alpha n_a}{\alpha n_a}\psi_a\right],$$ (3)

where E is the potential; E_e is the equilibrium potential; $(\Delta G_0^{\neq})_n$ is the chemical part of the standard free energy of activation (it includes, in particular, the standard free energy of adsorption at potential E); ψ_a is the potential in the plane where the transitional state is situated; n_a is the number of electrons transferred in the slow step; z is the charge of the initial substance; α is a constant which varies slowly with potential [3].

Equation (3) is based on the ideal that the standard free energy of the transitional state is a linear combination of the standard free energies of the original and final states [2];

$$G_0^{\neq} = \alpha G_{0,\,R} + (1 - \alpha)G_{0,\,O}.$$ (4)

Here the subscripts R and O denote values characteristic of the final (reduced) and initial states respectively. This idea originates from the Brønsted equation [4, 5]. When using Eq. (4) it is necessary to assign to the initial and final states the coordinates to which they actually correspond at the beginning and at the end respectively of the electron transfer process. In reduction of the hydrogen ion, for example, the initial state is located at the outer Helmholtz plane [4] or somewhat nearer to the electrode surface [6]. For reactions involving adsorbed substances the initial and final states are regarded as adsorption states [5, 7, 8]. The Brønsted relationship also makes it possible to calculate the value of

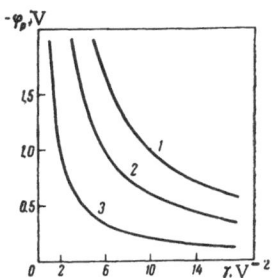

Fig. 1. Curves for Eq. (17) [21].
Values of αn_a: 1) 0.5; 2) 0.3;
3) 0.2.

$f^{\neq}$ [5]:

$$f^{\neq} = f_0^{(1-\alpha)} \cdot f_R^{\alpha}. \qquad (5)$$

Since the $f^{\neq}$, ψ_a, and $(\Delta G_0^{\neq})_n$ values enter into the kinetic equation, an answer to the question as to where in fact the transitional state is located is of considerable interest. However, after the values of these functions have been determined [from Eqs. (4) and (5) or by another method], the choice of point at which the original state is located in no way affects the value of the rate calculated from Eq. (3). In fact, in the final count, Eq. (3) contains the electrochemical potential of the original substance, which in the absence of concentration polarization does not depend on the coordinates of the particle. It is convenient [9] when using Eq. (3) to place the original state in the bulk of the solution. In [9] this state was denoted by I, and the states figuring in the right-hand side of Eqs. (4) and (5) were denoted by II and III.

Thus, the form of the kinetic equation depends substantially on the orientation of the transition state in relation to the electrode and particularly on whether the transition state is adsorbed or is located in the diffusion part of the double layer. In the light of what follows it will become clear that the first case holds in practice, but we will begin with the equation for reduction of unadsorbed particles. Only the case with z = 0 will be examined.

When considering a reaction involving unadsorbed particles the dependence of $(\Delta G_0^{\neq})_n$ on the potential can be disregarded. From Eq. (3) it is possible to obtain the Tafel equation [9] and, by taking concentration polarization into account, the equation for the irreversible polarographic wave [10]:

$$E = E_{1/2} - \frac{2,3RT}{\alpha n_a F} \log \frac{I}{I_d - I}, \qquad (6)$$

$$\frac{\alpha n_a F}{2.3RT}(E_{1/2} - \psi_a) = \log 0.81\, k_0 t^{1/2} D^{-1/2} + \log f_{01} - \log f^{\neq}. \qquad (7)$$

In order to determine the magnitude of $f^{\neq}$ use is made of Eq. (5). The original state (by stipulation) is not charged, and the final state carries a charge of n_a. The activity coefficient of molecules of type i may be represented in the form:

$$\log f_i = \log f_{z=0} + \log f_z, \tag{8}$$

where f_z only takes account of electrostatic interionic interactions. It can be evaluated from the Debye–Huckel equation assuming symmetry of the molecule. Hence:

$$\log f_{zR} = -\frac{A n_c{}^{n_a} \sqrt{\overline{\mu}}}{1 + K \sqrt{\overline{\mu}}},$$

where n_c is the charge carried by the cation of the indifferent electrolyte; μ is the ionic strength; A and K are constants.

It can be assumed [11] that the value of f_z does not change when the particles move from the bulk of the solution into the double layer. Since the radius of an organic molecule is usually large, its charge (according to Pleskov) should not affect the solvation energy. It can therefore be assumed that $f_{(z=0)R} = f_0$ [or at least $d(\log f_0) = d(\log f_{(z=0)R})$]. Then:

$$\log f^{\neq} = \log f_{0II} - \frac{\alpha A n_c{}^{n_a} \sqrt{\overline{\mu}}}{1 + K \sqrt{\overline{\mu}}}. \tag{9}$$

If state II is moved towards solution (is not adsorbed), it is practically identical with state I (z = 0), and:

$$\frac{\alpha n_a F}{2.3RT}(E_{1/2} - \psi_a) = \log 0.81 \, K_0 t^{1/2} D^{-1/2} + \frac{\alpha A n_c{}^{n_a} \sqrt{\overline{\mu}}}{1 + K \sqrt{\overline{\mu}}}. \tag{10}$$

According to Eq. (10), in the reduction of unadsorbed particles the half-wave potential, corrected for the value of the ψ_a potential, should move to the positive side with increase in the concentration of the electrolyte. The effect is, however, small (see Section 3).

2. REDUCTION OF ADSORBED PARTICLES

In deriving the equation for the polarization curve it is again necessary to start from Eq. (3) and place the initial state into the

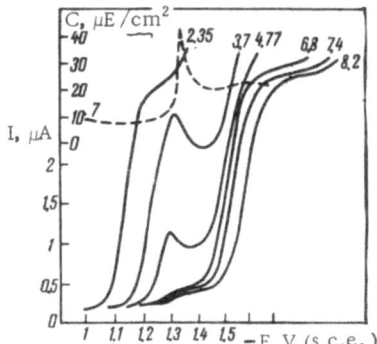

Fig. 2. Polarograms and differential capacity curve (dotted line) for 10^{-3} mole/liter solution of N-(2-methoxybenzoyl)caprolactam in 4% ethanol (0.2 N with respect to the sodium cation of the McIlvain buffer) [16]. The numbers on the curves respresent the pH values.

bulk of the solution. The result amounts to calculation of the $(G_0^{\neq})_n$ and $f^{\neq}$ values. The activity coefficient $f^{\neq}$ of the transition state must now reflect the competition of adsorbed species for sites at the surface and interaction between them. Therefore, according to Temkin, it should include the factor:

$$(1 - \sum_i \theta_i)^{-1} \exp\left(-2 \sum_i a_{\neq\, i} \theta_i\right).$$

In the region of small coverages this factor is equal to unity.

The adsorption energy of an organic substance at mercury and the standard free energy of an organic substance adsorbed at mercury are calculated from Eqs. (11) and (12):

$$W_{\mathrm{ads}} = W_{\mathrm{M}} - \gamma RT\,(E - E_{\mathrm{M}})^2, \tag{11}$$

$$G_{0\,\mathrm{ads}} = G_{0\,\mathrm{M}} + \gamma RT\,(E - E_{\mathrm{M}})^2, \tag{12}$$

where E_{M} is the potential of maximum adsorption,

$$\gamma = \frac{C_0 - C_1}{2RT\Gamma_{\infty}}, \tag{13}$$

and C_0 and C_1 are the integral capacities of the double electric layer with electrode coverages $\theta = 0$ and 1 respectively [12].

By using Eq. (12) to describe the characteristics of the transition state it is possible to obtain the following expression for the reduction rate of adsorbed organic substances:

$$i = \varkappa nF \left(\frac{kT}{h}\right) \frac{a_0}{f^{\neq}} \exp\left[-\frac{(\Delta G_0^{\neq}{}_{\mathrm{M}})_n}{RT} - \frac{\alpha n_a F}{RT}\,(E - E_{\mathrm{M}}^{\neq} - \psi_a) - \gamma^{\neq}\,(E - E_{\mathrm{M}}^{\neq})^2\right]. \tag{14}$$

In the region of small coverages, according to the above, we have:

$$i = \varkappa n F \left(\frac{kT}{h}\right) a_0 \exp\left[- \frac{(\Delta G_{0M}^{\neq})_n}{RT} - \frac{\alpha n_a F}{RT}(E - E_M^{\neq} - \psi_a) - \gamma^{\neq}(E - E_M^{\neq})^2 \right.$$
$$\left. + \frac{\alpha A n_c n_a \sqrt{\bar\mu}}{1 + B \sqrt{\bar\mu}} \right] \tag{14a}$$

and, disregarding the effect of the electrolyte concentration,

$$i = \text{const } c_0 \exp\left[- \frac{(\Delta G_{0M}^{\neq})_n}{RT} - \frac{\alpha n_a F}{RT}(E - E_M^{\neq}) - \gamma^{\neq}(E - E_M^{\neq})^2 \right]. \tag{14b}$$

Equation (14b) was derived by Frumkin [2]. In his opinion, for large molecules it differs little from the relationships obtained earlier [13, 14].

Tedoradze and Arakelyan [8] made use of Eq. (4) to describe the characteristics of the transition state and obtained for the region of small coverages:

$$2.3 \log i = 2.3 \log K_0 + 2.2 \log c - \alpha \gamma_R (E - E_{MR})^2 - (1 - \alpha) \gamma_O (E - E_{MO})^2$$
$$- \frac{\alpha n_a E}{RT} E. \tag{15}$$

With certain additional assumptions Eq. (15) is closely consistent with the experimental polarization curves for the discharge of the diphenylammonium ion. In particular, Eq. (15) takes account of the fact that the E_M values of the initial, transition, and final states do not coincide. As indicated by Krishtalik, Eqs. (14b) and (15) are equivalent. From Eq. (14b) it is not difficult to obtain the following relationships [14]:

$$\log i = \log i_p - \gamma^{\neq}(E - E_p)^2, \tag{16}$$

$$\varphi_p = E_p - E_M^{\neq} = - \frac{\alpha n_a F}{2\gamma^{\neq} RT}, \tag{17}$$

$$i_p = \varkappa n F \frac{kT}{h} c_0 \exp\left[- \frac{(\Delta C_{0M}^{\neq})_n}{RT} + \gamma^{\neq}(E_p - E_M^{\neq})^2 \right]. \tag{18}$$

As seen from Eqs. (16) and (17), the reduction rate of the adsorbed particles passes through a maximum with increase in negative potential. Substitution of plausible numerical values shows, however, that this maximum usually lies outside the limits of the region of

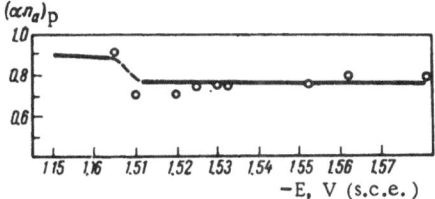

Fig. 3. Dependence of $(\alpha n_a)_p$ on $E_{1/2}$, calculated from Eq. (20) for 10^{-3} mole/liter solution of N-(2-methoxybenzoyl)caprolactam in 4% ethanol (McIlvain buffer). The points represent the experimental values (from [16]).

potentials accessible to measurement (Fig. 1); the φ values do not usually exceed 1-1.5 V, $\alpha n_a \geq 0.3$, and $\gamma \leq 10$. Furthermore, the fact that the maximum falls in an experimentally accessible region of potentials does not mean that it can be detected in practice. For this it is necessary that the i_p value [Eq. (18)] should be commensurable with the limiting diffusion current, otherwise either the maximum as a whole or the drop in current will not be noticeable. Additional complications appear when the orientation of the activated complex at the surface depends on the potential, and a simple formula of the type (12) no longer describes its properties [14].

If adsorption is strong the equation of the polarization curve becomes more complicated. It can, however, be shown [15] that there is a maximum on the curve at the potential:

$$E_p = E_M^{\neq} - \frac{\alpha n_a F}{2\gamma^{\neq} RT} \frac{1 - 2a(1 - \theta_p)\theta_p}{1 - \theta_p} . \tag{19}$$

if the Frumkin isotherm is correct. Here θ_p is the degree of the coverage at potential E_p. Consequently, with large positive values for the attraction constant a, a new maximum appears on the curve at a potential more positive than that calculated from Eq. (17). Such a case was evidently observed by Tsagareli et al., [15, 16] in the investigation of N-(2-methoxybenzoyl) caprolactam and certain related compounds.

On the polarogram (Fig. 2) there is a maximum at potentials corresponding to the desorption peak on the differential capacity

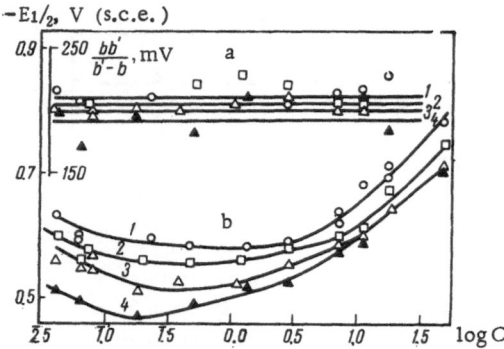

Fig. 4. Dependence of slope (a) and half-wave potential (b) of reduction wave of α-bromocaproic acid on its concentration [22] with various drop times: 1) 0.3 sec; 2) 0.6 sec; 3) 1.2 sec; 4) 2.4 sec.

curve. Along both branches of this maximum (both rising and descending branches) it seems that electron transfer is one of the slow steps [15]. The appearance of a dip in current in this case cannot therefore be attributed to limitations at the preceding chemical step. This dip must be connected with a decrease in the electron transfer rate resulting from decrease in the adsorption energy.

If $a = 0$ (i.e. Langmuir isotherm) Eq. (19) is practically identical to Eq. (17).

Equation (14a) can be transformed as follows:

$$-\frac{d \log i}{dE} = \frac{(\alpha n_a)_P F}{2.3RT} = \frac{\alpha n_a F}{2.3RT} + 2 \gamma^{\neq} (E - E_M^{\neq}). \qquad (20)$$

The left-hand side of this equation is a quantity which for irreversible processes can be determined from the slope of the polarographic curve or from the form of the relationship between half-wave potential or other reaction rate characteristic and pH. The function $(\alpha n_a)_P$ may be called the polarographic electron transfer coefficient. Mairanovskii, who writes it as $(b' - b)/bb'$ [17], has described several cases where this quantity has decreased with increase in the negative potential and has analyzed the results by means of equations equivalent to Eq. (20). Figure 3 shows the curve from [15].

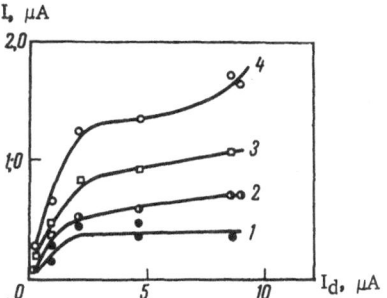

Fig. 5. Dependence of current passing through dropping electrode after 7 sec electrolysis at constant potential on concentration of ethyl α-bromobutyrate in 0.1 N sodium bromide solution at 27°C [20]. Values of $-\varphi$ (s.c.e.): 1) 0.22 V; 2) 0.30 V; 3) 0.35 V; 4) 0.40 V.

The straight line was computed from Eq. (20) on the basis of independent adsorption measurements, and the points represent the experimental data.

Until recently a polarographic wave equation which described the reduction of adsorbed particles was only known in the case of the Henry isotherm [18]. From the theory developed in [18] it follows that at constant potential the current density should pass through a maximum with increase in the period of electrolysis, but this effect was not observed in practice. The maximum evidently occurs after very short times.

The complex form of the theoretical current–time curve is explained by the fact that the degree of surface coverage with depolarizer molecules passes through a maximum. If $\psi_a = 0$, the following equation is valid for long periods [17–19]:

$$\frac{\alpha n_a F}{2,3RT} E_{1/2} = \log\left(\frac{7\pi}{3}\right)^{1/2} \frac{48}{169} + \log\frac{k_0 \Gamma_\infty}{D^{1/2}} + \log B + \frac{1}{2}\log t + \log\frac{f_{0.1}}{f^{\neq}};$$

where

$$B = B_M^{\neq} \exp\left[-\gamma^{\neq}(E_{1/2} - E_M^{\neq})^2\right]. \tag{21}$$

Most recently Filinovskii has solved the diffusion problem for the case of adsorption according to the Frumkin isotherm [19]. An expression for the half-wave potential for stationary electrolysis conditions and for $\theta = 1$ was obtained earlier by Feoktistov [20], Érshler, and others [15, 21]. It was shown that:

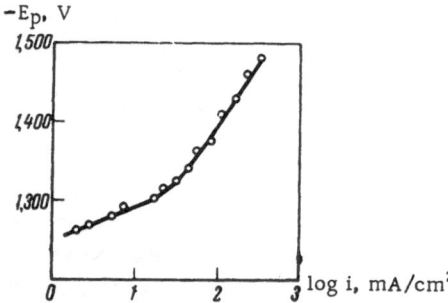

Fig. 6. Dependence of initial reaction potential E_p on current density in galvanostatic
reduction of 1,2-di(2-pyridyl)ethylene $(2 \cdot 10^{-4}$
mole/liter) [27]. Potential during preliminary period -0.900 V; supporting electrolyte
0.1 N phosphate buffer + 1.3 N potassium
chloride; pH 11.5. The potentials are given
with reference to a calomel electrode in the
same solution.

$$\frac{dE_{1/2}}{d \log c} = \frac{2,3 RT}{\alpha\, n_a F},$$

where c is the concentration of depolarizer in solution.

A displacement of the half-wave potential to the more negative
values or, what amounts to the same, tendency of the rate of the
process at constant potential to limit at higher depolarizer concentrations has been observed repeatedly in practice [20-26] (Figs. 4
and 5). This effect was first described and explained by Antropov
[78].

3. EFFECT OF SOLUTION COMPOSITION
ON REACTION RATE

The relationships given in the preceding section did not take
account of the effect of the ψ_a potential or treated this effect in a
purely formal manner. Consideration of the effect of the double
layer on the reduction of organic compounds is complicated for two
reasons: a) the diffuseness of the double layer at a given potential

depends on θ [9]; b) the quantitative adsorption theory [12], leading to Eq. (12), is based on the Frumkin model, which does not take into account the diffuse structure of the double layer.

Both these difficulties can be avoided by working with constant electrode surface coverage by depolarizer and reaction product molecules, preferably in the absence of the latter. Figure 6 gives the Tafel relationship, which Levinson [27] obtained under these very conditions by the method of charging curves. This problem can be solved also somewhat differently [15]: the effect of the solution composition on the process rate can be investigated at the potential corresponding to the desorption peak on the differential capacity curve; at this potential the total degree of coverage has a constant values, equal to 0.5 (with large positive values of a [28]).

In the region of small coverages the value of θ can only be determined by calculation, but as yet we have no theory which describes the effect of diffuseness of the double layer on the adsorption coefficient B, and this is essential for such a calculation. Since it is well known [29–31] that the desorption peak moves to the negative side when the electrolyte concentration is decreased, it can be supposed that the value of B increases with increase in the diffuseness of the double layer at a given potential. The same conclusion can be reached on the basis of model approximations. Formulas of the type (12) are thus unsuitable for the examination of effects connected with the influence of the ψ_a potential on the reduction of adsorbed particles. This fact makes the applicability of the well-known formula [17] to our case doubtful:

$$\Delta E_{1/_2} = \frac{z - \alpha n_a}{\alpha n_a} \Delta \psi_a.$$

In practice, during reduction of uncharged organic molecules ($z = 0$) in the region of small coverages, $\Delta E_{1/_2}/\Delta \psi' \leq 1$ (Table 1). The differences between experiment and the formula just given are related not only to the influence of the ψ_a potential on adsorption but also to the inadequacy of the Gouy–Stern theory for describing the properties of the ψ_a potential in the case of large molecules. In fact, in the region of large coverages, where changes in the value of B do not have such a significant effect on the coverage and consequently on the reaction rate, considerable deviations from the

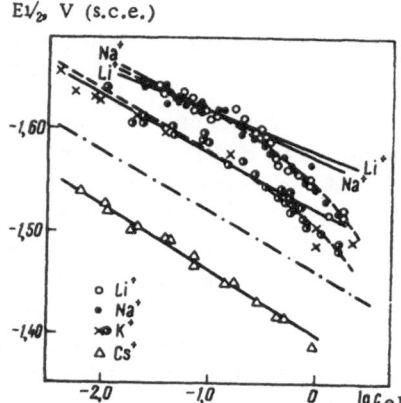

$E\frac{1}{2}$, V (s.c.e.)

Fig. 7. Polarographic reduction of iodobenzene in 2% aqueous ethanol at 27°C
with alkali metal acetates as supporting
electrolyte [21, 33].

formula are also observed
particularly in the reduction
of cations [15, 17, 32].

Next the effect of the
diffuseness of the double
layer on the reduction of adsorbed particles in the region of small coverages, and
with large electrode charges,
will be examined on the basis
of the following assumption:

$$\Delta E_{1/2} = x \Delta \log c_{el} \qquad (22)$$

This empirical formula
closely describes the experimental data for electrolyte
solutions which are not too
concentrated, e.g., as shown in Fig. 7. The values of the coefficients
x for a number of cases are given in Table 1.

When examining the effect of the electrolyte on the activity of
the organic depolarizer Eq. (21) was supplemented by the relationship (22). According to Eq. (14), in addition to the electrode potential and double layer structure, the reduction rate is affected by any
factors which change the activity of the initial and transition states.
We shall consider how the activity is affected by changes in the concentration of the indifferent electrolyte and in the composition of
the binary water – ethanol solvent. When the electrolyte concentration is increased in aqueous solution the activity of organic compounds in this solution varies according to the Sechenov formula
[39] ("salting out"):

$$\log S_0 - \log S_{el} = \log f_{el} - \log f_0 = - k_{el} c_{el}. \qquad (23)$$

where k_{el} is a constant which depends on the nature of the dissolved
substances; S is the solubility of the organic substance; the subscript 0 relates to pure water and the subscript el to a solution with
a concentration of c_{el} with respect to the electrolyte.

TABLE 1. Effect of Electrolyte Concentration on Half-Wave Potential

Depolarizer	Solvent	Electrolyte	$-E_{1/2}$, V (s.c.e.)	$\dfrac{\Delta E_{1/2}}{\Delta \log c_{el}}$	Reference
β-Bromopropionitrile	Water	0.1–1.0 N KCl	1.7	58	[24]
Iodobenzene	2% Aqueous ethanol	0.02–0.4 N CH₃COOLi	1.60	39±8	[21, 33]
	"	0.02–0.2 N CH₃COOK	1.57	52	[21, 33]
	"	0.02–0.7 N CH₃COOCs	1.45	67±5	[21, 33]
	87% Aqueous ethanol	0.1–2.0 N CH₃COOK	1.68	10–20	[24, 33]
2-Acetyl-5-bromothiophene	Water	0.1–2.0 N KCl + KOH	1.16	65	[34]
Benzaldehyde (second wave)	"	0.1 M CH₃COOH + 0.1 M CH₃COOK + x KCl	1.25	50	[35]
Acetophenone (second wave)	"	"	1.48	35	[35]
β-Acetyltetralin	"	—	1.6	36	[38]
N-(2-Methoxybenzoyl)caprolactam	4% Aqueous ethanol	0.02 N Na₂HPO₄ + NaClO₄ (up to 1 N)	1.5	20	[30, 36]
Benzyl chloride	Water	0.02–1.0 N KCl	1.0	50	[36]
Iso-butyl iodide	"	0.02–0.2 N KCl	1.81	38	[37]

TABLE 2. Salting out of Amyl Alcohol from Aqueous Solutions

Electrolyte	Concentration range, g-equiv /liter	$-E_{z=0}$, V (s.c.e.)	$-E_{des}$, V (s.c.e.)	ΔE_{exper}, V	ΔE_{calc}, V	ΔE_{sal}, V	k_{el}, liter/mole
KCl	1.0–0.1	0.55	1.09	−0.015	0.052	−0.067	0.21
Na₂SO₄	1.0–0.1	0.47	1.17	−0.040	0.042	−0.082	0.34
Na₂SO₄	0.1–0.01	0.47	1.14	+0.030	0.048	−0.018	0.71
H₂SO₄	0.1–0.01	0.47	1.04	+0.030	0.042	−0.012	0.39

TABLE 3. Variation of Half-Wave Potential Due to Effect of Electrolyte Concentration on Activity of Reaction Components

$\Delta E_{1/2}$, calculated from Eq.	k_{el}, liter /g-equiv	Range of electrolyte concentrations, g-equiv/liter			
		0–0.02	0.02–0.2	0.2–2.0	1.0–2.0
(9)	Any	0.002	0.003	0.015	0.003
(26)	0.10	0.002	0.006	0.040	0.015
(26)	0.20	0.003	0.011	0.090	0.039

Increased adsorbability due to salting out in tert.-amyl alcohol was observed by Frumkin [40]. Grigor'ev and Damaskin [29] report that owing to salting out the desorption peak of this compound is displaced towards greater charge density of the electrode with increase in the concentration of electrolyte in the solution. A similar effect was observed with N-(2-methoxybenzoyl)caprolactam [30]. Thus, with change in electrolyte concentration the desorption peak is displaced for two reasons, i.e., salting out and change of ψ_a potential. By using the Frumkin–Damaskin theory [12] it is possible to calculate the shift of the desorption peak caused by salting out for the case where $a > 1$. For this purpose the Frumkin isotherm must be written in the following form:

$$B_M c_0 f_{el} \exp[-\gamma(E - E_M)^2] = \frac{\theta}{1 - \theta} \exp(-2a\theta),$$

where the characteristic of the activity coefficient are given by Eq. (23).

By considering further that $\theta \simeq 0.5$ at the potential of the desorption peak [12], we obtain:

$$\left(\frac{dE_{des}}{dc_{el}}\right)_{sal} = \frac{2.3k_{el}}{2\gamma(E_{des} - E_M)}. \tag{24}$$

According to the data of Grigor'ev and Damaskin [29], it is possible to calculate the k_{el} values for amyl alcohol using Eq. (24) (Table 2). From these data [29] it was calculated that with sodium fluoride as supporting electrolyte $\gamma = 7.2$ V^{-2}, and the displacement of the zero-charge point relative to the maximum adsorption point is

$$E_M - E_{z=0} = -0.1 \text{ V}.$$

It was further assumed that these two quantities do not depend on the nature of the electrolyte. The shift of the potential due to salting out was calculated from the formula

$$\left(\frac{dE}{dc_{el}} \right)_{sal} = \frac{\Delta E_{exper} - \Delta E_{calc}}{\Delta c_{el}}.$$

Values for the shift of the potential in the absence of salting out (E_{calc}) were taken from [29], where use was made of the formula

$$\Delta E_{calc} = \frac{RT}{nF} \ln \frac{a_1}{a_2} \tag{25}$$

to compute the theoretical shift of potential on switching from a solution with supporting electrolyte activity a_1 to a solution with activity a_2. The k_{el} values obtained are reasonable in order of magnitude (Table 2). Disagreement in the k_{el} values for the two ranges of sodium sulfate concentrations show that the method of calculating the ΔE_{calc} value from Eq. (25) is not entirely correct. The k_{el} values for the two above-mentioned ranges of sodium sulfate concentrations coincide if it is assumed that the shift of the desorption peak in the absence of a salting out effect amounts to 38 mV when the concentration is changed by a factor of 10. For N-(2-methoxybenzoyl) caprolactam with sodium perchlorate as supporting electrolyte, when salting out does not take place, the potential of the desorption peak is shifted by 20 mV towards the positive side when the electrolyte concentration is increased by an order of magnitude, and this also is inconsistent with Eq. (25).

It is known that k_{el} depends little on changes in structure. For example, the k_{el} values for all the investigated monosubstituted benzene derivatives practically coincide [39]. It is therefore possible to write $k_{el\,O} \simeq k_{el\,R} \simeq k_{el}^{\neq}$ and to use Eq. (9) to calculate $f^{\neq}$. Accordingly $f_{0\,I} \simeq f_{0\,II}$ and Eq. (10) is valid in the case where the transition state is displaced towards the solution.

The situation is different if the transition state is located within the limits of the inner part of the double layer and the

coverages are small. As seen from Table 1, the half-wave potential is shifted by approximately 60 mV to the positive side when the
electrolyte concentration is increased by an order of magnitude.
Furthermore, according to [41], whereas the potential in the region
of large charge densities is displaced by 59 mV to a positive side
when the electrolyte concentration is increased by a factor of 10,
the structure of the inner part of the double layer does not change.
Thus, changes in the concentration of the electrolyte affect the
structure of the double layer at constant potential and the half-wave
potential, but the structure of the double layer at the half-wave
potential for reduction of the uncharged molecule remains constant.
The immediate environment of the adsorbed initial state (State II)
at the half-wave potential thus does not depend on the concentration
of the electrolyte, and it can be assumed that the activity coefficient
of this state $(f_{0\,II})$ does not depend on the concentration of the electrolyte.

The activity coefficient of the transition state can now be calculated from Eq. (9), taking $f_{0\,II}$ as constant. The activity coefficient
of the initial state in the bulk of the solution (State II) is calculated
from Eq. (23), and from Eqs. (9), (21), and (23) we can now obtain
[42]:

$$\frac{\alpha\,n_a F}{2,3\,RT}\,E_{1/2} = \log\left(\frac{7\pi}{3}\right)^{1/2}\frac{48}{13^2} + \lg\frac{k_0 B\Gamma_\infty t^{1/2}}{D^{1/2}} + k_{el}c_{el} + \frac{\alpha n_c A n_a\sqrt{\mu}}{1 + K\sqrt{\mu}}. \qquad (26)$$

The same expression, without the last term on the right, can be
obtained on the assumption that the reaction rate is proportional to
the concentration of depolarizer in the reaction layer.

The effect of electrolyte concentration on the reduction of unadsorbed particles is thus described by Eq. (10), and the effect on
the reduction of adsorbed particles is described by Eq. (26), supplemented by Eq. (22). The salting out constant only occurs in Eq.
(26). Both equations predict an effect of the same sign, but the
forms of the dependences of half-wave potential on supporting electrolyte concentrations and the magnitude of the effects at high
supporting electrolyte concentrations make it possible to distinguish
between reduction of adsorbed and unadsorbed particles (Table 3,
where calculated of $\Delta E_{1/2}$ was made for an interionic distance of
3–4 Å, and electron transfer coefficient of 0.5 and $n_c = n_a = 1$).

The effect of ethanol on the half-wave potential is largely equivalent to the effect of electrolyte concentration. In both cases variation in the composition of the solution affects the activity of the initial, transition, and final states, and also the structure of the double layer. In the region of large surface coverages with ethanol molecules there is a sharp decrease in the charge of the electrode and the diffuseness of the double layer [10].

In the region of small surface coverages with organic molecules the composition of the solvent which fills the inner part of the double layer changes very little with change in ethanol content of the solution.

It can be shown, on the basis of the idea that the reaction rate is proportional to the depolarizer concentration in the reaction layer, that with unchanged mechanism the addition of ethanol does not affect the reduction rate of the unadsorbed particles and reduces the reduction rate of adsorbed particles. In fact addition of ethanol does not alter the concentration of depolarizer in the bulk of the solution but reduces its adsorption. This can occur in two ways. First, addition of ethanol as a rule reduces the activity of organic substances in aqeuous solution, e.g., as seen in the increased solubility. The decrease in the adsorption of the depolarizer, due to its reduced activity in the bulk of the solution, must be particularly large in the region of low coverages, where the degree of coverage is proportional to the activity of the adsorbate (Henry isotherm). In the region of large converages the effect is negligibly small. Second, in the region of large coverages competition may take place between the ethanol and the depolarizer for a site on the surface, i.e., the surface may be blocked with ethanol molecules. On the other hand, this effect should not appear in the region of low coverages. It is thus possible to establish which of the two effects plays the principal part. In the case of N-2-(methoxybenzoyl) caprolactam a change from 4% ethanol to 50% shifts the half-wave potential by 20 mV in the region of large coverages and by 50 mV in the region of small coverages [15]. It is possible to indicate many other cases where addition of ethanol shifts the half-wave potential to the negative side in the region of small coverages [21, 43]. In these cases blocking of the surface with ethanol is out of the question, and the only conceivable reason is decreased activity of the depolarizer in the bulk of the solution.

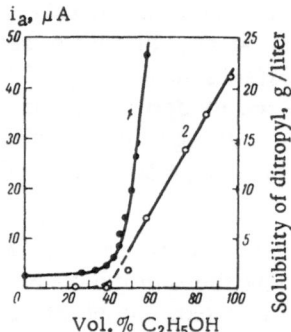

Fig. 8. Height of adsorption prewave on polarogram of tropylium perchlorate (supporting electrolyte 0.1 N chloric acid) (1) and solubility of ditropyl in aqueous ethanol (2) at 20°C [44].

For confirmation of this mechanism for the effect of ethanol one can refer to [44], where the effect of ethanol concentration on the retardation of the reduction of the tropylium ion by its reduction product (ditropyl) is examined. The retardation leads to the formation of an irreversible adsorption prewave; if the retardation is eliminated the limiting adsorption current increases to the level of diffusion current, as happens when ethanol is added to the solution (Fig. 8, Curve 1).

The adsorption current increases (Fig. 8) in proportion to the increase in solubility of ditropyl (Curve 2). Furthermore, in the absence of ditropyl, high degrees of surface coverage with ethanol at potentials corresponding to the adsorption current are attained even in a 20% solution with respect to ethanol, where the height of the adsorption wave has not yet begun to change. From this it is concluded [44] that the decrease in adsorption of ditropyl (detected from the increase in the height of the adsorption wave) arises from its reduced activity in the bulk of the solution and not from blockage of the electrode with ethanol.

We shall now attempt to examine the problem of the effect of ethanol on the kinetics of the electrode reaction in the region of small coverages by means of Eq. (3) using the same method as employed above in the investigation of the effect of electrolyte concentration. The change in activity under the influence of moderate additions of ethanol can be determined in the same way as in the abovementioned case:

$$\log S_0 - \log S_{et} = \log f_i - \log f_0 = -k_{et}c_{et}, \qquad (27)$$

where S_0 and S_{et} are the solubilities of the organic substance in the aqueous solution of the electrolyte and in a solution of the same electrolyte in the binary solvent respectively.

As already mentioned above, the composition of the solvent in this case changes at the Helmholtz layer; on one side, in the inner layer, water is present, and on the other side the binary solvent. Therefore, if a depolarizer molecule in State II is located in the inner part of the double layer in the region of small coverages, its activity does not change when ethanol is added to the solution; if, however, it is located on the other side of the Helmholtz layer, the activity of the State II decreases when ethanol is added. Consequently, if the transition state (like the initial state I) is located outside the limits of the inner part of the double layer, the $f_0 I/f^{\neq}$ function should not depend on the ethanol concentration, and addition of ethanol should not affect the half-wave potential. If, however, the transition state is located in the inner part of the double layer, its activity coefficient is practically independent of the ethanol concentration, and it follows from Eqs. (14) and (27) that

$$(\Delta E_{1/2})_{et} \simeq \frac{2.3\,RT}{\alpha\,n_a F}\,\Delta\,(\log f_{0\,I} - \log f^{\neq}) = \frac{2,3\,RT}{\alpha n_a F}\,\Delta \log S_{et}.$$

When this relationship was first obtained [43], it was also assumed that the activity coefficient of the transition state does not depend on the concentration of ethanol; however, for some incompletely understood reasons, the authors linked the formula obtained to the reduction of particles situated outside the limits of the inner part of the double layer.

Similarly as in [43], in examining the effect of the solvent we have above disregarded the fact that the final state carries a charge (n_a). However, according to Eq. (5), this means that the properties of $f^{\neq}$ must be described by equations characteristic of ions. In addition, when water is replaced by ethanol, the activity coefficient of the anion in an infinitely dilute solution increases by two to three orders of magnitude [45], which in our case should cause a shift of the half-wave potential by 0.1-0.2 V to the negative side if the water surrounding the transition state is replaced by ethanol (compare [46]). Therefore, the equation obtained above is not sufficient to provide a correct description of the experimental data, and the displacement of the half-wave potential to negative values when ethanol is added to the aqueous solution again does not signify that adsorbed particles are being reduced. In order to understand the role of adsorption in this system from its behavior in aqueous ethanolic solutions it is necessary to investigate the effect of ethanol at large and at small coverages of the electrode surface with depolarizer molecules. As already mentioned above, the displacement of the half-wave potential should be larger in the region of small coverages. If unadsorbed molecules are being reduced the displacement of the potential is not due to the magnitude of the adsorption energy and should not depend on the degree of coverage.

When the effect of temperature on the reduction of organic substances in the region of small coverages is examined the properties of the $f^{\neq}/f_0$ ratio should also be taken into account. It can be supposed [36] that this quantity will be more sensitive to changes of temperature during reduction of adsorbed particles. It is, however, necessary to take yet another effect into account; this is the displacement of the point of maximum adsorption with change in temperature. If, for example, the point of maximum adsorption moves appreciably to the positive potentials on heating, this will lead to a substantial decrease in the adsorption of the depolarizer at a constant potential lying in the region of small coverages to the right of the point of maximum adsorption and measured on an absolute scale; the reaction rate will consequently decrease at this potential. As a result it can be expected that on heating there would be a shift of the half-wave potential to negative values, corresponding to the reduction of adsorbed molecules in the region of small coverages. An analogous effect — decrease in reaction rate arising from shift of the zero-charge point to the positive potentials on heating — can take place in the reduction of anions, since the rate of this reaction is also substantially dependent on rational potential. However, changes in the position of the maximum adsorption point cannot substantially affect the reduction rate of a neutral unadsorbed particle [2]. The shift of the half-wave potential to negative values on heating can therefore be regarded as an adequate proof of reduction of adsorbed particles.

Above some relationships were reported which should describe the reduction of organic compounds at an electrode of the mercury type on the condition that, in the course of the electrode reaction, equilibrium is maintained between the adsorbed layer of the reacting

substance and the adjacent layer of solution. Here the question as
to whether the transition complex is adsorbed was to a considerable
degree found to be separate from the question as to how strong the
adsorption of the depolarizer is. The latter only interested us in so
far as the adsorption of the depolarizer affects the activity coefficient of the transition state. Accordingly this quantity was of no
interest to us in the region of small coverages. The situation may,
however, change at sufficiently negative potentials.

We will consider by way of an example the reduction of a
depolarizer, the surface concentration of which is related to potential in the following way:

$$\Gamma = B_M c f \, \Gamma_\infty \exp\left[-\gamma\,(E - E_M)^2\right] = 10^{-9} \exp\left[-\gamma\,(E - E_M)^2\right].$$

The magnitude of the preexponential term corresponds, for example,
to the following values: $B_M = 10^4$; $cf = 5 \cdot 10^{-4}$; $\Gamma_\infty = 2 \cdot 10^{-10}$ mole
/cm^2. It is easy to verify that if $\gamma = 2$ V^{-2} the surface concentration
at $E - E_M = -2.5$ V amounts to approximately 10^{14} mole/cm^2, which
is probably sufficient for the realization of an electrode reaction.
However, if $\gamma = 5$ V^{-2}, there is one molecule of adsorbate on a
square centimeter of surface; with $\gamma = 10$ V^{-2} concentrations as
small as this are obtained even at $E - E_M = -1.8$ V. It is clear that
reduction of particles characterized by large γ values should differ
by the appearance of limiting adsorption currents at negative potentials, and it should be possible to observe processes where slow
steps are strictly adsorption processes. In this case the adsorption
equilibrium is disturbed, and the current (like the adsorption rate)
depends on the potential.

The assumption can be made that the molecule of the depolarizer
approaches the electrode, replaces a solvent molecule of the surface (this operation requires the time τ_{ads}), and then leaves the
surface, while accepting n electrons; the rate of the electrode reaction is equal to $i\,\mu$A/cm^2, and the surface concentration of depolarizer
is Γ mole/cm^2. In order that such a reaction course should be realized in practice it is necessary that the time τ_{ads} should satisfy
the following inequality:

$$\tau_{ads} \ll \tau_{max} = \frac{nF\Gamma}{i}\sec.$$

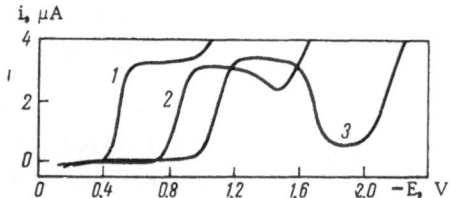

Fig. 9. Polarographic curves of 10^{-3} mole/liter
solution of phenolphthalein in 25% aqueous
ethanolic solution [46]. 1) 3.8 N hydrochloric
acid + 0.001% fuchsin; 2) biphthalate buffer at
pH 4.13; 3) phosphate buffer at pH 7.40.
Curves 2) and 3) are displaced by 0.1 and 0.2
V respectively in relation to curve 1.

If $\tau_{ads} \ll \tau_{max}$, the electrode reaction does not disturb the
adsorption equilibrium. When $\tau_{ads} \simeq \tau_{max}$ the adsorption equilibrium is disturbed, and the equations derived above cannot be used,
but the reaction can as before proceed through the adsorbed state.
If $\tau_{ads} > \tau_{max}$ in a reaction proceeding at rate i, adsorbed particles cannot participate. It is possible to verify that for the depolarizer just considered the τ_{ads} values corresponding to a reduction rate of $i/n = 100$ $\mu A/cm^2$, at a potential of $E - E_M = 2.5$ V,
amount to 10^{-14}, 10^{-10}, 10^{-6}, or 10^{-4} sec respectively with γ values
of 4.5, 3.3, 1.9, or 1.1 V^{-2}. According to Lorenz [47], in the vicinity
of the desorption peak the kinetic limitations at the stage of strictly
adsorption—desorption processes begin to appear at frequencies
of the order of a megahertz, i.e., $\tau_{ads} \simeq 10^{-5}$ sec. Since there is
no reason to expect that the adsorption rate increases on passing
to larger surface charges, it must be admitted that in the numerical
example considered above the reduction of adsorbed particles cannot in principle occur at the indicated rate of 100 $\mu A/cm^2$, normal
for polarography, if $\gamma > 2$ V^{-2}; for this the adsorption would have
to take place at an unlikely high rate. With $\gamma < 2$ V^{-2}, however,
under conditions where the adsorbed molecule occupies a small
volume in the dense part of the double layer the adsorption rate is
evidently sufficient for the reaction to proceed through the adsorbed
state at the most negative potentials.

Thus, at negative potentials, particles characterized by large
γ values can only be reduced at a limited rate, which decreases with

increase in the negative potential. Polarographic curves corresponding to such reactions should show dips resulting from the decreased
adsorption rate. Since such dips have never been observed it must
be agreed that in practice at such negative potentials the depolarizers
exist on the surface in states corresponding to small γ values or
that unadsorbed particles are reduced. In the general case we cannot make a choice between the two possibilities. However, on the
basis of the above-mentioned differences between the behaviors
of adsorbed and unadsorbed transition states, it is possible to
attempt a solution of this problem by comparing experimental data
for each specific case with the results predicted by theory.

4. LOCALIZATION OF TRANSITIONAL STATE

By definition, the adsorption energy depends on: a) work
against electric forces which prevent the entry of the organic molecule into the double layer; b) difference between the solvation energy of the adsorbate in the bulk of the solution and at the surface;
c) work involved in transfer of the organic solvent molecule, displaced during adsorption, from the surface into the volume of the
solution; d) reaction between the adsorbate and the electrode. After
the organic molecule, having displaced a group of solvent molecules,
has penetrated into the inner part of the double layer (the zone of
influence of the electrical forces) its free energy can be subject to
further changes on account of change in the solvate shell, reorientation, polarization and change of conformation through its
physical interaction with the electrode, and finally chemisorption
processes. If the latter effect is disregarded, however, it can be
assumed that the free energies of all these changes at the surface
are small. A decisive change in the properties of the particle takes
place at the moment when it crosses the Helmholtz plane, passing
from the bulk of the solution into the surface phase, and comes into
the electrical field of the inner part of the double layer; from this
moment the properties of the particle depend on the potential.

We will call a molecule which has entered the inner part of
the double layer an adsorbed molecule. Only those molecules which
are situated outside the inner part of the double layer will be
regarded as unadsorbed molecules.

We thus have to determine whether the transition state in the reduction of the organic substance is located within the limits of the dense part of the double layer or whether it can be displaced towards the solution and outside the limits of the dense layer. This problem was first posed by Kolthoff [48] while investigating the polarographic behavior of phenolphthalein in alkaline aqueous ethanolic solutions. On the polarogram the current passed through a maximum and then through a minimum and then began to rise again (Fig. 9).

It was shown [48] that the decrease can be explained by desorption of the depolarizer from the electrode surface at negative potentials. However, in order to explain the appearance of a second rising branch on the polarogram, it was necessary to make further assumptions, and Kolthoff and Lehmicke [48] put forward another explanation for the unusual form of the polarogram. They suggested that reduction was preceded by a surface chemical reaction; the rate of this reaction decreased with increase in the intensity of the field of the double layer, and a decrease appeared on the polarogram. Unadsorbed molecules took part in the charge transfer step (at least on the second rising branch).

This work [48] evidently remained unknown to Grabovskii and Frumkin, who discussed the same problem some years later. Frumkin [49] concluded that reduction of unadsorbed particles (in other words existence of an unadsorbed transition state) was unlikely. Grabovskii [50] defended the opposite point of view. He pointed out that the behavior of organic substances on the electrode was inconsistent with the predictions made by Antropov in the theoretical examination of the reduction of adsorbed organic substances. Antropov was evidently the first to point out that it was necessary to consider variation of the adsorption energy of an organic substance with potential when investigating the electrochemical changes of the substance; he then attempted to describe. the characteristics of such reactions [51, 78]. In particular, he supposed that as the electrode potential departed from the zero-charge point the reduction rate of the adsorbed organic substance necessarily decreased owing to a decrease in its equilibrium surface concentration.

On the basis of Antropov's papers, Grabovskii came to the conclusion that it was necessary to assume that unadsorbed molecules were reduced, since no decreases were found on the polarograms for reduction of organic substances. The analysis performed above of Eqs. (14)-(19) shows, however, that current decreases of this type can be observed in rare cases; if present, they should be regarded as sufficient indication of the reduction of adsorbed particles. (Incidentally, contrary to [51], these decreases should appear − if they appear at all − at potentials remote from the zero charge point). Today Grabovskii's arguments are inadequate because such decreases have been found in practice in a number of instances. However, before passing on to consider these experimental data, it is necessary to indicate other reasons [not allowed for in Eqs. (17) and (19)] for the appearance of decreases on polarograms.

In addition to decreased electron transfer rate on to the adsorbed particle current decreases on the polarogram may also be caused by: a) decrease in rate of antecedent chemical reaction [17]; b) change of reaction mechanism with increase in negative potential. This case has been investigated less than the preceding case but has also been described in the literature [52, 53]. Thus, the polarogram of nitropyrrolecarboxylic acid contains a minimum, near the zero-charge point [52], which is related to electrostatic desorption of the intermediately formed carbanion from the electrode. As a result of this desorption the consumption of electrons per molecule decreases− the reaction stops at the carbanion stage − and a drop is observed on the polarogram.

A unique type of decrease in current accompanied [54] the reduction of methyl bromide in a mixture of sodium and tetramethylammonium cations as supporting electrolyte. The potential of the tetramethylammonium desorption peak was situated exactly between the half-wave potentials of methyl bromide in the presence of sodium salts and those containing tetramethylammonium salts respectively as supporting electrolyte. In the presence of a mixture of these cations the reduction commences in the region of potentials where the double layer is composed of tetramethylammonium cations; the rate of the process then passes through a maximum, owing to displacement of the tetramethylammonium ions from the surface, and finally a new increase in current, coinciding with the wave recorded in solutions containing sodium salt as supporting electrolyte, is observed.

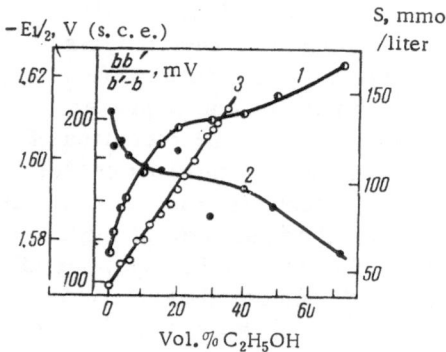

Fig. 10. Effect of ethanol on half-wave potential (1), slope of wave (2), and solubility (3) of methyl iodide [43]. The supporting electrolyte is 0.1 N potassium chloride.

The dip on the wave for benzyl chloride [13] will be discussed next. In this case the limiting diffusion current corresponds to addition of two electrons [55], while the current at the minimum on the polarogram in 60% ethanol is less than half the limiting diffusion current [56]. The assumption about electrostatic desorption of the carbanion [57] is therefore unsuitable in this case. Moreover the troughs on the polarograph curves of anions are observed in another potential range (near the zero-charge point). Tsagareli [36] anticipated a surface reaction which preceded the electron transfer onto the benzyl chloride molecule. It has been supposed earlier that reduction takes place by a chain mechanism [58]:

$$RX \xrightarrow[\text{slow}]{e} R^{\cdot} + X^{-} \xrightarrow{Hg} RHg^{\cdot} \xrightarrow{RX} RHgX + R^{\cdot} \xrightarrow[\text{fast}]{e,Hg} 2RHg^{\cdot} \xrightarrow{2RX}.$$

If this did in fact occur, the introduction of benzylmercury choride into the solution would substantially accelerate the reaction, since free radicals are generated in the one-electron reduction of this compound:

$$RHgX + e \xrightarrow[\text{fast}]{} RHg^{\cdot} + X^{-},$$

and their formation would represent the slow step of the chain mechanism. It was found, however, that additions of benzylmercury chloride did not affect the half-wave potential of benzyl chloride. It was further suggested that the reduction was preceded by reversible dissociation in the field of the electrode:

It was found, however, that addition of chloride ion to the solution did not retard the electrode reaction. Consequently, if a preceding reaction does take place, it does not lead to the formation of a free ion. It was thus not possible to find a preceding chemical step, and it can be supposed that the slow step in the reaction is electron transfer onto the adsorbed particle, as described in Eqs. (14)-(18).

Behavior similar to that of benzyl chloride is shown by benzyl chlorides and chloromethylated naphthalene substituted in the ring [57], certain N-benzoyllactams [16], and also possibly γ-hexachlorane [59], DDT [60], and phenyl vinyl ketone at pH 10.7 [61]. An example of such a reaction investigated in detail is the elimination of a bromine ion from 2-acetyl-5-bromothiophene [17, 34]. In the reduction of this compound in alkaline solutions the half-wave potential does not depend on the pH value and, with increase in the electrolyte concentration in the solution, is shifted to positive values; this and other evidence is characteristic of the reduction of neutral and (in this case) unprotonated particles (e.g., [4, 10, 17]). In alkaline solutions therefore the protolytic reaction no longer precedes the electron transfer, and the current decrease observed under these conditions (see Fig. 39 in the monograph [17]) probably results from a decrease in the electron transfer rate as described by Eq. (14). Application of Eq. (16) to these polarograms gave the value $\gamma \approx 30 \ V^{-2}$.

Thus, in a number of cases, it has been possible on polaro-graphic curves to detect dips predicted by the theory concerning reduction of adsorbed organic substances and located at fairly negative potentials beyond the desorption peaks of the depolarizers. It follows from this that it is possible for the reaction to pass through the adsorbed state under conditions where the adsorption is small. It remains unclear, however, as to whether this is the only reaction course, especially as in many cases the dip is followed by a new rise in current not predicted by the simple theory. A comparative investigation of the two rising branches on the polaro-graphic curve has been carried out in the case of N-(2-methoxy-benzoyl) caprolactam and benzyl chloride. In the first case only the second rising branch is in the region of small coverages. It has been shown that it corresponds to reduction of adsorbed particles [16, 30, 36]. For this purpose the effects of potential on the quantity $(\alpha n_a)_p$ and of salting out on the half-wave potential were considered.

In the case of benzyl chloride there were distinct differences
between the reaction characteristics on these two branches. Thus,
the first wave decreased in height when ethanol was added and could
not be detected in 70% ethanol. The limiting current of the second
wave retained its diffusion character in ethanolic solutions. Change
in the nature of the cation has a small effect on the rate of the process at the first wave and rapidly changes the rate at the second
wave [13].

In [13] the appearance of two rising branches on the polarographic curves was explained by the fact that benzyl chloride can
be oriented in two ways with respect to the electrode surface. The
first orientation corresponds to $\gamma_1^{\neq} = 9.5$ V^{-2} (from polarographic
data) and to greater adsorption energy. The second orientation
corresponds to considerably lower values of $B_{M_2}^{\neq}$ and $\gamma_2^{\neq}$, as a
result of which the proportion of molecules oriented in this way is
very small with small electrode charges and approximates to unity
with large charges. The two branches on the polarographic curve
correspond to reduction of the two different orientations. However,
the first and second branches of the polarographic curves of benzyl
chloride are separated from each other by a distance of the order
of a volt. The change in the half-wave potential resulting from
change in $B_M^{\neq}$ should be described by the simple formula, following
from Eq. (14) $(\gamma_1 \gg \gamma_2)$:

$$(\log B_{M1}^{\neq} - \log B_{M2}^{\neq}) = -(\gamma^{\neq}\varphi_{1/2}^{\bullet})_1 - \frac{\alpha\, n_a F}{RT}\, \Delta\varphi_{1/2}.$$

With $\alpha n_a = 0.25$, $\gamma_1^{\neq} = 8$ V^{-2}, and $(\varphi_{1/2}^{2})_1 = 0.5$ V displacement of the
half-wave potential by 0.7 V requires a change of approximately
four orders of magnitude in the value of B_M on reorientation, whereas B_M in general does not exceed 10^6. It was suggested [10] that
the second branch on the polarographic curve of benzyl chloride
corresponds to reduction of unadsorbed particles.

Other authors have also resorted to the hypothesis about
reduction of unadsorbed particles [48, 50, 62-66]. In the opinion of
Laviron [66], unadsorbed molecules are reduced at a higher rate
than adsorbed molecules. Mairanovskii [43] introduced the idea of
a "quasiunadsorbed" state. On the basis of [67], he supposes that
at strongly negative potentials the composition of the activated
complex includes a molecule of water located within the limits of

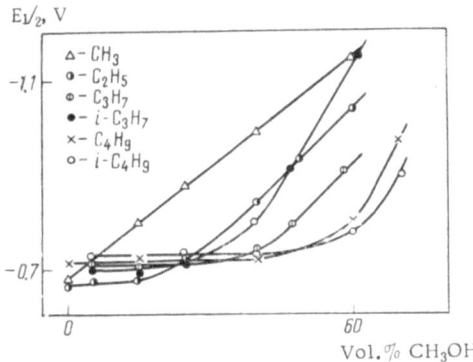

Fig. 11. Dependence of half-wave potentials for
α-hydroxy-β, β, β-trichloroethylphosphinic esters
on methanol concentrations in solution [62].

the inner double layer and that this secures contact between the
electrode and the activated complex lying with its largest portion
in the diffusion part of the double layer. In each of the above-mentioned cases, where the hypothesis concerning reduction of unadsorbed particles has been put forward, the observed phenomena can
be explained without resorting to this concept (e.g., [66] and [68]).
In the case of benzyl chloride account must be taken of the fact that
two orientations on the surface differ in their disposition with
respect to the supporting electrolyte cation; this is demonstrated
by the results of an investigation into the effect of the nature of the
cation on this reaction. It can be assumed that at more negative
potentials (second branch) the reaction becomes possible after accumulation of a sufficient quantity of adsorbed cations in the double
layer.

The active role of the supporting electrolyte cation in the
reduction of organohalogen compounds was mentioned by Levin as
early as 1952 [69]. Examples of how a small change in the
characteristics of the molecule brings about a sharp change in the
course of the reaction right up to a change in the final product can
be found with Feoktistov and Gol'din [70].

Yet another reaction of this type was discovered by Volke
[53]. He obtained a polarogram with a maximum in the reduction

of the nitrile of isonicotinic acid. By electrolysis at controlled potential it was shown that the reaction mechanism changed along the wave; aminomethylpyridine was formed on the first rising branch, and pyridine was formed on the descending part. Volke suggested that the change in the course of the reaction was due to reorientation of the depolarizer molecule at the surface.

The question of the possibility of reduction of unadsorbed molecules in the region of the second branch has thus not been excluded even in those numerous cases where a decrease in current connected with a decrease in the adsorption energy is observed on the polarogram. Other experimental methods by means of which it is possible to distinguish between reduction of adsorbed and unadsorbed particles are therefore necessary. One method is to investigate the dependence of $(\alpha n_a)_P$ on potential; it does not change in the reduction of unadsorbed particles but decreases according to Eq. (20) with increase in negative potential during reduction of adsorbed particles.

In using Eq. (20) it is necessary to take account of the fact that not only the polarographic but also the true electron transfer coefficient can decrease with increase in negative potential [3, 5]. At sufficiently positive potentials $\alpha = 1$ for any reaction. This is a barrier-free reduction, where the transition state coincides with the final state. Barrier-free reactions are convenient for kinetic investigations, since the structure of the transition state is known. With increase in overpotential a barrier to the course of the reaction is formed, and the value of α decreases [3, 5]. This effect was evidently observed by Levinson [71] in the reduction of dipyridylethylene on mercury in alkaline solutions (Fig. 6). The Tafel relationship shows a clearly defined inflection which corresponds to decrease of αn_a from 1 to 0.3 with increase in the negative potential. The reaction takes place in the region of large coverages, and Eq. (20) is not applicable. Moreover, this equation predicts a slow smooth change in the value of $(\alpha n_a)_P$, which is not consistent with experiment. Since $n_a = 1$ in the reduction of dipyridylethylene [71], the observed change in αn_a evidently signifies a decrease in α from 1 to 0.3, i.e., a transition from barrier-free reduction to normal reduction. The value of αn_a is close to unity in the reduction of nitromethane [72], iodobenzene in 80% ethanol [21, 33], certain benzoyllactams [15, 27, 36] (in this case $n_a = 1$ and consequently $\alpha = 1$; see [36]), dibutyl fumarate, and 2-acetyl-5-bromothiophene [27] on mercury in aqueous solutions. In considering the properties of the fumarate the authors of [27] rested particularly on the data obtained by Kargin [73], who showed that one electron per molecule was required in the first stage in the reduction of this compound (and thus $\alpha \simeq 1$, since $\alpha n_a \simeq n_a = 1$).

In continuing the discussion of Eq. (20), it should be pointed out that it is usual, instead of investigating the dependence of αn_a on potential, to investigate the dependence of this quantity on the half-wave potential, which is varied by changing the composition of the solution (e.g., Figs. 10 and 3). In the analysis it is then assumed that change in composition of the solution only affects the half-wave potential, and the observed change in αn_a is regarded as a secondary effect resulting from the dependence of $(\alpha n_a)_P$ on potential. In other words it is assumed that the true electron transfer coefficient does not depend on the composition of the solution. Such an approach is not entirely reliable. We will return, for example, to the data of Kargin [62] on the reduction of α-hydroxy-β,β,β-trichloroethylphosphinic esters in aqueous ethanolic solution. Figures 11 and 12 show the dependences of half-wave potential and polarographic electron transfer coefficient on the ethanol concentration. For the majority of compounds the potential in solutions which are dilute with respect to ethanol does not depend on the ethanol concentration, and the $(\alpha n_a)_P$ value rapidly decreases. On the other hand, the half-wave potential of the dimethyl ester moves relatively strongly to the negative value with increase in ethanol concentration, whereas the $(\alpha n_a)_P$ values changes little. Thus, in this case, it is clear that the change in slope of the wave with increase in ethanol concentration cannot be attributed to change in the half-wave potential. As supposed by Kargin, the effect which he observed is related to a decrease in the true electron transfer coefficient resulting from reorientation of the activated complex at the surface or from a change to reduction of unadsorbed particles. In practice both decrease (e.g., Fig. 3) and increases (Fig. 10) in the $(\alpha n_a)_P$ value are observed when the half-wave potential is shifted to the negative side. Instances have also been described where $(\alpha n_a)_P$ does not depend on the half-wave potential [43]. Use of Eq. (20) hardly helps us to progress in the desired direction.

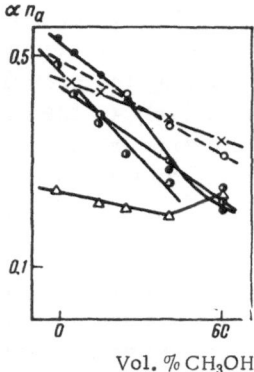

Fig. 12. Dependence of polarographic electron transfer coefficient
of α-hydroxy-β,β,β-trichloro-
ethylphosphinic esters on methanol
concentration in solution. The
legend is the same as in Fig. 11.

In principle it should be extremely fruitful to investigate the effect of the nature and concentration of the supporting electrolyte. The effect of the ψ_a potential has been investigated by very many authors, and the most interesting from our point of view is one of the first papers [74], the authors of which came to the conclusion that the transition state in the reduction of nitromethane is located at the inner Helmholtz plane (see [10, 17]). However, it is not possible to analyze the data which have been accumulated until a theory which takes account of the ψ_a potential has been formulated for the adsorption of organic molecules.

We will now consider experimental data on the effect of salting on the half-wave potential. The theory of this effect was examined above. It was first applied to the reduction of iodobenzene in 2% aqueous ethanol [21, 33] (Fig. 7). The continuous straight lines drawn on the graph represent the best lines through the experimental points at moderate concentrations of electrolyte in the solution; the slopes of these lines are given in Table 1. The deviation of the experimental points from these straight lines, which considerably exceeds the experimental error, was regarded as resulting from salting out and from the effect of the electrolyte concentration on the viscosity of the solution. The variation of viscosity and k_{el} values for iodobenzene were determined by independent methods, after which the dotted curves in Fig. 7 were calculated from Eq. (26). (For lithium and potassium acetates $k_{el} = 0.11$ liter/mole.) The appearance of a charge in the final state was not taken into account. In [21, 33] it was concluded that iodobenzene is reduced in the adsorbed state, since salting out affects the half-wave potential. The displacement of $E_{1/2}$ for iodobenzene resulting from salting out amounted to 25 mV in 2 N lithium, sodium, and potassium acetate solutions, and somewhat less in 2 N cesium acetate solution. The much greater effect, exceeding the potential shift, predicted by Eq. (10) was obtained with N-(2-methoxybenzoyl) caprolactam in sodium phosphate solution (McIlvain buffer pH 7.8) [30]. The constant for salting out by disubstituted phosphate is equal to 1.89, and the potential shifts amount to 50 mV even in 0.2 N solution. It can thus be concluded that salting out affects the half-wave potential and that benzoyl caprolactam is reduced in the adsorbed condition. Comparison of data on the effect of a number of other anions (chloride, bromide, iodide, sulfate, and perchlorate) on the reaction rate and adsorption of this compound leads to the assumption that salting out is not the only effect related to the nature of the anion. The half-wave potentials obtained with these anions as supporting electrolyte do not lie very close to the general line after the effect of salting out has been taken into account, and the desorption peaks in solutions of various sodium salts at equal concentrations and saturated with respect to the adsorbate do not coincide in respect of height and potential. Joint adsorption of the supporting electrolyte anion and the investigated caprolactam evidently occurs. In the region of large coverages the effect of the nature of the anion on adsorption has been observed before (see Table 2 and [75]). It is, however, surprising that in the case of N-(2-methoxybenzoyl) caprolactam this effect was retained at small coverages [36].

Investigation of the effect of ethanol on the half-wave potential is one of the best known and less complicated methods for detecting adsorption of a depolarizer. The effects of ethanol in regions of large and small coverages have been investigated simultaneously in rare instances (e.g., [16], where $E_{1/2}$ is greater in the region of small coverages). A displacement of the potential towards the negative side is nearly always observed when ethanol is added. Exceptions are dipyridylethylene [68] (where the potential shift to the positive side is due to the disappearance of the effect of large coverages), organic anions [22] (the activity of which in the solution increases with addition of ethanol), and certain chloro- and bromo-substituted allyl bromides [76]. In the last case no explanation has been found for the effect.

The effect of ethanol on the process rate has been examined above. When ethanol is added a change in mechanism, leading to very large potential shifts, can also occur. On switching from

ethanolic solutions of diiodobenzene to aqueous solutions [21], the
four-electron reduction wave of this compound splits into two,
where the second coincides with the wave of iodobenzene and the
first is situated at a potential approximately 1 V more positive.
Evidently in aqueous solutions diiodobenzene reacts with the electrode material, and the first wave represents the reduction of the
organomercury compound – the product of this reaction. The prewave on the polarograms of bromoalkanoic acids, corresponding to
reaction between these depolarizers and the electrode material,
also disappears when ethanol is added to the solution [22]. It is
possible that the effects discovered by Cisak [77] in the investigation of hexahalogenocyclohexanes are similar in nature.

5. CONCLUSIONS

Here we conclude the discussion of the problem concerned
with localization of the transition state. The experimental data
which have been considered and the equations which have been derived
are of importance in the region of small coverages, since the question of the location of the transition state does not arise in the region of large coverages. For the region of small coverages we have
at our disposal several experimental criteria (form of the polarization curve, relationship between the reaction rate and activity) which
can be used to solve this problem for each case. In all the reactions
which have been examined from this standpoint it can be concluded
that the transition state is adsorbed on the electrode. Especially
interesting in this sense are the data on N-(2-methoxybenzoyl) caprolactam, where it has been shown that reduction of the adsorbed particles takes place on the second rising branch of the polarogram.
In conclusion it is expedient to mention the point which was substantiated above but was not specially singled out, namely, that the
effect of adsorption of a depolarizer on the electrode reaction rate
is particularly large in the case where the adsorption energy and
the degree of coverage are small.

LITERATURE CITED

1.   N. M. Émanuél' and D. Knorre, Course in Chemical Kinetics [in Russian],
      Vysshaya Shkola, Moscow (1962).
2.   A. N. Frumkin, Élektrokhimiya, 1:394 (1965).
3.   L. I. Krishtalik, Usp. Khim., 34:1831 (1965).

4. A. N. Frumkin, V. S. Bagotskii, Z. A. Iofa, and B. N. Kabanov, Kinetics of Electrode Processes [in Russian], Moscow University Press, Moscow (1952).
5. L. I. Krishtalik, Thesis [in Russian], Inst. Élektrokhimii Akad. Nauk SSSR, Moscow (1967).
6. A. N. Frumkin, Advances in Electrochem. and Electrochem. Eng., 1:65 (1961).
7. D. M. Mohilner and P. Delahay, J. Phys. Chem., 67:588 (1963).
8. G. A. Tedoradze and R. A. Arakelyan, Élektrokhimiya, 4:62 (1968).
9. P. Delahay, Double Layer and Kinetics of Electrode Processes [Russian transla-tion], Mir, Moscow (1967).
10. P. Delahay, New Problems and Methods in Electrochemistry [Russian translation], IL, Moscow (1957).
11. H. Hurwitz, A. Sanfeld, and A. Steinchen-Sanfeld, in: Fundamental Problems in Modern Theoretical Electrochemistry [Russian translation], Mir, Moscow (1965), p. 198.
12. A. N. Frumkin and B. B. Damaskin, in: Modern Aspects of Electrochemistry (edited by J. Bockris and B. Conway) [Russian translation], Mir, Moscow (1967), p. 170.
13. G. A. Tedoradze, A. B. Érshler, and S. G. Mairanovskii, Izv. Akad. Nauk SSSR, Otd. Khim. Nauk, 235 (1963).
14. A. B. Érshler, G. A. Tedoradze, and S. G. Mairanovskii, Dokl. Akad. Nauk SSSR, 145:1324 (1962).
15. G. A. Tsagareli, G. A. Tedoradze, and A. B. Érshler, Élektrokhimiya, 4:418 (1968).
16. G. A. Tsagareli, A. M. Shkrob, and A. B. Érshler, Izv. Akad. Nauk SSSR, Ser. Khim., 1952 (1967).
17. S. G. Mairanovskii, Catalytic and Kinetic Waves in Polarography, Plenum Press, New York (1968).
18. V. G. Levich, B. I. Khaikin, and E. D. Belokolos, Élektrokhimiya, 1:1273 (1965).
19. V. Yu. Filinovskii, Report presented at Sixth Conference on Electrochemistry of Organic Compounds [in Russian], Inst. Élektrokhimii Akad. Nauk SSSR, Moscow (1968).
20. A. B. Érshler, G. A. Tedoradze, M. Fakhmi, and K. P. Butin, Élektrokhimiya, 2:319 (1966).
21. A. B. Érshler, Thesis [in Russian], Inst. Élektrokhimii Akad. Nauk SSSR, Moscow (1966).
22. A. D. Krasnova, Thesis [in Russian], Inst. Élektrokhimii Akad. Nauk SSSR, Moscow (1967).
23. S. G. Mairanovskii and A. P. Stradyn', Izv. Akad. Nauk SSSR, Otd. Khim. Nauk, 2239 (1961).
24. L. G. Feoktistov and S. I. Zhdanov, Electrochim. Acta, 10:657 (1965).
25. I. Rosenthal, C. A. Albright, and P. J. Elving, J. Electrochem. Soc., 99:227 (1952).
26. B. Kastening and L. Holleck, Talanta, 12:1259 (1965).

238 A. B. ÉRSHLER

27. I. M. Levinson, G. A. Tsagareli, G. A. Tedoradze, and A. B. Érshler, in: Advances in Electrochemistry of Organic Compounds [in Russian], Nauka, Moscow (1967), p. 65.
28. R. A. Arakelyan, G. A. Tedoradze, E. D. Belokolos, and Roland A. Arakelyan, in: Advances in Electrochemistry of Organic Compounds [in Russian], Nauka, Moscow (1967), p. 67.
29. N. B. Grigor'ev and B. B. Damaskin, in: Advances in Electrochemistry of Organic Compounds [in Russian], Nauka, Moscow (1967), p. 66.
30. A. B. Érshler, G. A. Tsagareli, and G. A. Tedoradze, in: Advances in Electrochemistry of Organic Compounds [in Russian], Nauka, Moscow (1967), p. 63.
31. B. K. Survilla, Thesis [in Russian], Moscow State University, Moscow (1968).
32. É. S. Levin and A. V. Yamshchikov, in this collection, p. 411.
33. A. B. Érshler, G. A. Tedoradze, E. A. Preic, and K. S. Korshunova, Reports at Fifth Conference on Electrochemistry of Organic Compounds [in Russian], Nauka, Moscow (1965), p. 20.
34. S. G. Mairanovskii, N. V. Barashkova, and Yu. B. Volkenshtein, Izv. Akad. Nauk SSSR, Ser. Khim., 1539 (1965).
35. S. G. Mairanovskii, J. Electroanalyt. Chem., 3:166 (1962).
36. G. A. Tsagareli, Thesis [in Russian], Inst. Élektrokhimii Akad. Nauk SSSR, Moscow (1968).
37. S. G. Mairanovskii, V. A. Ponomarenko, N. V. Barashkova, and A. D. Snegova, Dokl. Akad. Nauk SSSR, 134:387 (1960).
38. V. D. Bezuglyi, L. A. Mel'nik, and V. D. Dmitrieva, Zh. Obshch. Khim., 34:1048 (1964).
39. F. A. Long and W. F. McDevit, Chem. Revs., 51:119 (1952).
40. A. N. Frumkin, Electrocapillary Phenomena and Electrode Potentials [in Russian], Odessa (1919).
41. A. N. Frumkin, O. A. Petry, and N. V. Nikolaeva-Fedorovich, Electrochim. Acta, 8:177 (1963).
42. A. B. Érshler, E. D. Belokolos, and G. A. Tedoradze, Élektrokhimiya, 2:1429 (1965).
43. S. G. Mairanovskii and A. D. Filonova, Élektrokhimiya, 3:1397 (1967).
44. S. I. Zhdanov and A. M. Khopin, in: Advances in Electrochemistry of Organic Compounds [in Russian], Nauka, Moscow (1967), p. 75.
45. N. A. Izmailov, Electrochemistry of Solutions [in Russian], Khar'kov State University Press, Khar'kov (1959).
46. A. B. Lothe and L. B. Rogers, J. Electrochem. Soc., 101:1235 (1954).
47. W. Lorenz, Z. Phys. Chem. (N.F.), 65:433 (1959).
48. I. M. Kolthoff and D. J. Lehmicke, J. Am. Chem. Soc., 70:1879 (1948).
49. A. N. Frumkin, Proceedings of Fourth Conference on Electrochemistry [in Russian], Izd. Akad. Nauk SSSR, Moscow (1956), p. 237.
50. Z. Ch. Grabovskii, Proceedings of Fourth Conference on Electrochemistry, Izd. Akad. Nauk SSSR, Moscow (1959), p. 236.
51. L. I. Antropov, Trudy Erevansk, Politekhn. Inst., 1:97 (1946); Zh. Fiz. Khim., 24:1428 (1950).

52. M. Person, Bull. Soc. Chim. France, 1832(1966).
53. A. D. Kardős, P. Valenta, and J. Volke, J. Electroanalyt. Chem., 12:84 (1966).
54. I. Thamm. Analyt. Chem., 34:128 (1966).
55. R. Pasternak and H. V. Halban, Helv. Chim. Acta, 29:913 (1954).
56. A. N. Frumkin, in: Fundamental Problems in Modern Theoretical Electrochemistry [in Russian], Mir, Moscow (1965), p. 309.
57. L. W. Marple, L. E. I. Hummelstedt, and L. B. Rogers, J. Electrochem. Soc., 107:437 (1960).
58. N. S. Hush and K. B. Oldham, J. Electroanalyt. Chem., 6:34 (1963).
59. G. Draght, Analyt. Chem., 30:737 (1948).
60. M. Březina and P. Zuman, Die Polarographie in der Medizin. Pharmazie und Biochemie, Leipzig (1956), p. 197.
61. P. Zuman, Nature, 192:655 (1961).
62. Yu. M. Kargin, in: Advances in Electrochemistry of Organic Compounds [in Russian], Nauka, Moscow (1967), p. 40.
63. P. J. Elving, Rec. Chem. Progr., 14:99 (1953).
64. P. J. Elving and B. Pullman, Advances in Chem. Phys., 3:1 (1961).
65. A. M. Mirri and P. Favero, Ric. Sci., 28:2307 (1958).
66. E. Laviron, Bull. Soc. Chim. France, 418 (1962).
67. P. P. Schmidt Jr., and H. B. Mark, J. Chem. Phys., 43:329 (1965).
68. G. A. Tedoradze and A. B. Érshler, Usp. Khim., 34:1866 (1965).
69. É. S. Levin and Z. I. Fodiman, Zh. Fiz. Khim., 28:601 (1954).
70. L. G. Feoktistov, in this collection, p. 135.
71. I. M. Levinson, G. A. Tedoradze, and A. B. Érshler, Élektrokhimiya, 4:461 (1968).
72. M. Suzuki and P. Elving, Coll. Czechosl. Chem. Communs., 25:3202 (1960).
73. Yu. M. Kargin, Ya. A. Levin, V. Z. Kondranina, and R. T. Safin, in: Advances in Electrochemistry of Organic Compounds [in Russian], Nauka, Moscow (1967), p. 38.
74. M. Breiter, M. Kleinerman, and P. Delahay, J. Am. Chem. Soc., 80:5111 (1958).
75. A. L. Asatiani and G. A. Tedoradze, in: Advances in Electrochemistry of Organic Compounds [in Russian], Nauka, Moscow (1967), p. 27.
76. M. Kleine-Peter, Bull. Soc. Chim. France, 899 (1957).
77. A. Cisak, Roczn. Chem., 36:1895 (1962).
78. L. I. Antropov and N. T. Vagramyan, Zh. Fiz. Khim., 25:409 (1951).

Electrochemical Synthesis of Organometallic Compounds

A. P. Tomilov and I. N. Brago

The chemistry of organometallic compounds began to develop vigorously at the end of the nineteenth century and is now an important route in organic synthesis. Many such compounds are employed in industry and in agriculture. Thus, the use of the Grignard methods by Kipping [1] for synthesis of organosilicon compounds has eventually led to the creation of a new branch of the chemical industry, manufacturing the silicone polymers or silanes. The production of organosilicon products now totals more than 27,000 tons per annum [2]. The researches of Miugli [1], which showed that organic lead compounds provide an effective means for combatting the knock of fuels in motors, set the beginning to the industrial production of tetraethyllead, which has reached 227,000 tons per annum [3]. The volume of organotin compounds production has reached approximately 1360 tons per annum [4]. They are used as stabilizers for polyvinyl chloride, as antioxidants for rubbers, as polymerization catalysts for olefines, and as fungicides. Alkylaluminums are in demand to the extent of 2720 tons per annum [5]. Organic compounds of mercury, zinc, and magnesium, which are finding various applications, are produced in small amounts mainly on account of their high cost.

Instances are known where organometallic compounds are formed as a result of electrochemical processes. This interesting method for producing organometallic compounds has only comparatively recently begun to attract the attention of investigators. Nevertheless definite successes have been obtained in this field. Already a considerable amount of experimental material demonstrating the great possibilities of the electrochemical method has been accumulated. However, ideas on the mechanism of the process remain obscure in many details.

TABLE 1. Organometallic Compounds Synthesized by Electrolysis of Carbonyl Compounds

Cathode	Electrolyte (solvent)	Starting compound	Current density, A/cm²	Temperature, °C	Heteroorganic compound obtained	Yield, %	Reference
Mercury	40% H_2SO_4	Acetone	0.6	—	Diisopropylmercury	—	[7]
	H_2SO_4	Methyl ethyl ketone	—	—	Di-sec-butylmercury	—	[6]
	H_2SO_4	Menthone	—	80	Dimenthylmercury	—	[11,18]
	$HCl + CH_3COOH$	Phenylacetone	0.135	18	Di(2-phenylisopropyl)-mercury	25—30	[10]
	5% H_2SO_4	Cyclohexanone	0.3281	15—66	Dicyclohexylmercury	25—30	[10]
	3% H_2SO_4	2-Methylcyclohexanone	0.197	55	Di(2-methylcyclohexyl)-mercury	25—30	[10]
	" "	3-Methylcyclohexanone	0.197	55	Di(3-methylcyclohexyl)-mercury	25—30	[10]
	" "	4-Methylcyclohexanone	0.197	55	Di-(4-methylcyclohexyl)-mercury	25—30	[10]
	$HCl + CH_3COOH$	Cyclopentanone	0.197	20	Dicyclopentylmercury	25—30	[10]
	50% H_2SO_4	Benzaldehyde	0.197	25	Dibenzylmercury	8.5	[16]
	" "	Benzylmethyl ketone	0.197		Di(2-phenylisopropyl)-mercury	—	[10]
Lead	20% H_2SO_4	Acetone	0.05	20	Diisopropyllead, tetra-isopropyllead	—	[8]
	" "	Methyl ethyl ketone	0.05	20	Tetrabutyllead	—	[9]
	20% H_2SO_4	Diethyl ketone	0.05	20	Tetraamyllead	—	[9]
	" "	Methylisoamyl ketone	—	—	Not identified	—	[18]
	" "	Citral	—	—	" "	—	[15]
Cadmium	20% H_2SO_4	Acetone	—	—	" "	—	[12]
Germanium	H_2SO_4	" "	—	—	" "	—	[13]

TABLE 2. Organometallic Compounds Synthesized during Electrolysis of Unsaturated Compounds

Cathode material	Electrolyte (solvent)	Starting compound	Current density, A/cm^2	Temperature, °C	Organometallic compound obtained	Yield, %	Reference
Graphite	1.0 N Na_2SO_4	Acrylonitrile, selenium	0.016	45	Di(β-cyanoethyl)selenide	23	[20]
	1.0 N Na_2SO_4	Acrylonitrile, tellurium	0.016	50	Di(β-cyanoethyl)telluride	3.4	[20]
	1.0 N Na_2SO_4 (ethanol)	Acrylonitrile, sulfur	0.016	50	Di(β-cyanoethyl)sulfide	15.1	[19]
Tin	0.7 N NaOH	Acrylonitrile	0.02	15	Tetra(β-cyanoethyl)tin	60.0	[21]
	0.7 N NaOH	Methacrylonitrile	0.02	15	Tetra(β-cyanopropyl)tin	Low	[22]
	0.5 N K_2HPO_4	Acrylonitrile	0.03	20	Hexa(β-cyanoethyl)ditin	24	[23]
	0.7 N NaOH	1-Cyano-1,3-butadiene	0.005	30	Not identified	Low	22
Mercury	Not indicated	Methyl vinyl ketone	—	—	Di(2-butanone)mercury	—	[24]
	1.0% NaOH	Dimethylvinylcarbinol	0.0125	25	Not identified	—	[25]

TABLE 3. Organometallic Compounds Synthesized during Electrolysis of Halogen–Containing Compounds

Cathode material	Electrolyte (solvent)	Starting compound	Current density, A/cm²	Temperature, °C	Organometallic compound obtained	Yield, %	Reference
Lead	10% NaOH in ethanol + 2% lead acetate	Ethyl iodide	0.0115	55	Tetraethyllead	—	[26]
	10% NaOH in water	Ethyl iodide	0.0115	50	" "	—	[27]
	0.5 N H_2SO_4	Iodopropionitrile	0.035	15	Tetra(β-cyanoethyl)lead	13	[32]
Lead	$(CH_3)_4NBr$ (Acetonitrile)	Ethyl bromide	—	60	Tetraethyllead	70	[28]
Tin	0.5 N H_2SO_4	β-Chloropropionitrile	0.035	15	Hexa(β-cyanoethyl)ditin	12	[32]
Thallium	0.5 N H_2SO_4	Iodopropionitrile	0.035	10	Di(β-cyanoethyl)thallium iodide	—	[32]
Graphite	(Dimethylformamide)*	Butyl iodide, phosphorus	0.005	80	Butylphosphine Dibutylphosphine Tetrabutylcyclotetra-phosphine and tetrabutyl-diphosphine	2 6.3 9.1	[33] [33] [33]
Mercury	0.1 N $(C_2H_5)_4NI$ Ethanol:water = 1:1	Benzyl iodide	—	25	Dibenzylmercury†	—	[24, 30]
	0.5 N H_2SO_4	Iodopropionitrile	0.035	10	Di(β-cyanoethyl)mercury and cyanoethylmercury iodide	45	[32]

* Butyliodide forms an electrically conducting solution with dimethylformamide.
† The formation of dibenzylmercury was assumed on the basis of polarographic data.

The purpose of the present review is to attempt to indicate the applicability range of electrochemical methods and to examine possible mechanisms for the processes which lead to the formation of the carbon – metal bond. The carbon – metal bond can form both at the anode and at the cathode. Of course the mechanisms of the electrode processes are entirely different in these cases, and will therefore be examined separately.

1. CATHODIC PROCESSES

Formation of organometallic compounds at the cathode is observed during electrolysis of three classes of organic compounds: 1) carbonyl compounds; 2) unsaturated compounds; 3) alkyl halides.

Electrolysis of Carbonyl Compounds

Organometallic compounds are sometimes formed during electrolysis of acidic aqueous solutions of ketones. The overall form of this reaction can be represented by the following equation:

$$2\ \begin{matrix} R \\ \diagdown \\ \diagup \\ R \end{matrix} CO + M \xrightarrow{+6H^+,\ 6e} \begin{matrix} R \\ \diagdown \\ HCMCH \\ \diagup \quad \diagdown \\ R \quad\ \ R \end{matrix} + 2H_2O. \tag{1}$$

A reaction of this type was first performed by Tafel in 1906 [6] and was investigated in detail in later work [7–11]. On the basis of the available data it can be asserted that mostly aliphatic and alicyclic ketones are susceptible to organometallic compound formation.

Of the metals, mercury and lead enter into this type of reaction most readily. There are data on the formation of organometallic compounds during electrolysis of ketones at cadmium [12] and germanium [13] electrodes. The vigorous disintegration of a zinc cathode observed during electrochemical reduction of acetone is also attributed to the formation of an unstable organozinc compound [14].

In most cases the organometallic compounds are formed during electrolysis of sulfuric acid solutions of the ketones. It has been shown that organometallic compound formation is promoted by increase in temperature [6, 7, 10] and acid concentration [7, 10].

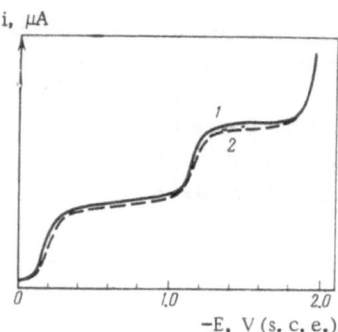

Fig. 1. Polarogram of benzyl iodide (1) and benzylmercury iodide (2).

There is much less information on the formation of organometallic compounds in electrochemical reduction of aldehydes. Low [15] observed the formation of an organolead compound in the electrochemical reduction of citral but did not study its structure. Organolead compounds which have been obtained by electrolysis of certain other aldehydes [16, 17] have also not been investigated.

The only identified organometallic compound formed in the electrochemical reduction of aldehydes is dibenzylmercury, obtained by reduction of benzaldehyde [16]:

$$2C_6H_5C\overset{O}{\underset{H}{\diagdown}} + Hg \xrightarrow{6H^+, 6e} C_6H_5CH_2HgCH_2C_6H_5 + 2H_2O. \qquad (2)$$

Attempts to carry out a similar reaction with vanillin, piperonal, p-methoxybenzaldehyde, furfural, and salicylaldehyde were unsuccessful.

The known cases where organometallic compounds are formed during electrolysis of carbonyl compounds are listed in Table 1.

Reduction of Unsaturated Compounds

During electrolysis of solutions of unsaturated compounds the carbon — metal bond forms at the point where the double bond breaks, and one carbon atom is protonated:

$$2RCH{=}CH_2 + M \xrightarrow{+2H^+, 2e} RCH_2CH_2MCH_2CH_2R. \qquad (3)$$

This type of reaction has been investigated in most detail for the electrochemical reduction of unsaturated nitriles [19–23].

The reaction is characteristic of a tin cathode. A similar process takes place if, for example, a graphite cathode is used and

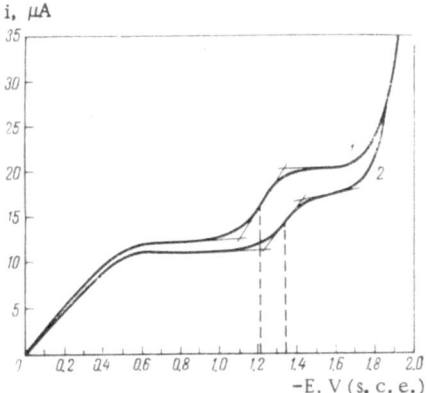

Fig. 2. Polarogram of saturated aqueous solution of β-cyanoethylmercury iodide. 1) 0.1 N lithium chloride; 2) 0.1 N lithium chloride + 0.001 mole/liter di(β-cyano-ethyl)mercury.

finely divided tellurium, selenium, or sulfur is added to the solution. Low current density and elevated temperature favors formation of the organometallic compounds [21, 22].

Formation of organometallic compounds has also been observed during electrolysis of methyl vinyl ketone with a mercury cathode [24]:

$$CH_2{=}CHCOCH_3$$

$$\rightarrow Hg\,(CH_2CH_2COCH_3)_2.$$

Dimethylvinylcarbinol evidently reacts by the same mechanism [25]. The nature of the compound obtained in the latter case was not investigated.

Information on the formation of organometallic compounds from unsaturated compounds is given in Table 2.

Electrolysis of Halogen-Containing Compounds

Symmetrical organometallic compounds can also be prepared by electrolysis of halogen-containing compounds:

$$2RX + M \xrightarrow{2e} RMR + 2X^-. \qquad (4)$$

This reaction was first used for preparation of tetraethyllead by electrolysis of a solution or suspension of ethyl iodide in ethanolic alkali [26, 27]. It was later found that the yield of tetraethyllead could be increased to 70% in an aprotic solvent, e.g., acetonitrile [28].

Formation of organomercury compounds has been observed in the electrochemical reduction of benzyl iodide [29, 30] and allyl bromide [31].

TABLE 4. Polarographic Behavior of Iodopropionitrile and β-Cyanoethylmercury Iodide*

Compound	Number of wave	Supporting electrolyte			
		0.1 N LiCl in water		0.1 N (CH₃)₄NI in 50% ethanol	
		$-E_{1/2}$, V†	i_d/c, mA /mole	$-E_{1/2}$, V†	i_d/c, mA /mole
ICH₂CH₂CN	I	1.03	1.78	0.895	1.63
	II	1.42	2.11	1.14	1.63
IHgCH₂CH₂CN	I	0.33	—	0.30	—
	II	1.13	—	0.885	—
ICH₂CH₂CN	I	0.390	—	0.320	—
+ IHgCH₂CH₂CN	II	0.975	1.83	0.895	1.61
mixture	III	1.335	2.30	1.120	1.48

* The measurements were made on an LP-60 polarograph; the capillary characteristics were $m^{2/3}\tau^{1/6} = 1.263$ mg$^{2/3}$sec$^{-1/2}$.

† The potentials were measured with reference to a saturated calomel cell.

Iodopropionitrile has proved very reactive. It was found [32] that during electrochemical reduction it readily forms cyanoethyl derivatives of mercury, lead, tin, and thallium. It is interesting to note that the principal product from the reduction of iodopropionitrile at a tin cathode was tetra(β-cyanoethyl)tin, whereas the product from electrochemical reduction of chloropropionitrile under the same conditions was mainly hexa(β-cyanoethyl)ditin.

Primary, secondary, and tertiary phosphines have been prepared by electrolysis of a solution of alkyl halide in dimethylformamide [33].

Finally the possibility of formation of organomercury compounds during electrolysis of certain thiocyanates [34] must be mentioned:

$$2RSCN + Hg \xrightarrow{2e} RSHgSR + 2CN^-. \tag{5}$$

The organometallic compounds which have been isolated and identified in the electrolysis of halogen-containing compounds are listed in Table 3.

Mechanism of Cathodic Process Leading to Formation of Carbon — Metal Bond

The mechanism of the formation of organometallic compounds in cathodic processes is at present under discussion. On its own the fact of "cathodic dissolution of metal" is paradoxical. Tafel, who first observed the dissolution of mercury during electrolysis of acetone solutions, called this phenomenon the "peculiar formation" of organomercury compounds [6].

Several hypotheses have been put forward to explain the formation of organometallic compounds in cathodic processes:

1) According to Reutov [35], Waters [36], and Walling [37], the formation of organometallic compounds results from the action of free radicals, formed initially at the cathode, on the cathode material;

2) According to Voitkevich [38], the initial product of electrolysis is an organic anion, which extracts a metal cation from the crystal lattice of the cathode;

3) Intermediate formation of a chemisorbed complex has been proposed [39]; upon electron transfer the coordination bond of the complex changes into a covalent bond;

4) It can be supposed that the first event is the formation of a metal hydride, which then reacts with the organic compound [40].

A critical examination of all the proposed mechanisms shows that the carbon – metal bond formation mechanism depends to a considerable extent on the nature of the initial reactants. Above all it should be noted that in all probability the radical mechanism does not hold in the formation of organometallic compounds during electrochemical reduction of unsaturated and halogen-containing compounds.

Cyanoethyl derivatives of tin, sulfur, selenium, and tellurium have been observed by electrolysis of acrylonitrile. Lead, mercury, and thallium do not enter into this reaction, but their cyanoethyl derivatives are obtained in satisfactory yield by electrochemical reduction of iodopropionitrile. If it is supposed that formation

TABLE 5. Effect of Acrylonitrile on Electrochemical Reduction of Tin, Sulfur, Selenium, and Tellurium

Substance	1.0 N sodium sulfate solution	1.0 N sodium sulfate solution + acrylonitrile	Electrolysis conditions		
	Current yield, %		temperature, °C	current density, A/cm^2	pH
	hydride	organo-metallic compound			
Tin	Traces	46.2	15	0.006	6-7
Sulfur	56.0	15.1	40	0.016	6-7
Selenium	83.4	23.0	46-50	0.016	7.5-8
Tellurium	~4.0	13.4	46-50	0.016	6-7

of organometallic compounds in all the above-mentioned cases takes place by the radical mechanism, it is then difficult to understand why the cyanoethyl radicals generated in the electrochemical reduction of acrylonitrile exhibit different properties from those formed from iodopropionitrile. It is clear that this difference must be explained by peculiarities in the mechanism of the electrode process.

Some halogen compounds, such as benzyl iodide and allyl bromide [31], give two waves (Fig. 1) during electrochemical reduction at a dropping mercury cathode, and each corresponds to a one-electron process. For the compounds mentioned it has been shown fairly convincingly [29, 30] that the first wave corresponds to the reduction of an intermediate organomercury compound. In this case formation of the organomercury compound occurs without the action of electrical current and results from reaction of the original iodide with the cathodically polarized mercury, where polarization of about −0.2 V (vs. s.c.e.) is required to initiate the reaction.

The reduction scheme for these compounds can be expressed in the following equations:

$$RI + Hg \rightarrow RHgI \tag{6}$$

$$RHgI + e \rightarrow {}^{\cdot}RHg + I^- \tag{7}$$

$${}^{\cdot}RHg + H^+ + e \rightarrow RH + Hg \tag{8}$$

TABLE 6. Electrical Conductivity and Dissociation Schemes for Some Complexes Used in Synthesis of Organometallic Compounds in Anodic Processes

Complex	Solvent	Proposed dissociation scheme	Equivalent electrical conductivity, mho·cm^2	Reference
NaF·2Al(C$_2$H$_5$)$_3$	Melt	Not considered, aluminum precipitated on anode	$4·10^{-2}$	[57]
KF·2Al(C$_2$H$_5$)$_3$	"	"	$7·10^{-2}$	[57]
NaAl(C$_2$H$_5$)$_4$	"	Na$^+$ + [(C$_2$H$_5$)$_4$Al]$^-$	$3.3·10^{-2}$	[57]
KAl(C$_2$H$_5$)$_4$	"	K$^+$ + [(C$_2$H$_5$)$_4$Al]$^-$	$8.0·10^{-2}$	[57]
NaB(C$_2$H$_5$)$_4$	Water (saturated solution)	Na$^+$ + [(C$_2$H$_5$)$_4$B]$^-$	$9.0·10^{-3}$	[58]
	Water (10-15% solution)	Na$^+$ + [(C$_2$H$_5$)$_4$B]$^-$	$5·10^{-2}$	[58]
C$_2$H$_5$MgBr	Ether (1 M solution)	MgBr$^+$ + [(C$_2$H$_5$)$_2$MgBr]$^-$	$3.2·10^{-2}$	[59]
C$_2$H$_5$MgI	" "	MgI$^+$ + [(C$_2$H$_5$)$_2$MgI]	$7·10^{-2}$	[59]

or, possibly,

$$2^{\cdot}RHg \rightarrow [RHgHgR] \rightarrow RHgR + Hg. \qquad (9)$$

A similar mechanism has been proposed in the discussion of the electrochemical reduction of disulfides. In the process it is suggested that the organomercury compound RSHgSR is formed initially, since the polarographic curve of this compound is identical with that of the original disulfide RSSR [41].

Compounds which have a sufficiently mobile halogen atom react according to the mechanism described above. Other halogen-containing compounds, e.g., iodopropionitrile, are reduced according to a substantially different mechanism. In aqueous and aqueous ethanolic solutions of iodopropionitrile [42] two waves are formed (Table 4) with $E'_{1/2} = -1.03$ to -0.895 and $E''_{1/2} = -1.42$ to -1.14 V (vs. s.c.e.), each of which corresponds to transfer of one electron. Under the same conditions β-cyanoethylmercury iodide forms two distinct waves with $E'_{1/2} = -0.40$ and $E''_{1/2} = -1.2$ V (vs. s.c.e.), and its reduction clearly takes place according to Eqs. (7) and (8).

Di(β-cyanoethyl)mercury is polarographically inactive, but its presence in a solution submitted to polarography substantially shifts the second wave of β-cyanoethylmercury iodide to more negative potentials. For example, the presence of 10^{-3} mole/liter di(β-cyanoethyl)mercury in the solution shifts the potential of the second wave of β-cyanoethylmercury iodide by almost 0.15 V (Fig. 2). It should further be noticed that addition of β-cyanoethylmercury iodide to a solution of iodopropionitrile gives rise to an increase in the second wave of iodopropionitrile, while the height of the first wave remains unchanged. The half-wave potential of the first wave of β-cyanoethylmercury iodide is substantially more positive than that of the first wave of iodopropionitrile, and when both compounds are present together the two waves are independent of each other. The above facts show that the first reduction wave of iodopropionitrile cannot result from reduction of β-cyanoethylmercury iodide, as could be expected by analogy with benzyl iodide, for example [30]. Since addition of β-cyanoethylmercury iodide causes an increase in the second wave of iodopropionitrile, it can be supposed that the second waves of both compounds correspond to a similar process. The fact that the half-wave potentials of the second wave of iodopropionitrile does not coincide with that of the pure mercury compound can be explained by the great sensitivity of the half-wave potential of β-cyanoethylmercury iodide to the presence of extraneous compounds, as shown in the case of electrochemical reduction of β-cyanoethylmercury iodide in the presence of di(β-cyanoethyl)mercury. At the same time the presence of di(β-cyanoethyl)mercury in the solution does not affect the half-wave potentials of iodopropionitrile.

The observed effects can be explained if it is assumed that iodopropionitrile forms an adsorption complex at the electrode surface. The first reduction wave corresponds to transfer of an electron to the molecule of the complex, followed by ionization of the iodine:

$$I - CH_2CH_2CN \xrightarrow{+e} \cdot HgCH_2CH_2CN + I^-. \qquad (10)$$
$$\underset{Hg}{\diagdown\diagup}$$

The fates of the free radicals formed in this process can be various. If the potential of the cathode is insufficient for their subsequent reduction, the radicals react with each other to form di(β-cyano-ethyl)mercury:

$$2\ \dot{H}gCH_2CH_2CN \rightarrow NCCH_2CH_2HgCH_2CH_2CN + Hg. \tag{11}$$

If the potential of the cathode is sufficiently negative, the radical acquires a second electron and forms an anion:

$$HgCH_2CH_2CN \xrightarrow{+e} [HgCH_2CH_2CN]^-, \tag{12}$$

which either interacts with another molecule of iodopropionitrile:

$$ICH_2CH_2CN + [HgCH_2CH_2CN]^- \rightarrow Hg(CH_2CH_2CN)_2 + I^- \tag{13}$$

or is protonated to propionitrile:

$$[HgCH_2CH_2CN]^- + H^+ \rightarrow Hg + CH_3CH_2CN. \tag{14}$$

It can be stated with some degree of certainty that cyanoethyl compounds of lead, tin, and thallium are formed from iodopropionitrile in accordance with the mechanism examined above.

In the electrolysis of ketones and unsaturated nitriles the organometallic compounds are formed by a different mechanism. At the dropping mercury cathode, for example, acrylonitrile (and other α,β-unsaturated nitriles) gives one wave, corresponding to transfer of two electrons [43–45].

It can be assumed that the anion which forms after addition of two electrons and one proton reacts with the cathode material, extracting a cation from the crystal lattice of the metal, according to the following equation:

$$Sn^{4+} + 4[CH_2CH_2CN]^- \rightarrow Sn(CH_2CH_2CN)_4, \tag{15}$$

i.e., the organometallic compound is formed by the mechanism proposed by Voitkevich [38]. It was also proposed in [38] that the formation of organometallic compounds in electrochemical reduction of aliphatic ketones also takes place by an ionic mechanism. In favor of this proposal is the fact that aromatic ketones do not form organometallic compounds, although there is no doubt that intermediate free radicals are formed in the electrochemical reduction of aromatic ketones.

In spite of the fact that the above mechanism for formation of organometallic compounds during electrolysis of nitriles appears most probable to us, it is impossible not to mention another possible explanation for this reaction. In electrochemical reduction of acrylonitrile organometallic compounds have been obtained from tin, sulfur, selenium, and tellurium, which are themselves reduced electrochemically.

For example, tin hydride is formed in the electrolysis of an aqueous solution of tin sulfate in sulfuric acid with 0.5% dextrin, gelatin, or agar-agar at a lead cathode [46]. Sulfur is electrochemically reduced to hydrogen sulfide [47, 48], and at a graphite cathode in alkaline and neutral solutions the current yield of hydrogen sulfide amounts to 55-57% [19]. Tellurium is electrochemically reduced to hydrogen telluride [49], and in alkaline solutions at a graphite cathode selenium is reduced to hydrogen selenide [19] with a current yield of 93-97%. It is also known that acrylonitrile, for example, readily reacts with hydrogen sulfide in alkaline solutions to form β,β'-dicyanodiethyl sulfide [50].

The question arises as to whether the formation of organometallic compounds during electrolysis of aqueous solutions of acrylonitrile is not due to cyanoethylation of hydrides formed initially at the electrode. In Table 5 data on the electrochemical reduction of tin, sulfur, selenium, and tellurium in aqueous solutions of sodium sulfate with and without acrylonitrile are compared [40].

The results show that the current consumed in the formation of organometallic compounds from tin and tellurium is considerably greater than the current consumed in the formation of the hydride under the same conditions. Consequently the cyanoethyl derivatives of the tin and tellurium cannot result from cyanoethylation of their hydrides. In the case of sulfur and selenium, on the other hand, formation of cyanoethyl derivatives through the hydride is quite likely.

There is one other observation which is not consistent with the hydride mechanism for formation of organotin compounds; a paper was recently published [51] in which the effect of electrolysis conditions on the production of tin hydride was discussed in detail. It was found that appreciable formation of tin hydride is observed at high current densities (10,000 A/m^2 and higher). At the same time the maximum yield of organotin compounds is observed at low current densities in the order of 100–200 A/m^2 [21], where formation of tin hydride is practically excluded.

2. ANODIC PROCESSES

As yet only one type of anodic process leading to the formation of organometallic compounds has been investigated in sufficient detail; this is the electrolysis of solutions or melts of organic complex electrolytes.

By themselves symmetrical organometallic compounds of the RMR type and their solutions in polar solvents possess insignificant electrical conductivity [52, 53] and cannot therefore be subjected to electrolysis. However, a mixture of such compounds with certain salts of the MX type, alkyl halides RX, metal hydrides MH, and finally other organometallic compounds sometimes gives electrically conducting solutions. This effect is explained by the formation of dissociating complexes.

The phenomenon was first discovered by Hein [54] in 1924. He found that a mixture of diethylzinc and ethylsodium forms a solution which is not inferior to 0.1 N aqueous potassium chloride solution in electrical conductivity. In this case the existence of electrical conductivity is explained by the formation of the dissociating complex:

$$NaZn(C_2H_5)_3 \rightleftarrows Na^+ + Zn(C_2H_5)_3^-. \tag{16}$$

A considerable number of complexes of similar composition have now been investigated, and most interesting from the practical standpoint are the complexes of magnesium, aluminium, and boron. Table 6 gives data on the electrical conductivity of some of the most widely investigated complexes. The reactions which may be realized during electrolysis of these complexes will be considered next.

TABLE 7. Electrolysis of Grignard Reagents

Anode	Cathode	Composition of solution		Current density, A/cm²	Temperature, °C	Organometallic compound obtained	Yield, %	Reference
		electrolyte	solvent					
Aluminium	Platinum	C_2H_5MgI	Diethyl ether	—	—	$Al(C_2H_5)_3$	—	[59]
Boron	Steel or platinum	CH_3MgCl	"	—	—	$B(CH_3)_3$	—	[60]
		C_2H_5MgCl	"	—	—	$B(C_2H_5)_3$	—	[60]
		C_3H_7MgCl	"	—	—	$B(C_3H_7)_3$	—	[60]
		$i\text{-}C_3H_7MgCl$	"	—	—	$B(i\text{-}C_3H_7)_3$	—	[60]
		C_4H_9MgCl	"	—	—	$B(C_4H_9)_3$	—	[60]
		$i\text{-}C_4H_9MgCl$	"	—	—	$B(i\text{-}C_4H_9)_2$	—	[60]
		C_6H_5MgCl	"	—	—	$B(C_6H_5)_3$	—	[60]
		$C_6H_{13}MgCl$	"	—	—	$B(C_6H_{13})_3$	—	[60]
Black phosphorus	Platinum	C_2H_5MgCl	"	0.1	—	$P(C_2H_5)_3$	—	[61]
		$AlkMgCl(C_1\text{-}C_8)$	"	0.1	—	$P(Alk)_3$	—	[61]
		C_6H_5MgCl	"	0.1	—	$P(C_6H_5)_3$	—	[61]
		$C_6H_5CH_2MgCl$	"	0.1	—	$P(C_6H_5CH_2)_3$	—	[61]
		$CH_5C_6H_4MgCl$	"	0.1	—	$P(CH_3C_6H_4)_3$	—	[61]
Lead	Lead	190 g 2 N CH_3MgCl 21.4 g tetrahydrofuran, 1.8 N CH_3MgCl	Dibutylcarbitol 40% tetrahydrofuran + 60% dibutylcarbitol	—	— 27	$Pb(CH_3)_4$ $Pb(CH_3)_4$	19.1 —	[62] [62]
Lead	Steel	4.5 mole benzene + 1.0 mole tetrahydrofuran + 0.95 g-atom Mg + 0.98 mole CH_3Cl	Hexylethylcarbitol (1.0 mole)	0.02	30	$Pb(CH_4)_3$	91.2	[63]

TABLE 7 (Continued)

| Anode | Cathode | Composition of solution | | Current density, A/cm^2 | Temperature, °C | Organometallic compound obtained | Yield, % | Reference |
		electrolyte	solvent					
Lead	Steel	1.3—1.4 M CH$_3$MgCl +0.5—2.5 M tetrahydrofuran(CH$_3$MgCl : : CH$_3$Cl = 0.45; tetrahydrofuran:benzene = 1:4)	Diethylene glycol in dibutyl ether (1000 g)	0.026	30	Pb(CH$_3$)$_4$	97.0	[64]
Lead	Lead	0.64 M C$_2$H$_5$MgBr	Diethyl ether	0.002	25	Pb(C$_2$H$_5$)$_4$	High	[65]
		3.16 M C$_2$H$_5$MgBr	1400 cc diethyl ether	0.01	—	Pb(C$_2$H$_5$)$_4$	86.5	[66]
Lead	Steel	1.27 M C$_2$H$_5$MgCl (C$_2$H$_5$Cl : C$_2$H$_5$MgCl = 9 : 10)	Dibutyl ether of diethylene glycol	0.003	35—40	Pb(C$_2$H$_5$)$_4$	100	[67, 68]
		0.9 N C$_2$H$_5$MgCl (C$_2$H$_5$Cl : C$_2$H$_5$MgCl = 2,5)	"	0.0015	55—60	Pb(C$_2$H$_5$)$_4$	80.5	[68]
		1.0 N C$_2$H$_5$MgCl (C$_2$H$_5$Cl : C$_2$H$_5$MgCl = 7)	"	0.009	65—85	Pb(C$_2$H$_5$)$_4$	73.5	[68]
		0.94 N C$_2$H$_5$MgCl (C$_2$H$_5$Cl : C$_2$H$_5$MgCl = 7,4)	"	0.003	33—38	Pb(C$_2$H$_5$)$_4$	84.0	[68]
		0.82 N C$_2$H$_5$MgCl (C$_2$H$_5$Cl : C$_2$H$_5$MgCl = 1,0)	Dimethyl ether of diethylene glycol	0.018	50—60	Pb(C$_2$H$_5$)$_4$	—	[68]
		2.79 N ethylmagnesium glycol	Dibutyl ether of diethylene glycol	—	25	Pb(C$_2$H$_5$)$_4$	98.0	[69]

Electrolysis of Organomagnesium Complexes

(Grignard Reagents)

While studying the electrolysis of ethereal solutions of a Grignard reagent, French and Drane [55] observed that anodes of aluminum, zinc, and cadmium dissolved and that these metals were deposited at the cathode. More recently Evans [56] proposed that triethylaluminum is formed when an aluminum anode dissolves in ethylmagnesium bromide:

$$6C_2H_5MgBr + 2Al \rightarrow 3Mg + 3MgBr_2 + 2Al(C_2H_5)_3. \tag{17}$$

The reaction has only comparatively recently attracted the attention of investigators as a method for preparation of organometallic compounds.

Electrolysis of Grignard reagents can be successfully applied to the synthesis of organic compounds of lead, boron, and phosphorus (Table 7). In view of the fact that work with ethyl ether solutions is fairly harzardous, higher ethers, e.g., dibutyl, are usually employed as solvent in the preparation of Grignard reagents intended for electrolysis.

Greatest attention has been paid to the development of a convenient method for the production of tetraethyllead. Evidently, one of the most successful methods for the realization of this reaction has been described in [70].

To reduce the fire risk in the process and simplify the technology for separation of the tetraethyllead a high-boiling ether with the following structure is used as solvent:

$$C_4H_8 \begin{matrix} O-C_2H_4-O \\ \diagup \qquad \diagdown \\ \diagdown \qquad \diagup \\ O-C_2H_4-O \end{matrix} C_4H_8$$

In this case electrolysis can be carried out at elevated temperatures; the tetraethyllead distills as it forms. During electrolysis an alkyl halide is added to the solution; the magnesium deposited at the cathode reforms the Grignard reagent and is thereby partially returned to the process. The reactions which take place at the electrode as a whole can be expressed in the following equations:

$$4RMgX + M \rightarrow MR_4 + MgX_2 + 2Mg, \tag{18}$$

$$RX + Mg \rightarrow RMgX. \tag{19}$$

It is recommended that dioxane is added to the solution to precipitate the magnesium halide formed.

Organoaluminum Complexes

With the development by Ziegler of an accessible method for the preparation of triethylaluminum [71, 72] interest in the use of this compound for the synthesis of other alkyl metallic compounds increased considerably. As already mentioned above, melts of mixtures of the $MX \cdot AlR_3$ or $MR \cdot AlR_3$ types (where M is a monovalent metal and X is halogen or hydrogen) possess sufficient electrical conductivity.

In 1955 Ziegler accomplished an electrochemical synthesis of tetraethyllead, using the complex $NaF \cdot 2Al(C_2H_5)_3$ as electrolyte [73]. During electrolysis of this complex metallic aluminum is liberated at the cathode and tetraethyllead is formed at the dissolving lead anode.

An electrolyte of this type has been used to synthesize alkyl derivatives of tin, zinc, antimony, indium, magnesium [74-77], silicon, and germanium [78] (Table 8). It is interesting to note that an attempt to synthesize organogermanium compounds on a germanium single crystal as anode was unsuccessful [86].

In spite of the initial successes in the use of organoaluminum complexes of the above-mentioned type, their application for preparative purposes encountered substantial difficulties in attempts to realize a continuous process. Thus, aluminum separated at the cathode in the form of a friable deposit, which is difficult to separate from the electrolyte. Moreover, aluminum suspended in the solution partly reacted with the metal alkyl. In order to avoid this undesirable reaction it was necessary to separate the anode and cathode compartments, and this in turn had undesirable features.

If electrolysis is continued over a fairly long period, the ratio between NaF and $Al(C_2H_5)_3$ approximates to unity and, together

TABLE 8. Electrochemical Synthesis of Organometallic Compounds by Electrolysis of Complexes

Anode	Cathode	Electrolyte	Current density, A/cm²	Temperature, °C	Synthesized compound	Yield, %	Reference
Aluminum	Copper	NaAl(C₂H₅)₄	0.04	140	Al(C₂H₅)₃	99	[79]
		NaAl(C₂H₅)₄ + KCl	—	—	Al(C₂H₅)₃	—	[80]
		NaAl(C₃H₇)₄	0.04	100	Al(C₃H₇)₃	94	[79]
		NaAl(C₄H₉)₄	0.04	—	Al(C₄H₉)₃	—	[79]
Magnesium	Copper	NaAl(C₂H₅)₄	0.04	100	Mg(C₂H₅)₂	95	[79]
		NaF·2Al(C₂H₅)₃	—	—	Mg(C₂H₅)₂	—	[76]
Magnesium	Steel	NaAl(CH₃)₄ + NaAl(C₂H₅)₄	0.25	100	Mg(C₂H₅)₂	95	[81]
		KAl(CH₃)₄ : KAl(i-C₃H₇)₄ = 1 : 8	0.25	100	Mg(i-C₃H₇)₂	—	[81]
		LiAl(CH₃)₄: NaAl(C₄H₉)₄ = 1 : 3	0.25	100	Mg(C₄H₉)₂	—	[81]
		RbAl(CH₃)₄: NaAl(CH₃)₄ = 1 : 5	0.25	140	Mg(CH₃)₂	—	[81]
Zinc	Copper	NaAl(CH₃)₄ + NaAl(C₂H₅)₄	0.35	105	Zn(C₂H₅)₂	80	[82]
		NaAl(CH₃)₄ : CsAl(C₂H₅)₄ = 1 : 2	0.35	80	Zn(C₂H₅)₂	—	[82]
		NaF·2Al(C₂H₅)₃	—	—	Zn(C₂H₅)₂	—	[76]
		3320 g NaAl(C₂H₅)₃ + 740 g KCl	—	100	Zn(C₂H₅)₂	—	[76]
		1050 g Na[Al(C₂H₅)₃(OC₄H₉)] + 830 g NaAl(C₂H₅)₄	0.04	100	Zn(C₂H₅)₂	91	[83]
		Na [Al(C₂H₅)₃(OC₂H₅)]	0.04	100	Zn(C₂H₅)₂	100	[83]
		NaF·2Al(C₃H₇)₃	—	—	Zn(C₃H₇)₂	—	[75]

TABLE 8 (Continued)

Anode	Cathode	Electrolyte	Current density, A/cm²	Temperature, °C	Synthesized compound	Yield, %	Reference
Zinc	Copper	$RbAl(CH_3)_4 : NaAl(CH_3)_4 = 5:1$	0,35	100	$Zn(CH_3)_2$	—	[82]
		$KAl(CH_3)_4 : KAl(C_3H_7)_4 = 8:1$	0,35	180	$Zn(C_3H_7)_2$	—	[82]
		$LiAl(CH_3)_4 : NaAl(C_{10}H_{21})_4 = 3:10$	0,36	40	$Zn(C_{10}H_{21})_2$	—	[82]
		$NaAl(CH_3)_4 : NaAl(C_6H_5)_4 = 1:10$	0,35	190	$Zn(C_6H_5)_2$	—	[82]
		$NaAl(C_4H_9)_4 + Na[Al(C_2H_5)_3(OC_{10}H_{21})]$	0,04	—	$Zn(C_4H_9)_2$	82	[83]
		$KAl(C_3H_7)_4 + NaAl(C_3H_7)_4$	0,04	—	$Zn(C_3H_7)_2$	88	[83, 84]
Tin	Mercury	$NaAl(CH_3)_4 + NaAl(C_2H_5)_4$	0,25	100	$Sn(C_2H_5)_4$	80	[83, 84]
		$KAl(CH_3)_4 + NaAl(C_2H_5)_4$	1,0	100	$Sn(C_2H_5)_4$	—	[85]
		$NaAl(CH_3)_4 : CsAl(C_2H_5)_4 = 2:1$	0,25	120	$Sn(C_2H_5)_4$	—	[85]
		$NaAl(CH_3)_4 + KAl(C_2H_5)_4 + LiAl(C_2H_5)_4$	0,25	—	$Sn(C_2H_5)_4$	—	[85]
Tin	Copper	3320 g $NaAl(C_2H_5)_4 + 740$ g KCl	—	100	$Sn(C_2H_5)_4$	—	[83]
		1050 g $Na[Al(C_2H_5)_3(OC_4H_9)]$ $+ 830$ g $NaAl(C_2H_5)_4$	0,04	100	$Sn(C_2H_5)_4$	85	[83]
		$NaF \cdot 2Al(C_2H_5)_3$	—	—	$Sn(C_2H_5)_4$	—	[75, 76]
Tin	Mercury	$RbAl(CH_3)_4 : NaAl(CH_3)_4 = 5:1$ in xylene	0,25	100	$Sn(CH_3)_4$	—	[85]
		$KAl(CH_3)_4 : KAl(i\text{-}C_3H_7)_4 = 8:1$	0,25	160	$Sn(i\text{-}C_3H_7)_4$	—	[85]
		$LiAl(CH_3)_4 : NaAl(C_8H_{17})_4 = 3:10$	0,25	40	$Sn(C_8H_{17})_4$	—	[85]
		$NaAl(CH_3)_4 : NaAl(C_6H_5)_4 = 1:1$ in biphenyl	0,25	180	$Sn(C_6H_5)_4$	—	[85]
Antimony	Copper	$NaF \cdot 2Al(C_2H_5)_3$	—	70—100	$Sb(C_2H_5)_3$	80—90	[76]
		3320 g $NaAl(C_2H_5)_4 + 740$ g KCl	—	—	$Sb(C_2H_5)_3$	—	[83]

TABLE 8 (Continued)

Anode	Cathode	Electrolyte	Current density, A/cm²	Temperature, °C	Synthesized compound	Yield, %	Reference
Antimony	Copper	1050 g Na[Al(C₂H₅)₃(OC₄H₉)] + 830 g NaAl(C₂H₅)₄	0.04	100	Sb(C₂H₅)₃	96	[83]
		NaAl(C₃H₇)₄ + Na[Al(C₃H₇)₃(OC₆H₁₁)]	0.04	100	Sb(C₃H₇)₃	89	[83, 84]
Mercury	Mercury	NaF·2Al(C₂H₅)₃	—	—	Hg(C₂H₅)₂	100	[57]
Mercury	Copper	1050 g Na[Al(C₂H₅)₃(OC₄H₉)] + 830 g NaAl(C₂H₅)₄	0.4	100	Hg(C₂H₅)₂	82	[83]
		3320 g NaAl(C₂H₅)₄ + 740 g KCl	—	—	Hg(C₂H₅)₂	—	[83]
Cadmium	Copper	3320 g NaAl(C₂H₅)₄ + 740 g KCl	—	100	Cd(C₂H₅)₂	—	[83]
		1050 g Na[Al(C₂H₅)₃(OC₄H₉)] + 830 g NaAl(C₂H₅)₄	0.04	100	Cd(C₂H₅)₂	81	[83]
Bismuth	Copper	3320 g NaAl(C₂H₅)₄ + 740 g KCl	—	—	Bi(C₂H₅)₃	—	[83]
		1050 g Na[Al(C₂H₅)₃(OC₄H₉)] + 830 g NaAl(C₂H₅)₄	0.04	100	Bi(C₂H₅)₃	95	[83]
Indium	Copper	NaF·2Al(C₂H₅)₃	—	—	In(C₂H₅)₃	—	[75, 76]

with aluminum, metallic sodium begins to separate at the cathode. It was suggested that high current densities should be used to deposit the aluminum in the form of readily separable dendrites, but even then it was not possible to go over to a continuous process.

Later, in 1959, Ziegler showed that if a complex of the NaAl $\cdot(C_2H_5)_4$ type was used as electrolyte, during electrolysis with an aluminum anode metallic sodium separated at the cathode and triethylaluminum at the anode [87]:

$$3NaAl(C_2H_5)_4 + Al \rightarrow 4Al(C_2H_5)_3 + 3Na. \tag{20}$$

In this case the question of regenerating the electrolyte is solved very easily, since if part of the triethylaluminum formed is returned to the cycle the original complex can again be obtained by its reaction with sodium, ethylene, and hydrogen [88]:

$$3Na + 3H + 3C_2H_4 + 3Al(C_2H_5)_3 \rightarrow 3NaAl(C_2H_5)_4. \tag{21}$$

In addition to ease of regeneration, a complex of the $NaAl(C_2H_5)_4$ type has the advantage that the sodium at the cathode can be obtained in the form of an amalgam or in liquid form and readily removed from the electrolyzer.

The investigations of Ziegler set the beginning to vigorous searches for continuous technological methods for electrochemical synthesis of metal alkyls, and tetraethyllead in particular. A detailed examination of the different variants of the processes suggested for production of tetraethyllead falls outside the scope of the present review; here we will only dwell on a few of the proposed solutions. However, to give a more general idea of the nature of the investigations, Table 9 shows the electrolytes recommended for preparation of tetraethyllead and Table 10 shows electrolytes recommended for preparation of other alkyl lead compounds.

In the original works of Ziegler [57] it was proposed that electrolysis should be carried out with a solid cathode under vacuum. In this case the tetraethyllead distills, which avoids possibility of its contact of sodium. Under vacuum the sodium is deposited as a film which readily wets the surface and is easily removed from the electrolyzer. However, in addition to tetraethyllead, a small amount of gaseous products is always formed at the

TABLE 9. Preparation of Tetraethyllead by Electrolysis of Aluminum Complexes (Lead Anode)

Cathode	Electrolyte	Current density, A/cm²	Temperature, °C	Yield of tetraethyllead, %	Reference
Platinum	$Al(C_2H_5)_3$ in diethyl ether	0.7	25	99	[89]
Mercury	$KAl(C_2H_5)_4$	0.465	100—110	100	[90]
	$NaAl(C_2H_5)_4$	0.8	—	92	[57, 79, 84, 91]
Steel	$NaAl(C_2H_5)_2(CH_3)_2$	—	—	—	[92]
Copper, steel	$NaAl(CH_3)_4 + NaAl(C_2H_5)_4$	0.25	100	77	[92, 93]
Copper	3 M $NaAl(C_2H_5)_4 + 1$ M $NaAl(CH_3)_4$	0.25	—	89	[93]
	$MAl(CH_3)_4 + MAl(C_2H_5)_4$ (M = 25% K, 75% Na)	0.5	100	96	[93]
	$NaAl(CH_3)_4 : CsAl(C_2H_5)_4 = 2:1$	0.25	80	—	[93]
	$NaAl(CH_3)_4 + KAl(C_2H_5)_4 + LiAl(C_2H_5)_4$	0.25	100	—	[93]
Steel	$NaAl(CH_3)_4 + NaAl(C_2H_5)_4 + CaAl(C_2H_5)_5$	0.25	100	—	[92]
Copper	$NaF \cdot 2Al(C_2H_5)_3$	0.022	28—45	87	[75, 94]
	3320 g $NaAl(C_2H_5)_4 + 740$ g KCl	—	100	—	[83]
Mercury	$NaF \cdot 2Al(C_2H_5)_3 + NaAl(C_2H_5)_4$	0.5—0.25	112—417	100	[57]

TABLE 9 (Continued)

Cathode	Electrode	Current density, A/cm²	Temperature, °C	Yield of tetraethyl-lead, %	Reference
Platinum	$NaF \cdot Al(C_2H_5)_3$	0.25	80	—	[89]
Iron, copper	$29,7$ kg $NaAl(C_2H_5)_4 + 27,7$ kg $NaF \cdot Al(C_2H_5)_3$ $+ 12,1$ kg $NaF \cdot 2Al(C_2H_5)_3$	0.08	70	98	[95]
Mercury	$1,753$ M $KAl(C_2H_5)_4 + 0,528$ M $KF \cdot Al(C_2H_5)_3$	0.185	85	98,5	[96]
Copper	$Na[Al(C_2H_5)_3(OC_{10}H_{21})]$	0.04	100	—	[83]
	$Na[Al(C_2H_5)_3(OC_4H_9)]$	0.04	100	93	[80]
	5 M $NaAl(C_2H_5)_4 + 5$ M $Na[Al(C_2H_5)_3(OC_4H_9)]$	0.053	100	95	[80, 83]
	$NaAl(OC_2H_5)_4$	0.3	—	—	[80]
	$Na[Al(C_2H_5)_3(i\text{-}C_6H_{11}O)]$	—	100	—	[80]
	$K[Al(C_2H_5)_3(OC_4H_9)]$	0.04	—	93	[83]
Copper	$NaAl(C_2H_5)_4 + Na[Al(C_2H_5)_3(OC_6H_5)]$	—	100	—	[80]
	$NaAl(C_2H_5)_4 + Al(C_2H_5)_2(OC_4H_9)$ $+ Na[Al(C_2H_5)_3(OC_4H_9)]$	—	100	—	[97]

TABLE 10. Preparation of Tetraalkyllead by Electrolysis of Aluminum Complexes (Lead Anode)

Cathode	Electrolyte	Current density, A/cm²	Temperature, °C	Synthesized compound	Yield, %	Reference
Mercury	200 g $NaAl(CH_3)_4$ + 400 g tetrahydrofuran	0.14	90	$Pb(CH_3)_4$	90	[98, 99]
	200 g $NaAl(CH_3)_4$ + 400 g tetrahydrofuran + $NaB(CH_3)_4$	0.15	—	$Pb(CH_3)_4$	94	[98, 99]
	200 g $NaAl(CH_3)_4$ + 500 g diglyme	0.30	100	$Pb(CH_3)_4$	96	[99]
	$Na[Al(CH_3)_3(OC_4H_9)]$	0.14	90	$Pb(CH_3)_4$	94	[99]
Copper	$RbAl(CH_3)_4 : NaAl(CH_3)_4 = 5:1$	0.25	100	$Pb(CH_3)_4$	—	[92, 93]
	$NaAl(C_2H_5)_4 + NaAl(C_3H_7)_4$	—	—	$Pb(C_3H_7)_4$	95	[83]
	$KAl(CH_3)_4 : KAl(i\text{-}C_3H_7)_4 = 8:1$	0.25	180	$Pb(i\text{-}C_3H_7)_4$	—	[92, 93]
	$NaF \cdot 2Al(i\text{-}C_3H_7)_3$	0.022	50	$Pb(i\text{-}C_3H_7)_4$	—	94
	$Na[Al(C_3H_7)_3(OC_9H_{11})] + NaAl(C_3H_7)_4$	0.04	100	$Pb(C_3H_7)_4$	90	[80, 83]
	$NaAl(C_2H_5)_4 + NaAl(C_4H_9)_4$	—	—	$Pb(C_4H_9)_4$	—	[83]
	$NaAl(C_4H_9)_4 + Na[Al(C_4H_9)_3(OC_{10}H_{21})]$	0.04	100	$Pb(C_4H_9)_4$	80	[80, 83]
Steel	$NaAl(i\text{-}C_4H_9)_4 + KAl(i\text{-}C_4H_9)_4$	0.2	—	$Pb(i\text{-}C_4H_9)_4$	—	[92]
Copper	$NaF \cdot 2Al(C_6H_5)_3$ in benzene	—	—	$Pb(C_6H_5)_4$	—	[94]
	$NaAl(CH_3)_4 : NaAl(C_6H_5)_4 = 1:10$	0.25	220	$Pb(C_6H_5)_4$	—	[94]
	$LiAl(CH_3)_4 : NaAl(C_{10}H_{21})_4 = 3:10$	0.25	40	$Pb(C_{10}H_{21})_4$	—	[93]
Steel	$NaB(C_6H_5)_4 : NaAl(C_2H_5)_4 = 2:3$	—	110	$Pb(C_6H_5)_4$, $Pb(C_2H_5)_4$	—	[92]
	$NaAl(C_6H_5CH_2)_4 + KAl(C_6H_5CH_2)_4 + 10\%$ biphenyl	—	—	$Pb(C_6H_5CH_2)_4$	—	[92]
	$NaAl(\beta C_6H_5C_2H_4)_4 + KAl(\beta C_6H_5C_2H_4)_4$	—	—	$Pb(\beta C_6H_5C_2H_4)_4$	—	[92]
	$NaAl(C_6H_{11})_4 + KAl(C_6H_{11})_4$	0.25	100	$Pb(C_6H_{11})_4$	—	[92]

anode and a powerful pump is required to maintain the vacuum, as a result of which the so-called "vacuum method" has not been developed further.

The "three cycle" method proved considerably more successful. In this process mercury is used as cathode. During electrolysis the sodium forms an amalgam, which does not interact with the metal alkyls; the need to use a diaphragm is thereby eliminated. The metallic sodium can be separated from the amalgam by repeated electrolysis [100, 101].

Since a potassium electrolyte has better conductivity but separation of metallic potassium from the amalgam is difficult, a method for regeneration of potassium-containing electrolytes has been developed on the basis of the reaction of the potassium amalgam with $NaAlR_4$ [57, 102]. Use of this double-decomposition reaction has made it possible to develop a scheme for the preparation of tetraethyllead with complete regeneration of the electrolyte:

$$KAl (C_2H_5)_4 + Pb \xrightarrow{\text{electrolysis}} Pb (C_2H_5)_4 + 4Al (C_2H_5)_3 + 4K (Hg)_x, \qquad (22)$$

$$Al (C_2H_5)_3 + NaH + C_2H_4 \rightarrow NaAl (C_2H_5)_4, \qquad (23)$$

$$NaAl (C_2H_5)_4 + K (Hg)_x \rightarrow KAl (C_2H_5)_4 + Na (Hg)_x, \qquad (24)$$

$$KOH + Na (Hg)_x \rightarrow NaOH + K (Hg)_x. \qquad (25)$$

Several methods have been proposed for separating the metal alkyl from the electrolyte. If the differences in boiling points are sufficiently large, vacuum distillation can be used [103]. It is sometimes recommended that the metal alkyl should first be precipitated by addition of sodiumalkoxyaluminum [104] or sodium hydride [105, 106]. In both cases the lower layer, which is richer in metal alkyl, is separated, and removal of the pure compound presents no great difficulty.

In some cases additives which form insoluble complexes with the metal alkyls are recommended. In particular, tributylamine [107], sodium azide [108], and potassium cyanide [109] are used for this purpose.

In spite of the fact that a large number of different metal alkyls can be synthesized by electrolysis of aluminum complexes,

TABLE 11. Electrolysis of Boron Complexes

Anode	Cathode	Composition of solution		Current density, A/cm^2	Temperature, °C	Synthesized compound	Yield, %	Reference
		complex	solvent					
Lead	Copper	$B(C_2H_5)_3 + KCN$	—	0.05	65	$Pb(C_2H_5)_4$	—	[111]
		0.75 part $B(C_2H_5)_3$ + 0.35 part 1-hexynyl-sodium	12 part toluene	—	Room	$Pb(C_2H_5)_4$	High	[111]
		1 part $NaB(C_2H_5)_4$	5 part diethylene-glycol dimethyl ether	—	—	$Pb(C_2H_5)_4$	100	[111]
			Diethyl ether and diethylene-glycol dimethyl ether	—	25	$Pb(C_2H_5)_4$	100	[111]
		$NaB(C_2H_5)_4 + NaOH$	Water and di-ethyleneglycol dimethyl ether	0.01—1.0	Room	$Pb(C_2H_5)_4$	88.2	[111]
		2 part $NaB(C_2H_5)(C_4H_9)_3$	10 part tetra-hydrofuran	—	60	$Pb(C_2H_5)_4$	High	[110]
		$NaC_2H_5 \cdot B(C_6H_5)_3$	Diethyleneglycol diethyl ether	—	100	$Pb(C_2H_5)_4$	—	[110]
		$NaC_2H_5 \cdot B(C_{18}H_{37})_3$	"	—	110	$Pb(C_2H_5)_4$	—	[110, 111]
		$NaB(C_2H_5)_4 + KB(C_2H_5)_4$	Diethyleneglycol dimethyl ether	—	100	$Pb(C_2H_5)_4$	High	[110, 111]
		$NaB(C_2H_5)_4 + 1\% KI$	Diethyleneglycol methyl ethyl ether	—	30	$Pb(C_2H_5)_4$	"	[110, 111]
		$NaB(C_2H_5)F_3$	Tetrahydrofuran	—	50	$Pb(C_2H_5)_4$	"	[110, 111]
		$LiB(C_2H_5)_4$	Diethyl ether	—	—	$Pb(C_2H_5)_4$	"	[110, 111]

TABLE 11 (Continued)

Anode	Cathode	Composition of solution		Current density, A/cm²	Temperature, °C	Synthesized compound	Yield, %	Reference
		complex	solvent					
Lead	Copper	$(C_2H_5)_4N \cdot B(C_2H_5)_4$	Triethylamine	—	—	$Pb(C_2H_5)_4$	Quantitative	[110, 111]
		$NaNH_2 \cdot B(C_2H_5)_3$	Pyridine	0.01	70	$Pb(C_2H_5)_4$	High	[110, 111]
		0.75 M $B(C_{18}H_{37})_3$ +1 M $NaC_2H_5 \cdot BH_3(C_{18}H_{37})$	—	—	110	$Pb(C_2H_5)_4$	"	[110, 111]
		0.354 part $B(C_2H_5)_3$ +0.1 part $TiCl_4$ + 1.86 part $NaB(C_2H_5)_4$	40 part diethylene glycol dimethyl ether	—	Room	$Pb(C_2H_5)_4$	"	[110, 111]
Lead	Mercury	$NaB(C_2H_5)_4$	Water	0.035—0.1	20	$Pb(C_2H_5)_4$	91	[58]
Lead	Copper	10% $NaB(C_2H_5)_4$ +90% $[NaAl(CH_3)_4 + NaAl(C_2H_5)_4]$	—	0.3	100	$Pb(C_2H_5)_4$	90	[112]
		1 part $NaB(C_2H_5)_4$ +3 part $NaF \cdot 2Al(C_2H_5)_3$	—	—	—	$Pb(C_2H_5)_4$	High	[112]
		$NaB(C_2H_5)_4$ + $NaAl(CH_3)_2(C_2H_5)_2$	—	—	110	$Pb(C_2H_5)_4$	"	[112]
		$KB(C_2H_5)_4$ + $K[Al(C_2H_5)_3(OC_2H_5)]$	—	—	110	$Pb(C_2H_5)_4$	"	[112]
		20% $NaB(C_2H_5)_4$ +10% $Na[Al(C_2H_5)_3 \cdot (OC_2H_5)]$ +70% $NaAl(C_2H_5)_4$	—	—	—	$Pb(C_2H_5)_4$	"	[112]

TABLE 11 (Continued)

Anode	Cathode	Composition of solution		Current density, A/cm²	Temperature, °C	Synthesized compound	Yield, %	Reference
		complex	solvent					
Lead	Copper	10% $NaB(C_6H_5CH_2)_4$ +40% $NaAl(C_6H_5CH_2)_4$ +50% $KAl(C_6H_5CH_2)_4$	—	0.004	—	$Pb(C_6H_5CH_2)_4$	High	[112]
		1 part $KB(C_6H_{11})_3$	20 part benzene	—	75	$Pb(C_6H_{11})_4$	"	[110, 111]
		$NaB(C_6H_5)_4$	Dioxane	—	70	Phenyllead	"	[110, 111]
		0.5 M $B(C_6H_{11})_4$ +1 M $KB(C_6H_{11})_4$	Cyclohexane	—	—	$Pb(C_6H_{11})_4$	"	[110, 111]
		$NaB(C_8H_{17})_4$	Diethylene-glycol dimethyl ether	—	Room	$Pb(C_8H_{17})_4$	"	[110, 111]
		5 part $B(C_2H_5)_3$ + 2 part Sodium methylcyclo-pentadienide	40 part diethylene-glycol dimethyl ether	—	"	Lead dimethyl-cyclopentadienide	"	[110, 111]
Mercury	Mercury	$NaB(C_2H_5)_4$	Water	0.13	20	$Hg(C_2H_5)_2$	81	[58]
Bismuth		$NaB(C_6H_5)_4$	"	0.035	20	$Bi(C_2H_5)_3$	98	[58]
Lead		$NaB(CH_3)_4$	"	0.035	20	$Pb(CH_3)_4$	78	[58]

the use of these electrolytes presents substantial difficulties on account of their high reactivity. It is sufficient to mention that tri-ethylaluminum self-ignites in air, which necessitates special pre-cautions.

Organoboron Complexes

Organoboron complexes of the $MR \cdot BR_3$ and $MBR_4 \cdot MAlR_4$ types merit attention from the safety standpoint. Numerous metal alkyls can be prepared by electrolysis of these compounds (Table 11). It is recommended that the dimethyl ether of diethylene glycol should be used as solvent. A small amount of water is added to increase the electrical conductivity of the solution [110]. Electrolysis of tetraalkylboron complexes in aqueous solution with a soluble anode has also been described [58]. In this last case alkyl derivatives of mercury, bismuth, and lead are obtained. Tin and antimony anodes become passivated, and the yields of the respective organometallic compounds are small.

Other Anodic Processes

Isolated reports are found in the literature on electrolysis of other complex compounds. The preparative significance of these experiments is as yet difficult to assess, but they are undoubtedly of theoretical interest.

During electrolysis of the $NaGe(C_6H_5)_3$ system in liquid am-monia [113], for example, hexaphenyldigermane and triphenylger-mane are obtained. It has been suggested that these products re-sult from two independent reactions:

$$2Ge(C_6H_5)_3^- \xrightarrow{-2e} [Ge(C_6H_5)_3]_2, \tag{26}$$

$$6Ge(C_6H_5)_3^- + 2NH_3 \xrightarrow{-6e} 6Ge(C_6H_5)_3H + N_2. \tag{27}$$

The content of the dimerization product varies between 10 and 35% depending on the anode material. The nature of the anode material affects the secondary chemical reactions leading to nitro-gen evolution.

Of considerable interest are reports on the preparation of complex compounds of manganese by electrochemical means [114]

TABLE 12. Preparation of Manganese Complexes (Manganese anode)

Cathode	Composition of electrolyzed solution		Current density, A/cm^2	Temperature °C	Synthesized organometallic compound	Yield, %	Reference
	organic compound	solvent					
Manganese	4 part methylcyclopentadiene dimer + 1.5 part iron pentacarbonyl	32 part N,N-dimethylformamide*	0.1	195	Methylcyclopentadienylmanganese tricarbonyl	7	[115]
	Methylcyclopentadiene + MnBr$_2$	N-methylpyrrolidone*	0.1	165	"	Good	[115]
Iron	Methylcyclopentadiene + Na$_4$Mn(CN)$_6$ + Fe$_2$(CO)$_9$ + NaCl	Ethyleneglycol dimethyl ether*	0.1	125	"	—	[115]
Manganese	Ethylcyclopentadiene + Fe$_3$(CO)$_{12}$, MnSO$_4$	N,N-dimethylformamide*	0.1	225	Ethylcyclopentadienylmanganese tricarbonyl	—	[115]
	Indene + Mn(CH$_3$COO)$_2$ + Co$_2$(CO)$_{12}$ + LiBr	Dicyclohexylamine*	0.1	200	Indylmanganese tricarbonyl	—	[115]
	Fluorene + MnCl$_2$ + Cr(CO)$_6$	Hexamethylphosphoramide*	0.1	85	Fluorenylmanganese tricarbonyl	Good	[115]
	Phenylcyclopentadiene + MnI$_2$ + Mn$_2$(CO)$_{10}$	N,N-dimethylformamide*	0.1	150	Phenylcyclopentadienylmanganese tricarbonyl	—	[115]
Copper	Sodium methylcyclopentadienide + 0.8 M methylcyclopentadiene	Tetrahydrofuran :dimethylformamide = 1:4	0.005-0.05	Room	Manganese bis-methylcyclopentadienide	25	[115]
	Sodium methylcyclopentadienide	Dimethyleneglycol dimethyl ether	0.005-0.05	"	"	Good	[115]

TABLE 12 (Continued)

Cathode	Composition of electrolyzed solution		Current density, A/cm²	Temperature, °C	Synthesized organometallic compound	Yield, %	Reference
	organic compound	solvent					
	Sodium methylcyclopentadienide + 10 vol. % $B(C_2H_5)_3$	" "	0.005–0.05	"		Good	[115]
	"	10% Diethyleneglycol dimethyl ether+90% pyridine	0.005–0.05	"		"	[115]
	3.36 part sodium methyl-cyclopentadienide +14.26 part methyl-cyclopentadiene	9.5 part diethyleneglycol dimethyl ether	0.005–0.05	170	Methylcyclo-pentadienylmanganese tricarbonyl	–	[116]
	Lithium methylcyclo-pentadienide + methyl-cyclopentadiene	Tetrahydrofuran :diethyleneglycol dimethyl ether = 1:4*	0.005–0.05	180	Methylcyclo-pentadienylmanganese tricarbonyl	Good	[116]
Graphite	Potassium ethylcyclo-pentadienide	Tetrahydrofuran*	0.1	100	Ethylcyclopentadienylmanganese tricarbonyl	"	[116]
Copper	50% Sodium n-decylcy-clopentadienide + 50% potassium n-decylcy-clopentadienide	Diethyleneglycol butyl ether	0.005–0.05	200	Manganese n-decyclo-pentadienide	–	[116]
Manganese	Magnesium bis(indenyl)-magnesium + indene	Diethyleneglycol dimethyl ether	0.005–0.05	200	bis(Indenyl)manganese	–	[116]
Aluminum	Fluorenylaluminum	Diethyleneglycol diethyl ether	0.005–0.05	–	bis(Fluorenyl)manganese	–	[116]

TABLE 12 (Continued)

| Cathode | Composition of electrolyzed solution | | Current density, A/cm² | Temperature, °C | Synthesized organometallic compound | Yield, % | Reference |
	organic compound	solvent					
Manganese	Copper phenylcyclo-pentadienide + $Al(C_2H_5)_3$ + $Fe(CO)_5$	10 vol. % benzene	0.001–0.03	70	Phenylcyclo-pentadienylmanganese tricarbonyl	Good	[116]
Platinum	bis(1,2-Diethylcyclo-pentadienyl)thallium + $Ni(CO)_4$	n-Butyrolactone	0.001–0.03	60	Diethylcyclopenta-dienylmanganese tricarbonyl	Good	
Iron	bis(Butylcyclopenta-dienyl)iron	N,N-Dimethyl-formamide*	0,003	165	Butylcyclopenta-dienylmanganese tricarbonyl	—	
Copper	Sodium methylcyclo-pentadienide: Na $Al(C_2H_5)_4$ = 2:1†	—	—	150	bis(Methylcyclo-pentadienyl)-manganese	Good	

* The experiments were performed under pressure in an atmosphere of carbon monoxide.
† The second component may be: $KAl(C_2H_5)_4$, $Al(C_2H_5)_3 \cdot 2NaF$, or $NaAl(C_{10}H_{21})_4$.

TABLE 13. Decomposition Potential of Grignard
Reagents Preparation of Manganese Complexes
(Manganese anode)

R in RMgBr compound	Decomposition potential, V	R in RMgBr compound	Decomposition potential, V
C_6H_5	2.17	$(CH_3)_2CH$	1.07
CH_3	1.94	$(CH_3)_3C$	0.07
C_3H_7	1.42	$CH_2{=}CH{-}CH_2$	0.86
C_4H_9	1.32		

(Table 12). For example, methylcyclopentadienyltricarbonylman-
ganese was prepared by electrolysis of a solution of methylcyclo-
pentadiene in dimethylformamide with a manganese cathode in the
presence of manganese chloride and iron pentacarbonyl.

A new process has recently been described for the preparation of organotin compounds [117] by electrolysis of alkyl halide
solutions in butyl acrylate (or other ester) with added bromine and
stannous bromide. The anode was tin and the cathode magnesium.
The authors noticed that the magnesium cathode disintegrated during
electrolysis, and the process probably takes place through intermediate formation of an organomagnesium compound.

Mechanism of Anodic Process Leading to
Formation of Metal − Carbon Bond

Practically all the investigators who have studied the anodic
dissolution of metals during electrolysis of organometallic complexes have expressed the view that the metal−carbon bond is
formed by the action of organic radicals, formed during electrolysis, on the electrode material. For example, in the electrolysis
of the complex between NaC_2H_5 and $Zn(C_2H_5)_2$, $Zn(C_2H_5)_3^-$ ions are
discharged at the anode:

$$Zn(C_2H_5)_3^- \xrightarrow{-e} Zn(C_2H_5)_2 + {^{\cdot}}C_2H_5, \qquad (28)$$

with the complex between $Al(C_2H_5)_3$ and KF, the $AlF(C_2H_5)_3^-$ anions
are discharged:

$$AlF(C_2H_5)_3^- \xrightarrow{-e} AlF(C_2H_5)_2 + {^{\cdot}}C_2H_5 \qquad (29)$$

and so forth.

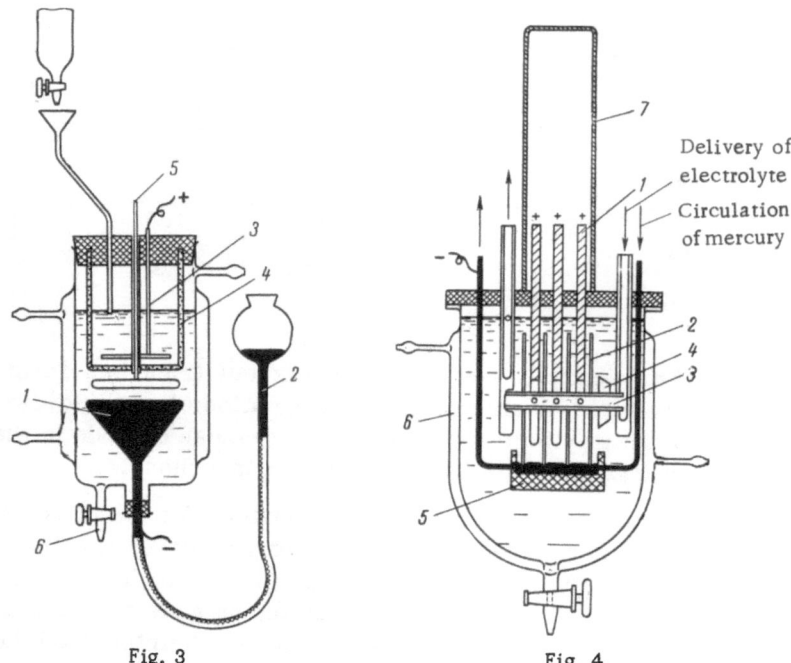

Fig. 3   Fig. 4

Fig. 3. Electrolyzer with mercury cathode for preparation of diisopropylmercury.
1) Mercury cathode; 2) levelling mercury container; 3) anode; 4) diaphragm; 5)
stirrer; 6) tap for removal of diisopropylmercury.

Fig. 4. Electrolyzer with rotating disk cathode. 1) Replaceable lead anodes; 2)
rotating cathode; 3) axis of rotating cathode; 4) pinion of cathode drive mechanism; 5) mercury container; 6) body of electrolyzer, 7) airtight cover.

The $\cdot C_2H_5$ radicals formed react with the anode material,
e.g., lead, as follows:

$$4\cdot C_2H_5 + Pb \rightarrow Pb\,(C_2H_5)_4. \tag{30}$$

The work of Evans on the electrolysis of ethereal solutions of
Grignard reagents [118] in general confirms the view of the anodic
process as a free radical process. Thus, he showed that electrolysis of Grignard reagents with an insoluble anode gives products,
the formation of which can be explained by disproportionation, dimerization of the initially formed radicals, and reaction with the

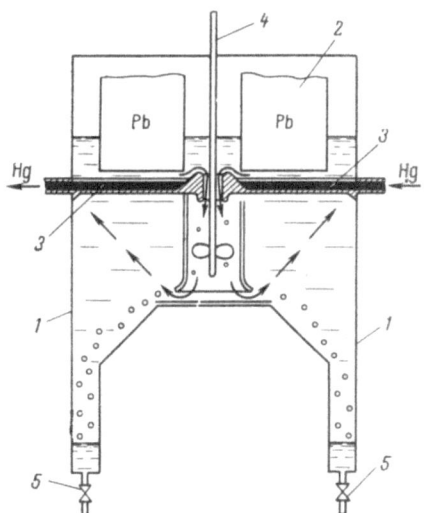

Fig. 5. Electrolyzer for preparation of tetra-ethyllead. 1) Body of electrolyzer 2) lead anode; 3) mercury cathode; 4) stirrer; 5) taps.

solvent. While studying the dependence of current on the applied potential, Evans found that at a specific potential (which he called the "decomposition potential") there is a sharp increase in current. The dependence of this potential for organo-magnesium compounds on the nature of the organic radical (Table 13) is consistent with the idea that the particle involved in the electrode process contains an organic free radical in its composition.

However, not all the phenomena can be explained in terms of the free-radical theory. Thus, there is now no doubt that the formation of hydrocarbons during electrolysis of carboxylic acids (the Kolbe synthesis) takes place by a radical mechanism [119]. In spite of this no case has been mentioned where metal alkyls have been formed during electrolysis of carboxylic acids.

Comparatively recently it was found that the composition of anodic gases formed during electrolysis of organic complexes is practically independent of current density but depends substantially on the anode material [120]. Thus, at a copper anode the hydrogen content of the anodic gas amounts to 93.5%. The greatest yield of ethane (68.7%) is observed at a palladium anode, while dimerization of radicals to form butane takes place most effectively (with 39.5% yield) at an iron anode. Just as in the electrolysis of $NaGe(C_6H_5)_3$ in liquid ammonia the composition of the electrolysis products depends substantially on the anode material [113].

These facts show that the free radicals formed at the anode do not exist in the form of kinetically independent particles but are in some way combined with the electrode material.

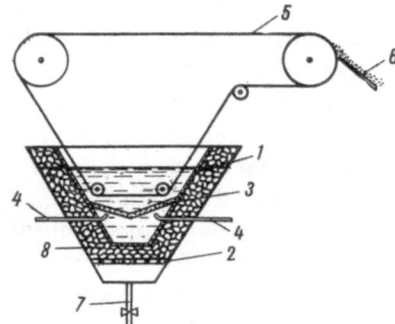

Fig. 6. Conical electrolyzer for preparation of tetraethyllead. 1) Porous screen;
2) perforated plate; 3) diaphragms; 4)
tubes for removal of gaseous products;
5) strip cathode; 6) knife for removal of
aluminum deposit; 7) emptying tube; 8)
anode compartment, filled with lead
granules.

3. ELECTROLYZERS FOR ELECTROCHEMICAL PREPARATION OF ORGANOMETALLIC COMPOUNDS

A special feature of the preparation of organometallic compounds by the electrochemical method is the fact that the electrode
dissolves during the reaction. Only comparatively short experiments can therefore be carried out on the laboratory apparatus
usually employed for electrosynthesis. Such apparatus cannot be
used for operations on a more-or-less large scale, and we will not
therefore dwell on it. We will only consider designs which permit
continuous processes.

The problem of continuous production of organometallic compounds is most easily solved by using liquid electrodes. Figure 3
shows an electrolyzer for the continuous production of diisopropylmercury. The electrolyser is provided with a mercury cathode.
The diisopropylmercury collects at the bottom of the electrolyzer
and is removed through a tube. The mercury level is kept constant
by moving the container 2.

A large-scale construction was undertaken in the development of a suitable electrolyzer for the production of tetraethyllead.
Figure 4 shows diagrammatically one model for such an electrolyzer, provided with replaceable lead anodes and rotating amalga-
mated copper disk cathodes [57]. It was constructed and tested for
a loading of 200 A. It is, however, very complicated.

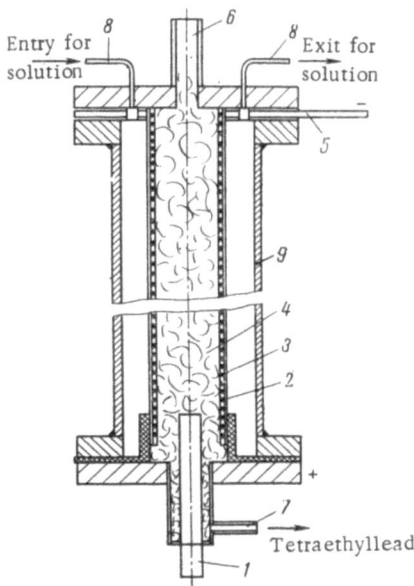

Entry for
solution

Exit for
solution

Tetraethyllead

Fig. 7. Electrolyzer of column type for preparation of tetraethyllead. 1) Graphite anode; 2) cathode; 3) screen; 4) anode compartment, filled with lead granules; 5) cathode strip; 6) pipe for filling with granulated lead; 7) removal of layer enriched with tetraethyllead; 8) tube for circulation of electrolyte; 9) body of electrolyzer.

Figure 5 shows an electrolyser with a mercury cathode, calculated for loadings up to 500 A [121]. The electrolyte is delivered to this electrolyzer continuously during the process, and a stirrer is used to keep the electrolyte continuously moving in the direction indicated by the arrow. The anodes are massive lead cylinders, and the cathode is a flat steel disk with a recess, in which mercury continuously circulates. The tetraethyllead is drawn by the stream of electrolyte into the central aperture by the stirrer and collects in the lower part of the vessel, from which it is periodically run off through the pipes at 5. By circulating the electrolyte it is possible to reduce the distance between the anode and cathode to a few millimeters without substantial losses of tetraethyllead through reaction with the amalgam on the cathode. With this type of electrolyzer the electrolysis can be carried out at temperatures in the region of 100°C, while the temperature in the lower part of the apparatus does not exceed 70°C; this reduces possible secondary chemical reactions to a minimum.

While considering electrolyzers for preparation of tetraethyllead the process in which use is made of liquid anode and

and liquid cathode should not be omitted [97]. In design the electrolyzer for this method differs little from the usual electrolyzer for the production of chlorine by the mercury method. Over the bottom of the electrolyzer passes lead amalgam, which becomes impoverished in lead after it has passed through the electrolyzer. The slightly inclined cathode is placed above the anode. The electrolyte contains a considerable amount of tetraethyllead, due to which its specific gravity is high; sodium floats in it and rises upwards along the inclined cathode.

For the electrolyzers described above to operate normally it is essential that the organometallic compounds formed should be liquid and should not dissolve in the electrolyte. On the other hand, if the organometallic compound is solid, it is more expedient to use electrolytes which dissolve it.

The problem of realizing a continuous process when working with solid electrodes is considerably more difficult to solve. It is perfectly clear that periodical replacement of electrodes is not suitable for an electrolyzer of comparatively large capacity.

In our opinion a very interesting solution to the problem is the use of a granulated anode. One possible design for such an electrolyzer, intended for the production of tetraethyllead, is shown in Fig. 6 [122]. The anode compartment is separated from the main bulk of the solution by a porous screen 1, made from an insulating material. A perforated plate is placed in the lower part of the electrolyzer. The space between the walls of the electrolyzer and the screen 1 is filled with lead granules, which serve as anode. A porous diaphragm 3 is located at the centre of the electrolyzer. Under the diaphragm there are tubes 4 for removal of the small amount of gases formed at the anode by side processes. The cathode is made in the form of a flexible continuously moving steel strip 5. The aluminum which is deposited on it is removed by the scraper 6. The bath is filled with an electrolyte in the form of a triethylaluminum complex. The tetraethyllead formed during electrolysis accumulates at the bottom of the electrolyzer and is periodically run off through the pipe 7.

A fairly original solution to the problem has been proposed in a patent [123], where an electrolyzer of column type is described (Fig. 7). This apparatus was successfully used for the electrochemical synthesis of tetraethyllead by electrolysis of a Grignard reagent. The anode is a graphite rod 1, surrounded by a porous screen of insulating material such as Teflon. This porous cylinder is surrounded by a metal gauze, which serves as cathode. The whole of the inner compartment 4, formed by the porous cylinder, is filled with lead balls through the pipe 6. Owing to the pressure created by the weight of the packing satisfactory contact is maintained between the individual balls, so that the whole load behaves as anode. The distance between the anode and cathode is determined by the thickness of the porous screen. This is a great advantage of this design, since it allows the distance between the electrodes to be reduced to a minimum, which is particularly important when working in nonaqueous solutions possessing low electrical conductivity. The tetraethyllead flows downwards and is removed through the pipe 7.

4. CONCLUSIONS

The information in this review shows that both anodic and cathodic processes can be used for synthesis of organometallic compounds. The mechanism of these processes has been far from adequately elucidated and clearly cannot be encompassed in a single common scheme. The metal — carbon bond formation mechanisms which have so far been proposed have only been verified in certain specific cases.

The overwhelming majority of electrochemical syntheses which have been investigated propose the use of alkyl halides as starting materials. If methods for production of organometallic compounds from alkyl halides are considered preference must be given to anodic processes (through Grignard reagents), where the greatest yields of the desired products are obtained. The anodic method for production of tetraethyllead has already found industrial use. However, in the anodic processes one metal in an organometallic compound is replaced by another, and the first stage of such a process is preparation of the more accessible Grignard type of compound or triethylaluminum by chemical means.

Reactions producing organometallic compounds from unsaturated substances and from ketones can only be realized at the cathode. Cathodic processes differ favorably from anodic processes in that it is possible to carry out a direct synthesis starting from metal and organic compound. Cathodic processes have not yet found applications for preparative purposes, but development of syntheses on this basis is more promising, since the organometallic compounds can be obtained without the intermediate stages.

LITERATURE CITED

1. Yu. Rokhov, D. Herd, and R. Lewis, Chemistry of Organometallic Compounds [Russian translation], IL, Moscow (1963).
2. Chem. Eng. News, 41(47):27 (1963).
3. Synthetic Organicals U. S. Production Sales, T. C. Publ. (1962), p. 114.
4. Chem. Week, 94(4):42 (1964).
5. Chem. Week, 93(8):62 (1963).
6. J. Tafel, Ber., 39:3626 (1906).
7. J. Haggarty, Trans. Electrochem. Soc., 56:421 (1929).
8. J. Tafel, Ber., 44:323 (1911).
9. G. Renger, Ber., 44:337 (1911).
10. T. Arai, Bull. Chem. Soc. Japan, 32:184 (1959).
11. C. Schall and W. Kirst, Z. Electrochem., 29:537 (1923).
12. T. Sekine, A. Iamura, and K. Sugino, Paper presented at the meeting of the Electrochemical Society (Sept. 30 to Oct. 3), New York (1963).
13. E. A. Efimov and I. G. Erusalimchik, Zh. Fiz. Khim., 38(12):2868 (1965).
14. V. G. Khomyakov and A. P. Tomilov, Zh. Prikl. Khim., 36:378 (1963).
15. H. Low, J. Chem. Soc., 101:1016, 1544 (1912); Proc. Chem. Soc., 28:98, 162.
16. T. Arai and T. Ogura, Bull. Chem. Soc. Japan, 33:1018 (1966).
17. W. Schepss, Ber., 46:2564 (1913).
18. J. Tafel, Ber., 42:3146 (1909).
19. A. P. Tomilov, L. V. Kaabak, and S. L. Varshavskii, Zh. Vses. Khim. Obshch., 8:703 (1963).
20. L. V. Kaabak, A. P. Tomilov, and S. L. Varshavskii, Zh. Vses. Khim. Obshch., 9:700 (1964).
21. A. P. Tomilov and L. V. Kaabak, Zh. Prikl. Khim., 32:2600 (1959).
22. L. V. Kaabak and A. P. Tomilov, Zh. Obshch. Khim., 33:2808 (1963).
23. A. P. Tomilov, L. V. Kaabak, and I. N. Brago, Zh. Vses. Khim. Obshch., 12:472 (1967).
24. L. Holleck and D. Marquarding, Naturwiss., 49:468 (1962).
25. A. I. Lebedeva, Zh. Obshch. Khim., 18:1161 (1948).
26. U. S. Patent No. 1539297 (1925); Chem. Abs., 19:2210 (1925).
27. U. S. Patent No. 1567159 (1925); Chem. Abs., 20:607 (1926).
28. British Patent No. 949925 (1964); Chem. Abs., 61:3935 (1965).
29. R. Benesch and R. Benesch, J. Am. Chem. Soc., 73:3391 (1951).
30. N. Hush, J. Electroanalyt. Chem., 6:34 (1963).
31. A. Kirrmann and M. Kleine-Peter, Bull. Soc. Chim. France, 894 (1957).

32. A. P. Tomilov and Yu. D. Smirnov, Zh. Obshch. Khim., 35:391 (1965).
33. L. V. Kaabak, M. I. Kabachnik, A. P. Tomilov, and S. L. Varshavskii, 36: 2060 (1966).
34. P. Lanza, L. Griggio, and G. Semerano, Ricere Sci., 26:230 (1956); 27:110 (1957).
35. O. A. Reutov, Theoretical Problems in Organic Chemistry [in Russian], Izd. MGU, Moscow (1956), p. 264.
36. W. A. Waters, Chemistry of Free Radicals [Russian translation], IL, Moscow (1948), p. 460.
37. C. Walling, Free Radicals in Solution [Russian translation], IL, Moscow (1960).
38. A. S. Voitkevich, Thesis [in Russian], All-Union Scientific-Research Institute of Synthetic and Natural Aromatic Materials, Moscow (1949).
39. L. G. Feoktistov, A. P. Tomilov, Yu. D. Smirnov, and M. M. Gol'din, Élektrokhimiya, 1:887 (1965).
40. A. P. Tomilov, Zh. Obshch. Khim., 28:214 (1968).
41. R. A. Bullerwell, J. Polarograph. Soc., 9:7 (1963).
42. A. P. Tomilov and Yu. D. Smirnov, Zh. Vses. Khim. Obshch., 10:101 (1965).
43. I. G. Sevast'yanova and A. P. Tomilov, Zh. Obshch. Khim., 33:2815 (1963).
44. L. G. Feoktistov, A. P. Tomilov, and I. G. Sevast'yanova, Élektrokhimiya, 1:1300 (1965).
45. M. Baizer, J. Electrochem. Soc., 111:215 (1964).
46. F. Penth, W. Haken, and E. Rabinowich, Ber., 57:1891 (1924).
47. E. Becqueral, Compt. Rend., 56:237 (1963).
48. A. Cossa, Ber., 1:117 (1868).
49. J. Mellor, Inorganic and Theoretical Chemistry, London—New York—Toronto, 11:36 (1954).
50. L. Gershbein and C. Hurd, J. Am. Chem. Soc., 69:241 (1947).
51. H. Salzberg and F. Mies, J. Electrochem. Soc., 105:64 (1958).
52. R. Dotzer, Chem. Ind. Tech., 36:616 (1964).
53. J. Gillet, J. Electrochem. Soc., 108:71 (1961).
54. F. Hein, Z. Anorg. Allg. Chem., 141:161 (1924).
55. H. French and M. Drane, J. Am. Chem. Soc., 52:4904 (1930).
56. W. Evans and F. Lee, J. Am. Chem. Soc., 56:654 (1934).
57. K. Ziegler and H. Lehmkuhl, Chem. Ing. Tech., 35:325 (1963).
58. K. Ziegler and D. Stendel, Ann. Chem., 652:1 (1962).
59. W. Evans and R. Pearson, J. Am. Chem. Soc., 64:2865 (1942).
60. U. S. Patent No. 3100181 (1963).
61. U. S. Patent No. 3079311 (1963).
62. U. S. Patent No. 3155602 (1967).
63. U. S. Patent No. 3234112 (1966).
64. U. S. Patent No. 3256161 (1966).
65. U. S. Patent No. 3007857 (1961).
66. British Patent No. 839172 (1958).
67. U. S. Patent No. 3007858 (1961).

68. British Patent No. 882005 (1960).
69. U.S. Patent No. 3118825 (1964).
70. U.S. Patent No. 3180810 (1965).
71. K. Ziegler, H. G. Gellert, K. Zosel, and W. Lehmkuhl, Angew. Chem., 67: 42 (1955).
72. U.S. Patent No. 2787626 (1957).
73. K. Ziegler and H. Lehmkuhl, Z. Anorg. Allg. Chem., 283:414 (1956).
74. Belgian Patent No. 543128 (1955).
75. British Patent No. 814609 (1959).
76. U.S. Patent No. 2985568 (1961).
77. British Patent No. 848364 (1960).
78. British Patent No. 797090 (1958).
79. German Patent No. 1150078 (1963).
80. British Patent No. 864393 (1961).
81. U.S. Patent No. 3028319 (1962).
82. British Patent No. 3028318 (1962).
83. U.S. Patent No. 3254009 (1966).
84. German Patent No. 1153754 (1963).
85. U.S. Patent No. 3028320 (1962).
86. R. Flannery, I. Thomas, and D. Trivich, J. Electrochem. Soc., 110:1054 (1962).
87. K. Ziegler, Brennstoff. Chem., 40:209 (1950).
88. German Patent No. 1127900 (1962).
89. British Patent No. 842090 (1960).
90. German Patent No. 1181200 (1965).
91. K. Ziegler, Angew. Chem., 71:628 (1959).
92. U.S. Patent No. 3088885 (1963).
93. U.S. Patent No. 3028322 (1962).
94. U.S. Patent No. 2944948 (1960).
95. U.S. Patent No. 3069334 (1963).
96. German Patent No. 1166196 (1964).
97. U.S. Patent No. 3159557 (1964).
98. Belgian Patent No. 617628 (1962).
99. U.S. Patent No. 3254008 (1966).
100. K. Ziegler, Angew. Chem., 72:565 (1960).
101. Belgian Patent No. 590573 (1963).
102. German Patent No. 1153371 (1960).
103. K. Ziegler, Organoaluminium Compounds, New York (1960), p. 251.
104. British Patent No. 864394 (1961).
105. German Patent No. 1120448 (1961).
106. U.S. Patent No. 3088957 (1963).
107. German Patent No. 1134672 (1962).
108. French Patent No. 1312426 (1963).
109. British Patent No. 923652 (1960).
110. U.S. Patent No. 3028352 (1962).

111. British Patent No. 895457 (1963).
112. U.S. Patent No. 3028323 (1962).
113. L. Foster and G. Hooper, J. Am. Chem. Soc., 57:76 (1935).
114. German Patent No. 1046617 (1958).
115. U.S. Patent No. 2915440 (1958).
116. U.S. Patent No. 2960450 (1960).
117. K. N. Korotaevskii, E. N. Lysenko, S. Z. Smolyan, L. N. Monastyrskii, and L. V. Armenskaya, Zh. Obshch. Khim., 1:167 (1966).
118. W. Evans, F. Lee, and C. Lee, J. Am. Chem. Soc., 57:489 (1935).
119. G. É. Svadkovskaya and S. A. Voitkevich, Usp. Khim., 29:364 (1960).
120. K. Ziegler, H. Lehmkuhl, and E. Lindner, Ber., 92:2320 (1959).
121. U.S. Patent No. 3164538 (1964).
122. U.S. Patent No. 2985568 (1961).
123. U.S. Patent No. 3180810 (1965).

Electrolysis of Organic Compounds with Use of Ion-Exchange Membranes

M. Ya. Fioshin

The effectiveness of many electrochemical processes is determined by the presence of a porous partition or diaphragm, which separates the electrolyzer into cathodic and anodic compartments. In processes of electrosynthesis of organic compounds the diaphragm has usually the role of separating the electrolysis products so as to prevent reverse oxidation or reduction and also to prevent mutual contamination of the products.

The porous diaphragm used to separate the liquid electrolysis products must consequently have high resistance to diffusion. At the same time the electrical resistance of a diaphragm, characterized by the ease with which the ions responsible for the electrical conductivity of the solution pass through it, must be as low as possible.

Many years experience in the use of industrial electrolyzers for various purposes has made it possible to work out fairly reliable recommendations for the selection of the most effective diaphragms for each specific case. Porous diaphragms do not, however, possess the ability to prevent reverse diffusion of ions or the penetration of the dissolved original or final organic compound molecules through the pores. In other words a porous diaphragm does not possess sufficient selectivity with respect to the ions and molecules present in the solution.

The imperfect nature of porous diaphragms has led to increased interest in ion-exchange membranes, which possess highly selective permeability [1-6]. This arises from the presence in the membrane phase of charged groups, which are loosely bound to fixed groups of opposite charge and are capable of exchanging with ions of like charge in the solution. Only ions which possess a

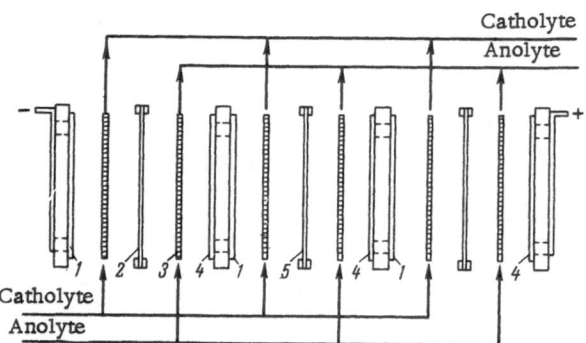

Fig. 1. Diagram of electrolyzer for hydrodimerization of nitriles. 1) Cathode; 2 and 5) membranes; 3) anode compartment; 4) anode.

charge opposite to that of the fixed group can pass through the membrane, i.e., the membranes are selectively permeable for oppositely charged ions. Consequently when the membrane is in an electric field the current will be carried through it exclusively by oppositely charged ions. The higher the concentration of the fixed ions in the membrane, the higher its exchange capacity and selectivity. Depending on the charge of the loosely bound ions the membrane can be cation-exchanging or anion-exchanging, i.e., permeable for cations or anions respectively.

There are a number of requirements for ion-exchange membranes [4]:

1) High selectivity, characterized by minimum permeability for ions with charge of the same sign as that of the fixed ion;

2) high electrical conductivity;

3) low free diffusion rate in electrolyte;

4) low osmotic permeability;

5) sufficient chemical resistance;

6) mechanical strength and dimensional stability.

Ion-exchange membranes have been used successfully in various branches of technology and in a number of cases have replaced porous diaphragms. They have found greatest use in electrochemical desalinization of water [2, 4]. The use of ion-exchange membranes or diaphragms in electrochemical synthesis or organic compounds is not yet very widespread. They have been very successfully used for more than 15 years in electrochemical demineralization processes and in the separation of organic compounds. These two fields of use will be examined below.

1. USE OF ION-EXCHANGE MEMBRANES IN ELECTROCHEMICAL SYNTHESIS OF ORGANIC COMPOUNDS

One of the reasons for the limited use of ion-exchange membranes in electrochemical synthesis of organic compounds is evidently the insufficient investigation of their behavior in the nonaqueous solvents which in many cases are used as the medium for these processes. There is little information, for example, on the swelling of ion-exchange resins in organic media [3], and the kinetics of ion exchange has been little investigated. (It is known, however, that the rate of ion exchange is considerably lower than in aqueous solutions.) It is clearly essential to take account of the change in chemical stability of the membrane in the nonaqueous solvents, particularly at the elevated temperatures which in many cases are most favorable for electrochemical synthesis. In spite of the fact that many difficulties involved in the use of ion-exchange membranes in electrosynthesis of organic compounds have not yet been overcome, a number of examples can be named where these diaphragms have been successfully used.

Recently great attention has been paid to the electrochemical hydrodimerization of nitriles, by means of which it is possible to obtain at the cathode a large number of bifunctional compounds with high, and in some cases practically quantitative, current yields [7-13]. Such compounds are of great interest in certain branches of polymer chemistry. In general form the electrochemical hydrodimerization reaction can be represented by the following equation:

$$2 \underset{|}{\overset{|}{C}}{=}\underset{|}{\overset{|}{C}}{-}CN + 2e + 2H_2O \rightarrow CN{-}\underset{H}{\overset{|}{C}}{-}\underset{|}{\overset{|}{C}}{-}\underset{H}{\overset{|}{C}}{-}\underset{|}{\overset{|}{C}}{-}CN + 2OH^-. \tag{1}$$

Greatest attention has been paid to the development of an effective method for electrochemical hydrodimerization of acrylonitrile into adiponitrile [14-18], which is a valuable intermediate in

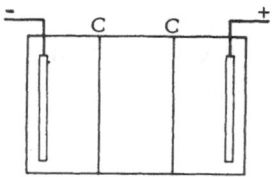

Fig. 2. Diagram of three-compartment electrodialyzer for production of organic acids from their salts. C = cation-exchange membrane.

the production of polyamide resins. A special feature of the electrochemical synthesis of adiponitrile and other dinitrides is their behavior in aqueous solutions of the hydrotropic salts which serve as electrolytes.

McKee and Brockman [19] showed that salts of p-toluenesulfonic acid increase the solubility of an organic compound in water. A positive feature of hydrotropic salts is their high solubility in water. By the use of acrylonitrile solutions in water containing the tetraethylammonium salt of p-toluenesulfonic acid, $p-CH_3C_6H_4SO_3N(C_2H_5)_4$, the firm Monsanto (U. S. A.) realized electrochemical hydrodimerization of acrylonitrile on an industrial scale with an adiponitrile yield amounting to 99% [16, 20]. The high yield is possible on account of the use of the p-toluenesulfonate salt, which secures high solubility of the acrylonitrile. (With an acrylonitrile content of less than 10% in the solution the reaction leads to the formation of large quantities of propionitrile as side product.)

A cation-exchange membrane of sulfonated polystyrene, with a working area of 0.93 m², has been used to prevent oxidation of acrylonitrile and p-toluenesulfonate anion at the anode in an industrial pressure-filter bipolar electrolyzer consisting of 24 cells and calculated for a loading of 2000 A. The membrane was reinforced with glass fiber to give it sufficient mechanical strength, which is particularly necessary with such large dimensions. The positioning of this membrane in the electrolyzer is shown in Fig. 1. In this process the cation-exchange membrane only admits hydrogen ions into the cathode compartment from the anode compartment, and they are converted into water by combining with hydroxyl ions formed by Reaction (1) in the cathode compartment. The p-toluenesulfonate ions, $CH_3C_6H_4SO_3^-$, and acrylonitrile molecules are unable to enter the anode compartment, which is filled with sulfuric acid solution, or take part in the electrochemical reaction at the platinum –titanium anode.

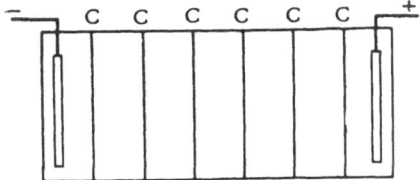

Fig. 3. Diagram of seven-compartment electrodialyzer for production of organic acids from their salts. C = cation-exchange membrane.

Consequently oxygen must be evolved at the anode, and the overall reaction is described by the following equation:

$$2CH_2=CHCN + H_2O$$

$$\rightarrow CN(CH_2)_4CN + \frac{1}{2}O_2. \quad (2)$$

Baizer, the author of this method for the production of adiponitrile, proposed that electrochemical dimerization in solutions of tetraethylammonium salts of sulfonic acids with cation-exchange membranes as diaphragm should also be employed for production of other bifunctional compounds. For example, by hydro-dimerization of 1-cyano-1,3-butadiene at the cathode it is possible to obtain 1,8-dicyano-2,6-octadiene with high yield (up to 88%), and by catalytic hydrogenation this can be converted to the dinitrile of sebacic acid [12].

Another example of the use of ion-exchange membranes in electrosynthesis is cathodic dehalogenation. Dichloroacetic acid has been obtained with almost quantitative current yields from trichloroacetic acid by elimination of chlorine at controlled potential [21]. The overall reaction which takes place in the electrolyzer can be written as follows:

$$CCl_3COOH + H_2O \rightarrow \frac{1}{2}O_2 + HCl + CHCl_2COOH. \quad (3)$$

Electrolysis at controlled potential possesses definite advantages over chemical methods, and over electrochemical dehalogenation without potential control owing to the high selectivity, which leads to a good yield and eliminates the problem of separating a mixture of reaction products. A second reason for the production of a high yield of dichloroacetic acid is the use of a cation-exchange diaphragm (Amberplex-C1) of the sulfonated polystyrene type. This diaphragm prevents trichloroacetate and dichloroacetate ions from entering the anode compartment and prevents losses due to their participation in electrochemical reactions at the platinum anode [22].

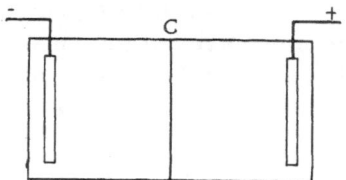

Fig. 4. Diagram of two-compartment electrodialyzer for production of sebacic acid from sodium sebacate. C = cation-exchange membrane.

Certain organometallic compounds which find use as antiknock additives to motor fuels, catalysts, insecticides, and fungicides can be obtained by electrolysis of a solution of alkylating agent (ethyl bromide, diethyl sulfate, ethyl iodide, ethyl chloride, ethyl acetate, etc.) in an organic solvent (acetonitrile, dimethylformamide, methylene chloride, etc.) containing an electrolyte such as tetraethylammonium, lithium, or calcium bromides, sodium or potassium iodides, lithium perchlorate, etc., [23]. The nature of the organometallic compound and its yield depend on the nature of the cathode material, the electrolyte, and the solvent. (The yield varies between 10 and 100%.) Owing to participation in electron transfer the ions which cause the electrical conductivity of the solutions employed will migrate into the anode compartment, which is filled with an aqueous solution of sodium carbonate. Migration of Cl^-, Br^-, CNS^-, and ClO_4^- anions from the cathode compartment to the anode compartment gives rise to undesirable changes in the composition of the catholyte and can lead to losses due to participation of these ions in electrochemical reactions at the platinum anode.

Only a few examples of the use of ion-exchange membranes in electrochemical synthesis of organic compounds have been discussed above. Such diaphragms will clearly be used more widely in the future [24]. The use of a cation-exchange membrane is the most convenient method for preventing transfer of electrolyte and other components of the solution from one compartment of the electrolyzer to another.

2. USE OF ION-EXCHANGE MEMBRANES IN ELECTRODIALYSIS PROCESSES INVOLVING ORGANIC COMPOUNDS

Ion-exchange membranes have become considerably more widespread in this field than in electrochemical synthesis. They are employed for demineralization of solutions of organic compounds, i.e., for removal of mineral acids or salts. Considerable interest is attached to the separation of organic compounds by means of electrodialysis, e.g., fractionation of amino acids on the basis of the difference in their isoelectric points and, consequently, their ionic composition.

During electrolysis it is necessary to strive towards conditions where the electrochemical reactions take place at the electrode surfaces without participation of the organic compounds. In the general case hydrogen must be liberated at the cathode and oxygen at the anode, i.e., electrolysis of water must occur. In some cases, when organic hydrochlorides are subjected to electrodialysis, chlorine may be obtained simultaneously at the anode.

We will consider some regions where ion-exchange membranes are used in electrodialysis of organic compounds.

Preparation of Free Organic Acids from Solutions of Their Salts

A free organic acid can be obtained from its salts by treatment with strong mineral acid. In this case, however, mineral salts (usually sulfates or chlorides) which are difficult to utilize are formed as a by-product, and the organic acid of an adequate degree of purity cannot be obtained without special additional operations.

It is also possible to realize electrodialysis with an inert porous diaphragm, which (as mentioned above) is, however, not capable of preventing reverse diffusion and, consequently, mixing of the organic acid and strong base.

Several variants of electrodialysis with ion-exchange membranes have been proposed for the production of free organic acids from solutions of their salts. For example, Bodamer [25] suggested that free acetic acid could be obtained from sodium acetate by electrodialysis in three-compartment (Fig. 2) and seven-compartment (Fig. 3) cells divided by heterogeneous cation-exchange membranes made from a sulfonated copolymer of styrene and divinylbenzene (Amberlite IR-120).

TABLE 1. Electrodialysis of Sodium Sebacate with
Cation-Exchange and Asbestos Diaphragms [26]:
The Anolyte is a 4.5% Solution of Sodium Sebacate,
the Catholyte is 0.1 N Caustic Soda Solution, and
the Anode is Platinum

Electrolysis conditions	Diaphragm	
	cation-exchange	asbestos
Current density, A/dm²	10	10
Length of experiment, min	58	124
Final pH value of anolyte	1.5	1.7
Current yield of sebacic acid, %	95	90
Electricity consumed per 1 kg sebacic acid, kWh	8.25	16
Energy yield, %	77.3	30.9

It should be noted that regeneration of acetic acid from sodium
acetate is of certain practical interest, since sodium acetate is
formed as a by-product in the manufacture of synthetic fibres. A
10% solution of sodium acetate is placed into the central compartments of the electrodialyzers. The anode compartment is filled
with 0.1 N sulfuric acid solution and the cathode compartment with
0.1 N caustic soda solution. The electrodes are made of platinum,
and the current density is 6.5 A/dm².

By electrodialysis with cation-exchange membranes it is
possible to achieve practically complete conversion of the acetate
into acetic acid and simultaneously to obtain alkali in the cathode
compartment; this can be reused for manufacturing purposes. The
electrochemical reaction products here will be exclusively hydrogen
at the cathode and oxygen at the anode, i.e., participation of the anion
in the electrochemical reactions is completely eliminated. If, however, an attempt is made to carry out the electrodialysis in a two-
compartment cell, with the acetate solution in the anode compartment, about 75% of the acetic acid is oxidized at the anode and clearly forms mainly ethane and carbon dioxide.

In some cases, however, it is possible to produce free acid
by electrodialysis in a two-compartment cell, divided by a cation-
exchange membrane (Fig. 4). One such example is the electrolysis

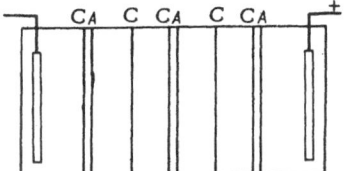

Fig. 5. Diagram of electrodialyzer with bipolar membranes. C = cation-exchange membrane; A = anion-exchange membrane.

of sodium sebacate solution [26]. The object of this process is to obtain sebacic acid, which is a valuable starting material for the production of polyesters, polyamides, lubricating oils, and perfumes. Sodium sebacate is formed as an intermediate product in the processing of castor oil, which is a starting material in one of the industrial methods for synthesis of sebacic acid [27], or in the production of sebacic acid by anodic condensation of adipic monoester [28]. In the last case the direct anodic condensation product is sebacic diester:

$$2CH_3OOC(CH_2)_4COOH - 2e \rightarrow CH_3OOC(CH_2)_8COOCH_3 + 2CO_2 + 2H^+. \qquad (4)$$

The diester is then saponified by boiling with an aqueous solution of caustic soda to form sodium sebacate:

$$CH_3OOC(CH_2)_4COOCH_3 + 2NaOH \rightarrow NaOOC(CH_2)_8COONa + 2CH_3OH. \qquad (5)$$

The sodium salt is neutralized by sulfuric acid:

$$NaOOC(CH_2)_8COONa + H_2SO_4 \rightarrow HOOC(CH_2)_8COOH + Na_2SO_4. \qquad (6)$$

Here 1 ton of sebacic acid gives about 1 ton sodium sulfate, and the essentially irreversible consumption of caustic soda and sulfuric acid is unavoidable.

The proposed method of electrodialysis provides for electrolysis of a comparatively dilute solution of sodium sebacate (3.7-4.5%), and this clearly prevents oxidation of the sebacate anions in spite of the fact that platinum is recommended as anode [26]. The cathode compartment of the electrodialyzer is filled with 1.07 N caustic soda solution. The diaphragm is a cation-exchange membrane made from sulfonated polystyrene and divinylbenzene, reinforced with polyethylene film. As the sodium ions migrate from the anode to the cathode compartment the free sebacic acid, being poorly soluble, precipitates in the form of a white deposit in the anode compartment and falls to the bottom of the electrolyzer, which prevents it from being oxidized at the anode.

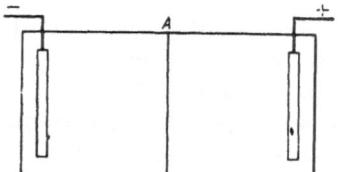

Fig. 6.   Diagram of electrodialyzer
for preparation of ethylenediamine.
A = anion -exchange membrane.

The effectiveness of electrodialysis with a cation-exchange
membrane can be judged by comparison with the results of experiments carried out with an asbestos diaphragm. Comparative data
are given in Table 1.

The advantages of electrodialysis with a cation-exchange
membrane are perfectly obvious, as seen from Table 1. The greater
duration of the process and the larger consumption of electricity
with the asbestos membrane arise from the impossibility of preventing reverse diffusion of alkali from the cathode compartment into
the anode compartment. It should also be noticed that for each
kilogramme sebacic acid electrolysis with cation-exchange
membrane gives 0.7 kg caustic soda (which can be recycled
and used for saponification) in the cathode compartment and the
corresponding quantities of hydrogen and oxygen at the electrodes.

For electrodialysis of salts of organic acids Oda and Murakosi
employed double "bipolar" membranes, manufactured by compressing cation-exchange and anion-exchange membranes made from IR-
120 and IRA-400 resins respectively [29]. The intermediate
compartments, formed by the double diaphragms, were in turn
partitioned off by cation-exchange membranes (Fig. 5). A 1 N sodium acetate solution was continuously delivered to the compartment
formed by the cation-exchange side of the double membrane and the
cation-exchange membrane (8 × 12 cm in size), and water was delivered to the adjacent compartment. The cathode and anode
compartments were filled with 0.5 N common salt solution. Electrolysis was performed with a current density of about 2 A/dm^2 at the
diaphragm. A 0.76 N solution of acetic acid, containing 0.17 g-
equiv/liter sodium acetate, was obtained in the compartment into
which the sodium acetate solution was delivered. A 0.35 N solution
of caustic soda, containing 0.06 g-equiv/liter sodium acetate (which
had clearly passed through the membrane by diffusion), was obtained
from the adjacent compartment. Hydrogen was evolved at the
cathode and chlorine at the anode.

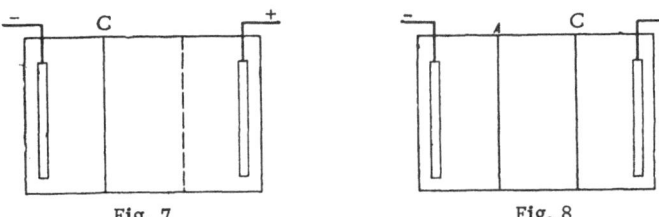

Fig. 7   Fig. 8

Fig. 7. Diagram of electrodialyzer for production of tetrabutylammonium hydroxide. C = cation-exchange membrane.

Fig. 8. Diagram of electrodialyzer for production of quaternary
ammonium bases and quaternary ammonium salts. C = cation-
exchange membrane; A = anion-exchange membrane.

Preparation of Amines from Solutions
of Their Salts

The greatest attention has been devoted to the use of ion-
exchange membranes to obtain ethylenediamine from solutions of
its chloride. Ethylenediamine is usually obtained by amination of
dichloroethane, and the efficacy of the whole technological process
therefore depends to a considerable degree on a method for conversion of the hydrochloride into the free base.

The chemical method, which comprises neutralization of the
solution with caustic soda, possesses the disadvantage that it
involves formation of sodium chloride as a by-product. The presence
of sodium chloride is undesirable since it involves considerable
difficulties in subsequent distillation to isolate the free base.

Electrodialysis was proposed by Bodamer [30] for the production of free ethylenediamine from an aqueous solution of its hydrochloride. A 40% solution of ethylenediamine hydrochloride was
placed in the cathode compartment which was separated from the
anode compartment by an anion-exchange membrane made from
Amberlite IRA-400 (70%) and polyethylene (30%). A diagram of the
electrodialyzer is shown in Fig. 6. The anolyte was a 0.1 N sulfuric acid solution, and the electrodes were platinum. The products
from the cathodic and anodic electrochemical reactions were hydrogen and oxygen respectively.

According to Bodamer, it is advisable to continue electrolysis until 85.8% of the hydrochloride has been converted to the free diamine. The amount of electrical energy consumed is 8.2 kWh/kg. If the anode compartment is filled with hydrochloric acid solution, chlorine can be obtained instead of oxygen at the anode together with the hydrogen evolved at the cathode. This process is undoubtedly of industrial interest since it does not involve use of caustic soda and permits the production of ethylene diamine with a high degree of purity. The chlorine formed at the anode is a valuable product and can be used for the chlorination of ethylene in the first stage of the process.

There is some information in the literature on the use of electrodialysis for production of other diamines, e.g., propylene-diamine [31], from solutions of their salts.

Preparation of Quaternary Ammonium Bases from Solutions of Their Salts

Electrodialysis will only give quaternary ammonium bases with a high degree of purity when ion-exchange membranes are used. In fact it has not been possible to obtain quaternary ammonium bases completely free from their salts in electrodialysis with porous ceramic diaphragms [32].

Dzbanovskii, Tsodikov, Borkhi, and Khleborodova [33] investigated the electrodialysis of tetrabutylammonium iodide in the electrolyzer, shown in Fig. 7. A characteristic feature of this three-compartment electrolyzer is the presence of a cation-exchange membrane which separates the cathode from the central compartment and a porous ceramic diaphragm which confines the anode compartment.

The heterogeneous cation-exchange membrane MK-40 was made by addition of polyethylene to the cation-exchange KU-2 resin. A saturated solution of tetrabutylammonium iodide was poured into the central compartment, and this was changed three times during electrodialysis as it became impoverished with $[N(C_4H_9)_4]^+$ ions migrating through the membrane; distilled water slightly acidified and sulfuric acid was initially poured into the cathode compartment.

TABLE 2. Losses of Amino Acids from Central Compartment of Electrodialysis Cell with Ion-Exchange Membranes

Electrolyte composition		Losses of amino acids, %, with isoelectric point in parentheses			
Catholyte	Anolyte	Aspartic (3.0)	Alanine (6.1)	Lysine (9.7)	Arginine (10.8)
$0,2N$ H$_2$SO$_4$	$0,02N$ NaOH	38.0	30.0	50.0	48.0
$0,2N$ Na$_2$SO$_4$	$0,2N$ Na$_2$SO$_4$	10.0	4.0	30.0	24.0
$0,2N$ NaOH	$0,02N$ H$_2$SO$_4$	1.0	0.0	5.0	14.0

The electrolysis was stopped after a 7.5% solution of tetrabutyl-ammonium hydroxide had been obtained in the cathode compartment. The yield of the compound amounted to 96-99% and the current yield to 15%. The current density did not exceed 3 A/dm^2, and the temperature of the solution amounted to 25-30°C. Hydrogen was formed at the steel cathode, and iodine was liberated at the graphite anode, which was immersed in 2% sulfuric acid solution. The authors showed that tetramethyl- and tetraethylammonium hydroxides, and also butylhexylamine hydroxide, can be obtained by a similar method.

Krokhv [34] suggested a rather different method of electrodialysis to produce quaternary ammonium bases. The electrolyzer was divided into three compartments by ion-exchange membranes. The anode compartment was separated from the central compartment by an MK-40 cation-exchange membrane, and the cathode compartment was separated by an MA-40 anion-exchange membrane (Fig. 8). A solution of a salt of the quaternary ammonium base was placed in the anode compartment, distilled water was placed in the central compartment, and alkali was placed in the cathode compartment. The base can be obtained as the result of migration of the quaternary base cation from the anolyte into the central compartment and migration of OH$^-$ anion from the catholyte. Tetramethyl-ammonium, tetrabutylammonium, and triethylbenzylammonium hydroxides were obtained in this way. If the cathode compartment is filled with a solution of a salt, the corresponding salt of the quaternary ammonium base can be obtained in the central compartment. In this way tetramethylammonium chloride was obtained from the iodide. It should, however, be mentioned that the advisability of

placing the salt of the ammonium base in the anode compartment is
doubtful owing to the risk of oxidation at the anode.

Demineralization and Separation of Amino Acids

Electrolysis of amino acid solutions is largely carried out to
purify them from mineral salts and acids and to isolate the individual amino acids from their mixtures. The need for separate
examination of electrolysis of amino acids with ion-exchange membranes arises from the fact that these compounds are amphoteric
in nature [35]. In acidic solution the amino acids exist in the form
of $H_3N^+ - R - COOH$ cations, and when a direct electric current is
passed through the solution they should migrate to the cathode. In
alkaline solution they are present in the form of $H_2N - R - COO^-$
anions and migrate to the anode. If the pH of the solution
corresponds to the isoelectric point of the given acid, it exists in
the form of a bipolar ion and should not migrate to either electrode.

a) Purification of Amino Acids from Mineral Salts. Naturally
most complete separation of inorganic anions from solutions of
amino acids can only be attained at a pH value of the solutions which
is numerically equal to the isoelectric point. Because of the considerable difference between the transport numbers of the inorganic
anions and the amino acid anions, effective demineralization can
also be secured at conditions differing somewhat from those mentioned above.

In work by Bobrovnik and Litvak it was shown that the migration rate of anions decreases in the following order: chloride
> oxalate > sulfate > citrate > lactate > tartrate > glutamate, foll-
owed by aspargate and betainate, i.e., the amino acids have the
lowest migration rate among all the anions examined [36]. Electrodialysis with ion-exchange membranes for removal of salts from
amino acids was first performed by Astrup and Stage [37]. Quantitatively the role of correspondence between pH and isoelectric
point in the migration of amino acids was evidently first evaluated
by Di Benedetto and Lightfoot [38], who showed that at pH values
close to the isoelectric point migration of glycine was not greater
than 6.1% and that migration of glycine through membranes amounted
to 15.7% at pH 9 and 51% at pH 11.3.

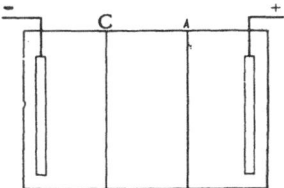

Fig. 9. Diagram of electrolyzer for removal of salts
from amino acids. C = ca-
tion-exchange membrane;
A = anion-exchange mem-
branes.

Yamabe, Seno, and Takai [39] investigated the permeability of a monocarboxylic amino acid (glutamic) and a diamino acid (lysine), solutions of which mixed with sodium chloride (0.1 N) were placed in the central cell of a five-compartment electrodialyzer. In the adjacent cells, which were separated from the central compartment by a cation-exchange membrane (on the cathodic side) and an anion-exchange membrane (on the anodic side), was placed 0.1 N sodium chloride solution. The changes in concentration of chloride and amino acid were determined as a function of the pH value in the central cell, which was varied by addition of alkali or hydrochloric acid.

It was found that appreciable penetration of the amino acids through the cation-exchange membrane takes place at pH values below the isoelectric point (pl), i.e., at pH < pl, and through the anion-exchange membrane at pH > pl. The smallest losses of amino acids from the central compartment take place at pH values equal to the isoelectric point, i.e., pH = pl. Separation of chloride ions, i.e., demineralization of amino acids, can also be achieved on this basis.

These results were further supported by investigation of three different amino acids with isoelectric points in various pH regions: glycine with pl = 6.1; glutamic acid with pl = 3.2; lysine with pl = 9.7 [35]. The conclusions reached are consequently general in nature.

These results were fully confirmed in [40], where electrodialysis of solutions of glycine (70 g/liter) and methionine (25 g/liter) in the presence of sodium formate and chloride (20 g/liter) was investigated. Selemion CSG cation-exchange membranes and AST anion-exchange membranes were used in the investigation.

In this case, too, migration of amino acids from the central compartment were lowest at the isoelectric point, and more than 99% of the sodium formate and chloride could be removed with loss of not more than 1.5% glycine and 1% methionine.

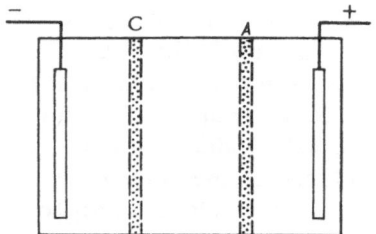

Fig. 10. Diagram of electrodialyzer with "filled" membranes. C = cation-exchange membrane; A = anion-exchange membrane.

During investigation of the electrochemical removal of salts from numerous amino acids (glycine, alanine, valine, lysine, arginine, glutamic acid, and aspartic acid) with varying isoelectric points, considerable losses were found from the central compartment of the cell, shown in Fig. 9 [41]. It was found that not only the pH of the amino acid solution in the central compartment but also the nature of the electrolyte in the electrode compartments was important [42]. Some data, illustrating this conclusion, are given in Table 2 [42].

As seen from the table, low pH values in the catholyte correspond to large losses of amino acids, as also shown in other work [41]. This is clearly explained by the fact that an acidic solution promotes conversion of amino acids into the cationic form and consequently increases their migration into the cathode compartment; the effect is manifest to a greater degree in amino acids with a high isoelectric point. If the pH value of the catholyte is high, the charge of the amino acid changes sign when it reacts with the alkali; consequently the direction of migration also changes.

A particularly significant effect in replacement of an alkaline anolyte by an acidic anolyte is observed with amino acids having high isoelectric points. This is explained by the fact that the principal amino acids exist as negative ions in the neutral solution of the central compartment, and their electromigration increases in the case of an alkaline anolyte. If the anolyte is acidic, the ions change their charge and their losses from the central compartment decrease.

Naturally alanine, the isoelectric point of which is near the pH of the neutral solution, participates to the lowest degree in the carrying of current, and its losses therefore depend less of all on the medium in the electrode compartments.

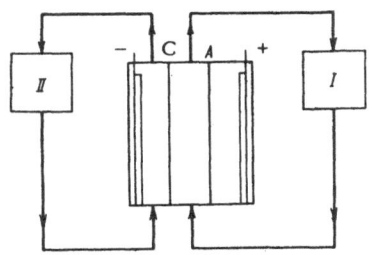

Fig. 11. Diagram of electrodialyzer
for production of L-lysine.

General questions relating to the separation of amphoteric and nonamphoteric electrolytes are discussed in the literature, e.g., in the review by Isaev and Shaposhnik [43].

It should be noted that, in spite of the low concentrations of mineral salts in the initial solutions, their separation by electrodialysis with ion-exchange membranes can be achieved with high yields. For example, demineralization of 9-aminopelargonic acid with Permaplex C-20 and A-20 membranes gives yields up to 93%, in spite of the fact that the concentration of the chloride in the initial solution does not exceed 1.5 g/liter [44].

In place of sheet ion-exchange membranes Wakasa, Saotome, and Kawamoto have used filled diaphragms for the demineralization of 7-aminoheptanoic acid [45]. A diagram of the electrolyzer is shown in Fig. 10. The diaphragms are in the form of two porous partitions between which the appropriate resin (cation-exchange or anion-exchange) is placed. Such an arrangement evidently allows the ohmic potential drop in the diaphragm to be reduced. A neutral solution containing 85 g/liter heptanoic acid and 32 g/liter ammonium chloride was demineralized.

After electrodialysis for 140 min (carbon electrodes with a current density of $1 A/dm^2$) the ammonium chloride content in the central compartment of the cells was decreased to 5.3 g/liter. The extraction of aminoheptanoic acid as a result of demineralization amounted to 99%.

In the works considered above demineralization was realized by migration of inorganic ions through ion-exchange membranes. The conditions were selected so that the amino acid remained in the central compartment, i.e., electrolysis was performed at a pH value equal to the isoelectric point.

Khleborodova and Ryazanov arrived at a different method in an investigation of the electrodialysis of L-lysine [46], which belongs

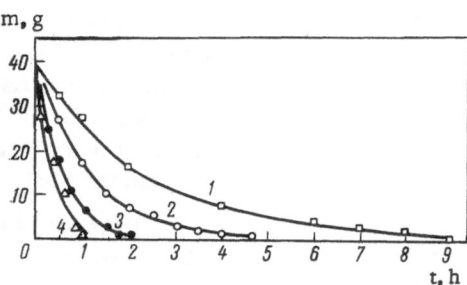

Fig. 12. Dependence of migration of L-lysine on time at various current densities: 1) 0.1 A /dm^2; 2) 0.2 A/dm^2; 3) 0.5 A/dm^2; 4) 1.0 A /dm^2. The initial concentration of L-lysine was 10 g/liter.

to a group of diamino acids with high isoelectric points (pl = 9.7). They electrolyzed a hydrochloric acid solution of lysine at pH 5, i.e., under conditions where the lysine existed as a cation and consequently migrated through the cation-exchange membrane, while the chloride ion passed through the anion-exchange membrane into the anode compartment. Investigation of the migration of lysine through membranes is of interest in the development of electrodialysis methods for production of this material from hydrolyzates of albuminous material and cultural liquids.

A diagram of the electrodialyzer is shown in Fig. 11. The initial solutions contained various concentrations of lysine hydrochloride (5, 10, 20, and 30 g/litre) at pH 5.0. These solutions were placed in vessel I and pumped continuously through the central compartment of the dialyzer. The anode compartment was filled with distilled water, and distilled water from vessel II also passed through the cathode compartment. Heterogeneous MA-40-2s and MK-40-2s ion-exchange membranes were used in the electrodialyzer. Lysine in cation form migrated into the cathode compartment through the membrane and was withdrawn into vessel II.

The process was continued until the lysine had been completely removed from the central compartment. The anode was made of graphite and the cathode of stainless steel.

The principal results from the investigation of the migration of L-lysine through a cation-exchange membrane are shown in Figs. 12 and 13. It can be concluded from the results that the migration rate of lysine through the cation-exchange membrane increases with decrease in the current density and in the concentration of this amino acid.

The dependence of the migration of lysine on time, as shown by curves 2 and 3 in Figs. 12 and 13, bears an exponential character and is described by the equation:

$$m = m_0 \cdot e^{-kt},$$

where m is the amount of L-lysine (g) remaining in the vessel I; m_0 is the original amount of L-lysine (g) in vessel I; t is the length of electrodialysis (h); k is a constant which characterizes the migration of the lysine ions.

The k value can be obtained from the following equation:

$$k = \frac{i \cdot S}{Q \cdot c},$$

where i is current density (A/dm^2); S is the area of the membrane (dm^2); c is the initial concentration of L-lysine in the solution (g/liter); G is a quantity which characterizes the electrochemical characteristics of the membrane. By substituting the k value in the equation given above, Khleborodova and Ryazanov obtained an equation by means of which they were able to calculate the amount of L-lysine remaining in vessel I after time t:

$$m = m_0 \cdot e^{-\frac{iS \cdot t}{cQ}}.$$

The amount of L-lysine which has passed into vessel II at time t as a result of migration can then be determined from the equation:

$$\Delta m = m_0 - m = m_0 \left(1 - e^{-\frac{iSt}{cQ}}\right).$$

Khleborodova and Ryazanov also evaluated the effectiveness of the electrodialysis process from the standpoint of current yield η:

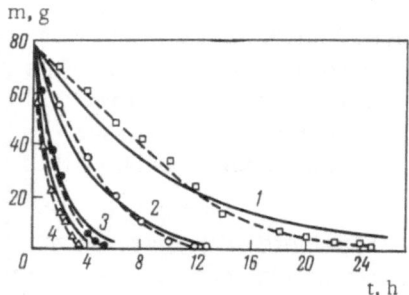

Fig. 13.   Dependence of migration of L-lysine on time at various current densities:
1) 0.1 A/dm^2; 2) 0.2 A/dm^2; 3) 0.5 A/dm^2;
4) 1.0 A/dm^2. The initial concentration of L-lysine was 20 g/liter. The dotted line represents the calculated curves.

$$\eta = \frac{Q_{theor}}{Q_{cons}} \cdot 100\%,$$

$$Q_{theor} = \frac{cV}{M} F,$$

where c is the concentration of L-lysine (g/liter); V is the volume of dialyzed solution (liter); M is molecular weight; F is the Faraday number;

$$Q_{cons} = It,$$

where I is the current strength (A); t is time (sec).

The following expression can then be obtained for the current yield:

$$\eta = \frac{cVF}{MIt} \cdot 100.$$

This equation was used to calculate the current yield for various current densities and various initial concentrations of L-lysine. It was found that the current yield decreased with increased current density. This is evidently related to decreased selectivity of the membrane owing to participation of hydrogen ions in migration. The decrease in current yield with increased initial concentration of L-lysine was explained by decreased selectivity of the membrane and reverse diffusion of lysine from vessel II into vessel I.

Khleborodova and Ryazanov also determined the effectiveness of the process from the standpoint of electricity consumption, the dependence of which on current density and initial L-lysine concentration is shown in Fig. 14. They found that the conditions for electrodialysis which gave the greatest migration of L-lysine through the cation-exchange membrane (30 g/h) comprised the following: initial concentration c = 10 g/litre; electricity consumption W ≈ 35 kWh/kg; i = 0.7 A/dm^2. The current yield η here amounted to 60%.

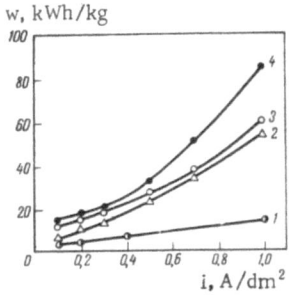

w, kWh/kg

i, A/dm²

Fig. 14. Dependence of relative flow rate of electric energy on current density for various original concentrations of L-lysine: 1) 5; 2) 10; 3) 20; 4) 30 g/liter.

b) Separation of Amino Acids from Their Mixtures. Amino acids play an important part in various biochemical processes and can be obtained by hydrolysis of proteins. The hydrolyzate contains up to 20 different α-amino acids, which can be subdivided into six groups:

1) Monocarboxylic amino acids (glycine, alanine, valine, leucine, isoleucine, phenylalanine, pl = 6.0-6.1);

2) dicarboxylic amino acids (aspartic and glutamic acids, pl = 3.0-3.2);

3) hydroxyamino acids (serine, threonine, tyrosine, pl = 5.7);

4) thioamino acids (cysteine, methionine, pl = 5.1);

5) diamino acids (ornithine, lysine, δ-hydroxylysine, pl = 9.7);

6) heterocyclic amino acids (proline, hydroxyproline, histidine, tryptophan, pl = 5.9-7.6).

Quantitative separation of amino acids contained in protein hydrolyzates is a fairly complex problem, a substantial part of which can be solved by electrodialysis processes. The isoelectric points show that it is possible to separate a mixture of amino acids into at least three groups by making use of their ability to form ions of different charge depending on the pH of the solution. These three groups include monoaminocarboxylic acids, the isoelectric points of which are close to the pH of neutral solutions; dicarboxylic acids, the isoelectric points of which lie in the acidic region; and finally diamino acids, which are basic in nature.

Consequently, by placing a neutral solution of the protein hydrolyzate in the central compartment of a multicompartment cell, it can be expected that during electrolysis the monoamino acids will remain in the central compartment owing to the fact that they exist in the form of dipolar ions; dicarboxylic amino acids will migrate into the anode compartment and diamino acids into the cathode compartment. Electrodialysis or electrolytic fractionation of amino acids is based on this principle.

The first attempt at electrodialysis of amino acids was made by Sadikov, who used three-compartment and five-compartment electrolyzers with porous inert diaphragms for this purpose [47]. However, the method did not find widespread use owing to the low selectivity and mechanical strength of the diaphragms. Effective separation of amino acids by electrodialysis was only possible with the use of ion-exchange membranes.

The first attempts to use ion-exchange membranes for the electrodialysis of amino acids were made by Sautsch and co-workers [48]. It took several more years for the development of preparative methods of separation. Figure 15 shows a diagram of a five-compartment electrodialyzer with ion-exchange membranes, intended for separation of amino acids into groups [49].

The mixture of amino acids was placed in the central compart-ment of the electrolyzer, the pH in which was maintained close to the isoelectric point of the neutral monoaminocarboxylic acids. During electrodialysis the basic amino acids migrated, in the form of cations, through the cation-exchange membrane into compartment 2, which was filled with 0.025 N hydrochloric acid solution. The amino acids with low isoelectric points passed, in the form of anions, through the anion-exchange membrane into compartment 4, filled with 0.025 N caustic soda solution. The anion-exchange membrane between compart-ments 1 and 2 and the cation-exchange membrane between compart-ments 4 and 5 were used to prevent the basic amino acids from enter-ing the cathode compartment, which was filled with 0.5 N HCl acid sol-ution, and to prevent the acidic amino acids from entering the anode compartment, which was filled with 0.5 N caustic soda solution.

Hara [50, 51] employed membranes of styrene—butadiene copolymer for electrodialysis. He showed that the best separation effect could be obtained by using cation- and anion-exchange mem-branes with specific resistance of 100 and 120 $\Omega \cdot$ cm respectively and with an initial current density of 0.8 A/dm^2. Under these condi-tions it was possible within 2 hours to remove completely acidic and basic amino acids from the initial solution, in the central compart-ment. However, 10-25% of the neutral acids also left the central compartment due to migration [51].

Considerable successes in the separation of amino acids by elec-trodialysis methods were achieved by Ryazanov and co-workers [52, 53].

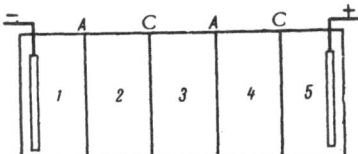

Fig. 15. Diagram of electrodialyzer for separation of amino acids into groups. A = anion-exchange membrane; C = cation-exchange membrane.

Investigations into the electrodialysis of gelatin hydrolyzate (pH 5-7), circulating through the central compartment of a cell (separated from cathode compartment by an MK-40 cation-exchange membrane and from the anode compartment by MA-40 anion-exchange membrane) showed that it is possible to isolate in the central compartment neutral monoamino acids (proline, hydroxyproline, α-alanine, leucine, valine, and glycine) free from basic and acidic amino acids [52]. However, a certain amount of the neutral acids passed into the cathode compartment with the arginine, lysine, and histidine and into the anode compartment with the aspartic and glutamic acids. Consequently with this method it is only possible to obtain the neutral monoaminocarboxylic acids as a pure fraction. The transport of some of the neutral acids from the central fraction was explained in the following way [53]: firstly, the pH value in the boundary layers of the membranes differs from that in the bulk of the electrolyte. This means that the neutral acids become negatively charged when they enter the boundary layers. As a result the amino acids contribute to the transport of electric current and enter both the cathode and anode compartments even when the pH of the solution in the central compartment is close to the isoelectric point. Secondly, these acids can diffuse through both the cation-exchange and the anion-exchange membranes [53]. This source of losses cannot, however, be substantial [4].

The problem of obtaining pure fractions of acidic, basic, and neutral amino acids was overcome by using the so-called "cycling" process, which is shown in Fig. 16. The protein hydrolyzate is delivered into the central compartment of a five-compartment electrolyzer I [53]. Here, as mentioned above, a certain amount of monoamino acids passes into the adjacent compartment through anion-exchange membrane together with the dicarboxylic amino acids. The same acids together with the main fraction of diamino acids also pass through the cation-exchange membrane into the compartment adjoining the cathode compartment.

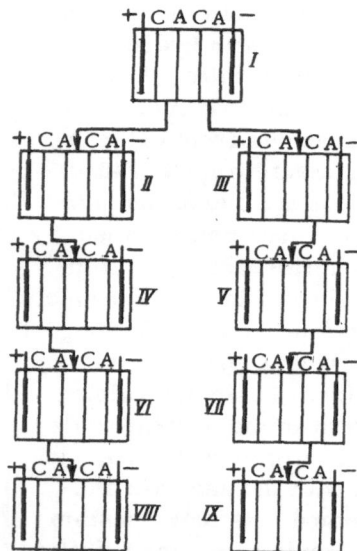

Fig. 16. "Cycling" process for production of acidic, basic and neutral amino acid fractions.

To separate the acidic and basic amino acids from the neutral amino acids, solutions from the compartments adjacent to the central compartment are delivered to the central compartments of dialyzers II and III. The solution, enriched in dicarboxylic amino acid, which passes through the anion-exchange membrane of electrodialyzer I does not contain basic acids. Further treatment in electrodialyzers II, IV, VI, and VII of the left branch of the cascade thus leads to separation of the neutral and acidic fractions by passing them successively through the central compartments of dialyzers, as seen in Fig. 16.

The solution leaving the compartment adjacent to the anode compartment of electrodialyzer I contains the basic amino acids together with a certain amount of the neutral fraction. They are separated by passing successively through the central compartments of electrodialyzers III, V, VII, and IX on the right branch of the cascade. Pure neutral fraction should be obtained from the central compartments. For the production of the pure acidic and basic fractions from the compartment adjoining the anode compartment of electrolyzer VIII and the compartment adjoining the cathode compartment of electrodialyzer IX respectively the cycling process should be repeated four to five times.

Demineralization of Glucose, Glycerol, Sugar Solutions, and Certain Other Products

Electrolytic desalinization of glucose is fairly effective and allows the content of mineral salts to be considerably reduced with small losses of the product. This was shown in the work by Shchegrov and Koblyanskii [54], who employed a four-compartment

TABLE 3. Demineralization of Glucose with Various Sodium Chloride Contents

Electrodialysis conditions	Chloride concentration in initial solution	
	5 g/liter	15 g/liter
Final content of sodium chloride, g/liter	0.2	0.24
Degree of desalinization, %	95.2	98.7
Current yield, %	36.0	44.0
Electricity consumed per 1 kg chloride removed, kWh/kg	9.1	6.5
Losses of glucose into adjoining compartments, %	0.33	0.99

electrolyzer divided up by DÉU-2 cation-exchange membranes and DÉN-2 anion-exchange membranes. A dilute sulfuric acid solution passed through the electrode compartments. The anode was of magnetite and the cathode of steel. A 20% solution of glucose, containing chlorides, was passed continuously into the electrolyzer. During electrodialysis it was possible to remove up to 80% of the chlorides with losses of only 1% glucose, arising from migration into the adjacent compartments.

Demineralization of glucose was studied in a multicompartment cell [55], which was developed by Partridge and Peers [56] for desalinization of amino acids. The cell was divided into compartments by cation- and anion-exchange membranes. The electrode compartments were filled with 0.5 N sodium sulfate solution. The anode was made from platinum and the cathode from steel. Water was circulated through the even compartments and 30% glucose solution, containing sodium chloride, through the odd compartments. Some of the results are given in Table 3.

The effectiveness of glucose demineralization decreases with decrease in the initial and final chloride concentrations. (The current yield decreases, and the amount of electricity used for desalinization increases.)

Demineralization of glycerol solutions has been studied [55] in an analogous way (Table 4). Losses of glycerol are fairly significant.

TABLE 4. Demineralization of Glycerol with Various Contents of the Solution

Electrodialysis conditions	Initial glycerol concentration, %	
	30	45
Chloride content, g/liter:		
Initial	30	30
Final	0.1	0.1
Degree of desalinization, %	99.7	99.7
Current yield, %	66.0	61.0
Electricity consumed per 1 kg chloride removed, kWh/kg	2.15	2.4
Losses of glycerol in adjoining compartments, %	5.45	3.8
Loss of glycerol per 1 kg chloride removed, g	0.58	0.64

One of the patents gives conditions for electrochemical desalinization of glycerol, by means of which it is possible substantially to reduce losses arising from migration through the ion-exchange membranes [57]. A special feature of this method is the creation of a pressure difference in the compartment to which the desalted solution passes and in the compartment enriched with electrolyte. With change in the pressure difference from 0 to 1200 mm H_2O the losses of glycerol were reduced from 21 to 0.6% during electrodialysis of solution which initially contained 10% glycerol and 13% chloride. The membranes employed were 0.75 mm thick and consisted of 70% ion-exchange resin and 30% inert thermoplastic material.

Attempts have been made to determine the effectiveness of glycerol demineralization from the economic standpoint [58]. It was found that the total cost passed through a minimum with increase in current density. (The amount of electricity consumed on dialysis increases with increased current density, but the working area of the ion-exchange membranes decreases.)

Considerable attention has been paid recently to the use of ion-exchange membranes for the electrodialysis of sugar solutions formed in the processing of sugar beet or sugar cane [59-72]. The sugar from chopped beet is leached out with hot water in diffusers. To remove nonsugar materials from the diffusion liquor it is treated with milk of lime (defecation). The calcium saccharate remains in solution, and the impurities form insoluble compounds with the calcium and are filtered off. The calcium saccharate solution is saturated with carbon dioxide, which converts the calcium into the insoluble carbonate. After removal of the latter the sugar solution is submitted to a second defecation and a second saturation for more complete removal of impurities. Nevertheless, certain impurities (mineral salts, oxalic and acetic acids, amino acids, amides, bases of the betaine type) remain. On further evaporation of the sugar solutions the nonsugar impurities are converted into a syrup. These operations lead to the loss, with the defecation and saturation residues and with the syrup, of up to 10-12% of the sugar contained in the diffusion liquor.

Ion-exchange resins can be used for demineralization of sugar liquors [73]. However, electrodialysis with ion-exchange membranes has the advantage that it avoids inversion of the sucrose, use of reagents for regenerating the resin, and dilution of the liquors [62].

Sugar solutions can be submitted to electrochemical demineralization at various stages of treatment: the green syrup diluted with liquor from the second saturation [62]; the liquor from the second saturation [67]; the molasses solution (residue after fractional recrystallization of the sugar) [59, 71, 72]; the liquor and syrups formed in the treatment of sugar cane [66].

In most cases multicompartment electrodialyzers divided by cation-exchange and anion-exchange membranes are used for demineralization of sugar solutions. Cation-exchange membranes of the Permaplex C-20 [64, 71], MK-40, MK-102 [69], and other types and anion-exchange membranes of the Permaplex A-20 [64, 71], MD-40, MA-41, MA-100, MA-101, and other types [69] have been used. The sugar solution was passed through the even compartments of the electrodialyzer, and an electrolyte solution through the odd compartments (water compartments). The effectiveness varies between 50 [67] and 85% [66] nonsugar materials removed.

Calculations have shown that treatment of the liquor obtained from 1000 ton beet requires the consumption of ~235 kW electrical energy and 650 m² useful surface area of cation-exchange and anion-exchange membranes [65]. The process can be intensified by increasing the temperature to 70°C and reducing the area of the membranes by nearly a half [65]. The consumption of electrical energy in dialysis is fairly considerable (up to 11 kWh per 1 m³ liquor [67]); this arises for a number of reasons, including an increase in the resistance of the cation-exchange membranes during the process due to sorption of amino acids, which then block the electronegative functional groups.

One of the methods for reducing the electricity consumption in dialysis is to fill the space between the membranes with ion-exchange resins [70]. By this method it is possible to reduce the resistance in the solution to approximately a third and, consequently, to reduce the voltage applied to the electrodialyzer and the consumption of electrical energy.

The examples considered above do not exhaust the fields of application of ion-exchange membranes in electrodialysis processes involving organic compounds. Tests have been carried out, for example, on the demineralization of milk with a view to removing the larger part of the sodium and potassium ions while retaining a specific calcium content. It was found that by electrodialysis with ion-exchange membranes it was possible to remove up to 99% of the potassium ions, 95% of the sodium ions and only 33-35% of the calcium ions [74, 75]. The possibility of avoiding complete removal of calcium evidently arises from its combination with betaine, in spite of the fact that in order of selective permeability of the membrane (according to data from Yamabe and Tanaka [76]) calcium stands to the left of sodium, as follows:

$$H^+ > K^+ > NH_4^+ > Ca^{2+} > Hg^{2+} > Na^+.$$

Wingerd and Block also accomplished demineralization of whey [77]. Among other examples may be mentioned the use of ion-exchange membranes in electrodialysis of acidified aqueous solutions of cinchona bark extract [78] and demineralization of polar solvents (acetone, acetonitrile, nitrobenzene) in high-voltage electrodialyzers [79, 80]. Removal of cations by electrodialysis with ion-exchange membranes has the advantage of selectivity, whereas demineralization by means of ion-exchange resins removes all the ions.

3. CONCLUSIONS

1. In electrochemical synthesis higher yields of the final organic products can be obtained with ion-exchange membranes than with porous inert diaphragms. The latter are not selective with respect to the ions participating in the transfer of electricity and do not have adequate resistance to diffusion.

2. With ion-exchange membranes it is possible to prevent the occurrence of side electrochemical reactions involving costly electrolytes such as salts of arylsulfonic acids.

3. With ion-exchange membranes it is possible to accomplish demineralization of organic compounds with high selectivity and to separate amphoteric compounds, such as amino acids.

LITERATURE CITED

1. K. M. Saldadze, A. B. Pashkov, and V. S. Titov, Ion-Exchange High-Molecular Compounds [in Russian], Goskhimizdat, Moscow (1960).
2. B. N. Laskorin, N. M. Smirnova, and M. N. Gitman, Ion-Exchange Membranes and Their Use [in Russian], Gosatomizdat, Moscow (1961).
3. F. Helferich, Ionites [Russian translation], IL, Moscow (1962).
4. J. R. Wilson (ed.) Demineralization by Electrodialysis [in Russian], Gosatomizdat, Moscow (1963).
5. G. Osborne, Synthetic Ion-Exchangers [Russian translation], Mir, Moscow (1964).
6. B. Tremillon, Separation on Ion-Exchange Resins [Russian translation], Mir, Moscow (1967).
7. Belgian Patent No. 623657;623691 (1963).
8. French Patent No. 1401175 (1965).
9. M. Baizer, J. Org. Chem., 29:1670 (1964).
10. M. Baizer and J. Anderson, J. Electrochem. Soc., 111:215, 223 (1964).
11. M. Baizer, U. S. Patent No. 3250690 (1966).
12. L. V. Kaabak, A. P. Tomilov, and S. L. Varshavskii, Zh. Obshch. Khim., 34:2107 (1964).
13. M. Baizer and J. Anderson, J. Org. Chem., 30:1357 (1965).
14. I. L. Knunyants and N. S. Vyazankin, Izv. Akad. Nauk SSSR, Otd. Khim. Nauk, 238 (1957).
15. M. Baizer, Tetrahedron Letters, 15:973 (1963).

16. I. Prescott, Chem. Eng., 72:238 (1965).
17. A. P. Tomilov, V. A. Klimov and S. L. Varshavskii, Khim. Prom., 12:892 (1967).
18. J. Riordan, Nitrogen, 40:43 (1967).
19. R. McKee and K. Brockman, Trans. Electrochem. Soc., 39:441 (1921).
20. G. Cauquis, Bull. Soc. Chim. France, 459 (1966).
21. M. Urabe, T. Sejama, and W. Sakai, J. Electrochem. Soc., Japan, 28:E69 (1960).
22. S. Okada and T. Eguchi, Japanese Patent No. 208257 (1954).
23. E. Silversmith and W. Sloan, U. S. Patent No. 3197392 (1965).
24. C. Mantel, Chem. Engin., 74:128 (1967).
25. G. Bodamer, U. S. Patent No. 2921005 (1960).
26. German Patent No. 1002750 (1960).
27. G. Khonff, A. Myuller and F. Venger, Polyamides, Goskhimizdat, Moscow (1958).
28. M. Ya. Fioshin, A. I. Kamneva, Sh. M. Itenberg, L. I. Kazakova, and Yu. A. Ershova, Khim. Prom., 4:263 (1963).
29. K. Oda and M. Murakosi, Japanese Patent No. 2023 (1958).
30. G. Bodamer, U. S. Patent No. 2737486 (1956).
31. D. Kato, J. Arimoti, I. Arai, and T. Arakava, Japanese Patent No. 6517 (1964).
32. G. S. Supin, Zh. Fiz. Khim., 34:4 (1960).
33. N. A. Dzbanovskii, V. V. Todikov, L. D. Borkhi, and R. T. Khleborodova, Trudy IREA, No. 25, 427 (1963).
34. V. V. Krokhv, Zh. Prikl. Khim., 37:2321 (1964).
35. K. D. Nenitzesku, Organic Chemistry [Russian translation], Vol. 2, IL, Moscow (1963), p. 376.
36. L. D. Bobrovnik and I. M. Litvak, Trudy Kievsk. Tekhnol. Inst. Pishch. Prom., No. 27, 31 (1963).
37. T. Astrup and A. Stage, Acta Chem. Scand., 6:1302 (1952).
38. A. Di Benedetto and E. Lightfoot, Ind. Eng. Chem., 50:691 (1958).
39. T. Yamabe, M. Seno, and H. Takai, J. Chem. Soc. Japan, Industr. Chem. Soc., 64:556 (1961).
40. S. Itoi and T. Ucunomie, Repts. Res. Lab., Asahi Glass Co., 15:171 (1965).
41. J. Bliney, H. Yardley, and J. Dumber, Nature, 177:83 (1956).
42. A. Peers, J. Appl. Chem., 8:59 (1958).
43. I. I. Isaev and V. A. Shaposhnik, Trudy Laboratorii Ionoobmennykh Protsessov i Sorbtsii, No. 1, Nauchn. Fiz. Khim., Inst. Izd. Voronezhsk. Univ. (1966), p. 70.
44. E. Tszin-shen, Shen' Tszy-Min, and U. Tun-li, Gaofen'tsza tunsyun', 6:389 (1964).
45. P. Wakasa, K. Saotome, and H. Kawamoto, Japanese Patent No. 3813 (1964).
46. R. T. Khleborodova and A. I. Ryazanova, in: Advances in Electrochemistry of Organic Compounds [in Russian], Nauka, Moscow (1968), p. 73.
47. V. S. Sadikov, Practical Work on Proteins [in Russian], LGU (1938), p. 194.
48. W. Sautsch, G. Manecke, and W. Broser, Nature, 8B:232 (1953).

49. J. Traxler, U. S. Patent No. 3051640 (1962).
50. Y. Hara, J. Chem. Soc. Japan, Ind. Chem. Sec., 65:885 (1962).
51. Y. Hara, Bull. Chem. Soc. Japan, 36:1373 (1963).
52. A. I. Ryazanov, Z. V. Gol'tseva, R. T. Khleborodova, and L. M. Butorina, Trudy IREA, No. 30, 69 (1967).
53. A. I. Ryazanov, Z. V. Gol'tseva, R. T. Khleborodova, and L. M. Butorina, Trudy IREA, No. 30, 77 (1967).
54. M. I. Shchegrov and A. G. Koblyanskii, Izv. Vuzov, Pishchevaya Tekhnologiya, 6:49 (1961).
55. D. Corming, Proc. Sympos. Less. Common Means Separat., Birmingham, 1963, London, Inst. Chem. Engrs. (1964), p. 48.
56. S. Partrige and A. Peers, J. Appl. Chem., 8:49 (1958).
57. L. Tye, British Patent No. 845938 (1960).
58. H. Sanders, and J. Parsi, Proc. Sympos. Less. Common Means Separat., Birmingham, 1963, London, Inst. Chem. Engrs. (1964), p. 16.
59. A. Warshaw, U. S. Patent No. 2688572 (1954).
60. H. Barger, Zucker, 7:80 (1954).
61. A. Anderson and C. Wylam, Chem. a. Inc. (Rev.), 191 (1956).
62. I. Buriánek and D. Šlechtová, Listy Cukravam, 75:62, 66 (1959).
63. I. Buriánek and D. Šlechtová, Listy Cukravam, 75:82 (1959).
64. I. Buriánek and D. Šlechtová, Listy Cukravam, 76:193 (1960).
65. I. Buriánek, D. Slechtova, and O. Lisy, Listy Cukravam, 76:217 (1960).
66. M. Leszko, Gaz. Cukrown., 62:246 (1960).
67. M. Leszko, Gaz. Cukrown., 64:38 (1962).
68. L. D. Bobrovnik and I. M. Litvak, Sakharnaya Prom., 11:18 (1962).
69. D. M. Leibovich and I. F. Zelikman, Sakharnaya Prom., 9:30 (1963).
70. I. Buriánek and P. Kadlec, Listy Cukrovam., 81:260 (1965).
71. P. Kadlec, Listy Cukrovam., 82:37 (1966).
72. R. Bretschneider, I. Buriánek, and P. Kadles, Czech. Patent No. 117128 (1965).
73. D. I. Ryabchikov and I. K. Tsitovich, Ion-Exchange Resins and Their Use [in Russian], Izd. Akad. Nauk SSSR, Moscow (1962), p. 118.
74. R. Block and W. Wingerd, U. S. Patent No. 2758965 (1958).
75. R. Block and W. Wingerd, U. S. Patent No. 2830905 (1958).
76. T. Yamabe and I. Tanaka, J. Chem. Soc., Japan, Ind. Chem. Sec., 63:1342 (1960).
77. W. Wingerd and R. Block, Dairy Sci., 27:932 (1954).
78. I. Kamii, T. Tanaka, and T. Yamabe, J. Pharmac. Soc. Japan, 81:931 (1961).
79. G. Briere and N. Felici, C. r. Acad. Sci., 259:3237 (1964).
80. G. Briere, N. Felici, and J. Filippini, C. r. Acad. Sci., 261:5097 (1965).

Electrochemical Oxidation of Organic Compounds at Metals of the Platinum Group

O. A. Petrii

If the electrochemistry of organic compounds as a whole is a rapidly developing field of modern electrochemistry, the electrochemical oxidation of organic compounds at metals of the platinum group is a section in which a particularly large amount of experimental material has been accumulated in recent years. This is due primarily to attempts to utilize oxidation reactions of organic compounds for direct conversion of chemical into electrical energy.

The present review briefly examines the most important work on electrochemical oxidation of simple organic compounds of various classes* (saturated alcohols, aldehydes, acids, carbon monoxide, hydrocarbons) and also the directions of searches for catalytically active materials for these reactions.

I. ADSORPTION OF ORGANIC COMPOUNDS ON PLATINUM METALS

The participation of adsorbed organic compounds in anodic processes had already been proposed in the early work on electrochemical oxidation by Müller and co-workers [2-8]. Owing to the development of experimental methods for determining adsorption at solid electrodes (reviews of these methods are given in [1, 9-12]) it has become possible in recent years to compare quantitatively the results from adsorption and kinetic measurements under identical conditions, and this has enabled the mechanism of

---

* A short historical outline of the development of research into electrochemical oxidation of organic compounds is given in [1].

oxidation reactions to be investigated in more detail. The first such comparison was made by Gilman and Breiter [13] for methanol and Gilman [14] for carbon dioxide.

Direct evidence for the effect of adsorption of organic compounds on the stationary process of their anodic oxidation at a platinum electrode was obtained by Bockris and co-workers [15, 16], who found a negative reaction order in the electrochemical oxidation of ethylene; this could only be explained on the assumption that ethylene was adsorbed on the electrode.

The role of chemisorption of organic compounds at the electrode surface during anodic oxidation has been investigated in detail by various Soviet investigators [17-26]. Before going on to discuss modern ideas on the mechanism of electrochemical oxidation it will be useful to dwell briefly on the characteristics and relationships concerned in the adsorption of organic compounds on platinum metals.

1. Adsorption Mechanism of Organic

Compounds

As far as can be judged from an analysis of the experimental results, in the initial investigations on the adsorption of organic compounds on platinum metals it was assumed that, as on the mercury electrode, organic molecules are physically adsorbed or reversibly chemisorbed [27-31].*

Accumulation of further experimental data led to the conclusion that adsorption on platinum electrodes from electrolyte solutions may be partly accompanied by destruction of the organic molecules [36-42]. This point of view is being ever more widely corroborated and expanded and is at present held by the majority of investigators concerned with the problems of adsorption on platinum metals. The idea of the dissociative nature of adsorption has been confirmed by results from experimental investigations into chemisorption from the gas phase onto catalyst surfaces [43-49].

---

* A detailed account of the early work on adsorption of organic compounds on catalysts was given in the monograph by Sokol'skii [35].

a) Adsorption Mechanism of Methanol. By comparing the amounts of electricity consumed in the oxidation of adsorbed hydrogen and organic compound during contact between a Pt/Pt electrode and methanol solutions in open circuit, Podlovchenko and Gorgonova [39] proposed that particles with stoichiometric composition HCO were chemisorbed as a result of dissociative adsorption of methanol according to the following equation:

$$CH_3OH \rightarrow HCO_{ads} + 3H_{ads}.$$

Bagotskii, Vasil'ev, and co-workers [12, 22, 26] confirmed this conclusion by measurements on a smooth platinum electrode using pulse methods. Recently Biegler and Koch [50], using similar procedures, came to the conclusion that carbon monoxide was adsorbed on smooth platinum in methanol solutions. However, this conclusion is insufficiently convincing, since in [50] use was made of adsorption values obtained by other investigators [17, 33], and the electrodes were differently pretreated in [17, 33, 50].

The adsorption mechanism of methanol has also been investigated by certain new methods. In addition to electrochemical measurements, Podlovchenko, Frumkin, and Stenin [51] analyzed the liquid phase for methanol, formaldehyde, and formic acid. Formation of HCOOH (or HCOOK) and CH_2O was not detected during electrochemical oxidation of a substance chemisorbed in methanol solution at the potentials of the double-layer region on a Pt/Pt electrode in 1 N sulfuric acid and 1 N potassium hydroxide. The strongly chemisorbed substance is thus practically completely oxidized to carbon dioxide. Under open circuit conditions the decrease in the amount of methanol in the solution during adsorption, established by direct analytical means, is found to be greater than the amount of strongly adsorbed substance with composition corresponding to HCO. This result indicates that on a platinum surface in methanol solution there are (in addition to strongly bound HCO particles) less strongly adsorbed particles, which are removed by washing and may have a different composition and a different type of bond with the surface.

To determine the nature of the species involved in adsorption of methanol in acidic solution Breiter [52, 53] used a combination of anodic charging curves at electrodes with an extended surface and gas-chromatographic carbon dioxide determination. Since the anodic oxidation reaction of an adsorbed substance with composition $C_sH_pO_q$ can be represented in the following form:

$$C_sH_pO_q + (2s - q)\, H_2O = sCO_2 + (4s + p - 2q)\, H^+ + (4s + p - 2q)\, \bar{e},$$

the following equation can be written for the number of electrons R required to produce a carbon dioxide molecule:

$$R = 4 + (p - 2q)/s. \qquad (1)$$

The R value for the formation of each carbon dioxide molecule from adsorbed methanol was found to be equal to ~2. This corresponds to an average composition of $H_2C_2O_3$ in the adsorbed particles or to simultaneous adsorption of HCO and COOH on the electrode.*

A comparison of the amount of hydrogen formed during dehydrogenation of methanol and the amount of organic compound adsorbed was made on platinum − ruthenium [54] and rhodium [55] electrodes. Since these values agreed to a first approximation, it was proposed that particles with composition HCO were chemisorbed on both electrodes. Evolution of methane was observed at rhodium and platinum − ruthenium electrodes [54, 55] at 20°C and at a Pt/Pt electrode at 80°C [56] in concentrated acidic solutions of methanol, which indicates that the methanol is hydrogenated by adsorbed hydrogen.

b) Identification of Adsorption Products from Saturated Alcohols, Ethylene Glycol, Aldehydes, and Acids. Other saturated alcohols react with the surface of platinum metals in an even more complex manner than methanol [20, 21, 38, 55, 57, 58]. In these systems particles of various composition can be formed at electrode surfaces as a result of dehydrogenation, hydrogenation, and self-hydrogenation of alcohols and of the products from their decomposition at the C − C bonds.

In [59] experiments were carried out on cathodic reduction of chemisorption products from alcohols.† Chromatographic

---

* Experiments with C^{14} showed, however, that the stoichiometry of the adsorbed particle corresponds to the formula HCO [250].

---

† This method for determining the nature of chemisorption products from organic compounds was proposed by Niedrach [42].

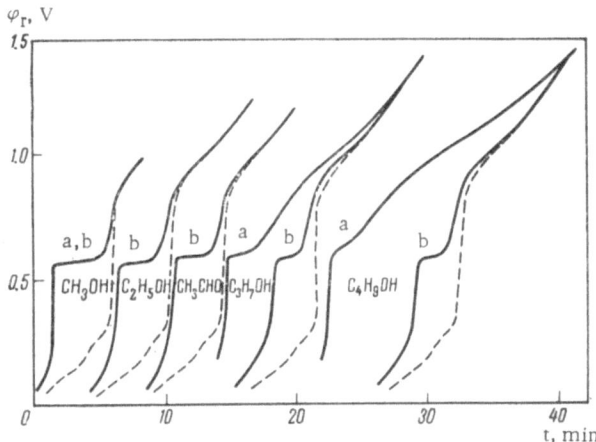

Fig. 1. Anodic charging curves for Pt/Pt electrode in 1 N sulfuric acid solution with chemisorbed organic substances before (a) and after (b) cathodic polarization of the electrode. The methanol was adsorbed at 480 mV, the other compounds were adsorbed at an open circuit (starting potential 500 mV). The concentrations were 0.5 mole/liter. The dotted lines represent the curves for the indifferent electrolyte (from data in [59]).

analysis of the gases formed during cathodic polarization showed that main hydrogenation products were ethane from chemisorbed ethanol and ethane and propane (more ethane than propane), and also small traces of butane and isobutane, from propanol. The comparison, in Fig. 1, of oxidation curves for the adsorbed substances before and after cathodic polarization show that the hydrocarbons are formed from less easily oxidized species. The oxidation curve for the product remaining at the electrode after cathodic polarization was close to the oxidation curve for chemisorbed methanol, which evidently means that the same HCO particles are formed during adsorption of various alcohols.

The chemisorption of ethylene glycol on platinum has been investigated [59, 60] and the results obtained by pulse methods were interpreted [60] as due to chemisorption process on smooth platinum at low temperatures in the following equation:

$$CH_2OHCH_2OH \rightarrow HO\underset{..}{C} \text{———} \underset{..}{C}OH + 4H_{ads}.$$

At 70°C rupture of the $C-C$ bond occurs:

$$CH_2OHCH_2OH \rightarrow 2\underset{...}{C}OH + 4H_{ads}.$$

Measurements on a Pt/Pt electrode in 1 N sulfuric acid in open circuit [59] showed that ethylene glycol could be reduced by adsorbed hydrogen as well as dehydrogenated. Chemisorbed ethylene glycol is partly reduced by ethane during cathodic polarization. The curve for electrochemical oxidation of the product which remains after reduction of chemisorbed ethylene glycol is close to the curve for electrochemical oxidation of chemisorbed methanol. Thus, the adsorption products from ethylene glycol evidently consist of both HCO particles and particles containing two carbon atoms capable of being reduced.

The chemisorbed substance which forms on platinum in formaldehyde solutions behaves like the chemisorption product from methanol [20, 21, 52, 53].* Dehydrogenation and destructive hydrogenation take place during chemisorption of acetaldehyde, and the rates of these processes are higher than the rates of the analogous processes in saturated alcohols [20, 38, 55, 57, 59]. The adsorption product from acetaldehyde probably contains both HCO particles (Fig. 1) and particles with two carbon atoms. During cathodic polarization the latter are removed as ethane [59].

In the literature great attention has been paid to investigations on chemisorption of formic acid on platinum [20, 34, 36, 52, 53, 61, 63-76]. Dehydrogenation and hydrogenation of formic acid take place during adsorption on a Pt/Pt electrode [20], and this can lead to the formation of HCO and COOH particles at the surface. According to adsorption measurements on smooth platinum [75], COOH particles are formed during adsorption of formic acid in the region with $\varphi_r \geqslant 0.4$ V.† At lower potentials the amount of hydrogen formed during adsorption of formic acid is found to be considerably less than the amount of chemisorbed compound [70, 75].

---

* In [61, 62] it was mentioned that a polymeric product can be formed when concentrated solutions of formaldehyde are in contact with platinum.

† Here and henceforth φ_r denotes the potential measured with reference to the reversible hydrogen electrode in the same solution, and φ denotes the potential with reference to the normal hydrogen electrode.

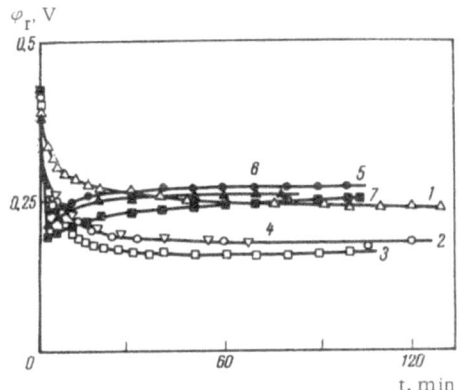

Fig. 2. Curves for displacement of potential for platinum black electrode in presence of methane (1), ethane (2), propane (3), isobutane (4), ethylene (5), propylene (6), and cyclopropane (7) at 25°C [42]. The indifferent electrolyte was 5 N sulfuric acid.

This result is explained by the possibility of either chemisorption of formic acid with cleavage of the $C = O$ bond or hydrogenation of formic acid leading to nonelectrochemical removal of hydrogen from the surface [75].

Adsorption of formic acid on palladium takes place in the same way, with decomposition of the molecules into CO_2 and H_{ads} or H_2 [77–80] and is thus an example of an irreversible chemisorption process. Analysis of the isotopic composition of the gas during decomposition of DCOOH in light water and HCOOD in heavy water [79], showed that the deuterium content of the gas in the first case and the protium content in the second case considerably exceeds the equilibrium values. This means that the hydrogen formed during decomposition of formic acid first forms molecules and is only then adsorbed at the electrode surface:

$$2HCOOH \rightarrow H_2 + 2HCOO_{ads},$$
$$2HCOO_{ads} \rightarrow H_2 + 2CO_2 \qquad \text{or} \qquad HCOOH \rightarrow H_2 + CO_2.$$

c) Adsorption of Carbon Monoxide and Carbon Dioxide. Adsorption of carbon monoxide on electrodes has been investigated by galvanostatic [81–87] and potentiostatic [88–92] methods. According to data obtained by Sokol'skii, Fasman, and co-workers [82, 85–87], the processes which take place in the presence of carbon monoxide depend largely on the temperature and the nature of the catalyst (platinum, palladium, and ruthenium). The strongest chemisorption reaction of carbon monoxide with the metal surface is observed at a palladium electrode. Analysis of the gas phase with the Pt/Pt electrode saturated with carbon monoxide indicates

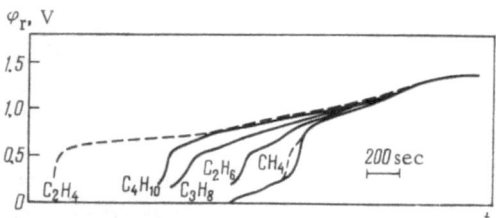

Fig. 3. Galvanostatic oxidation curves for hydro-carbons adsorbed by platinum black in an open circuit at 25°C [42]. The indifferent electrolyte was 5 N sulfuric acid solution.

formation of carbon dioxide and hydrogen in approximately equi-molar quantities, which can be explained by conversion of carbon monoxide, as follows:

$$CO + H_2O \rightarrow CO_2 + H_2.$$

It is suggested in [85] that formation of platinum dicarbonyl $Pt(CO)_2$ predominates at 50–70°C. It is, however, possible that carbon monoxide reacts with adsorbed hydrogen under these conditions [38], i.e., that processes of the Fischer–Tropsch type [93] take place.

Comparing the amounts of electricity consumed in the oxidation of adsorbed carbon monoxide and in the formation of a mono-layer of oxygen on platinum, Warner and Schuldiner [83] proposed the existence of physically adsorbed carbon monoxide on a mono-layer of chemisorbed carbon monoxide. It is indicated that com-paratively weakly bound particles, which are desorbed when car-bon monoxide is removed from the solution, are present in addi-tion to strongly adsorbed particles [84, 88].*

Gilman [88, 89, 91] proposed the existence of two structures in adsorbed carbon monoxide — a linear or one-site structure and a bridging or two-site structure. Two electrons are required for each occupied adsorption site during oxidation of the first form and

---

* In quantitative respects the results from [83, 84, 88, 89, 91] do not agree well with each other. The reasons for the discrepancies have been discussed [91, 94, 95] and evidently arise from special features of the methods used for the determination and for analysis of the results [91].

one electron during oxidation of the second. The amount of electricity Q_{CO}^t involved in the oxidation of the adsorbed layer is given by the following equation:

$$Q_{CO}^t = Q_{CO}^B + Q_{CO}^L = 2\Delta_S Q_H - Q_{CO}^B, \qquad (2)$$

where $\Delta_S Q_H$ is the decrease in the amount of adsorbed hydrogen during adsorption of carbon monoxide on the platinum surface; Q_{CO}^B and Q_{CO}^L are the amounts of electricity required for two-site and one-site adsorption.

Giner [36] found adsorption of carbon dioxide on platinum covered with adsorbed hydrogen. The adsorbed particles were removed during anodic polarization of the electrode to potentials of the double layer region. The adsorption product was called "reduced carbon dioxide" or $(CO_2)_r$. It was later suggested that $(CO_2)_r$ comprised formate radicals [96], carbon monoxide (or similar substance requiring two electrons for complete oxidation) [61], or "dermasorbed" (dissolved in the surface layer) hydrogen formed in the presence of CO_2 or HCO_3^- [97]. Breiter [98] compared the amount of electricity obtained during oxidation of $(CO_2)_r$ under galvanostatic conditions with the amount of carbon dioxide evolved and showed that on average two electrons are required for the formation of a carbon dioxide molecule from the adsorbed particle. This result, and also the coincidence of the charging curves in the presence of $(CO_2)_r$ and in the presence of carbon monoxide, led to the conclusion that $(CO_2)_r$ is identical with CO_{ads}.

Podlovchenko, Ekibaeva, and Stenin [99] observed a similarity between the behavior of $(CO_2)_r$ and that of species which are adsorbed during anodic polarization of the electrode in solutions of methanol, formaldehyde, and formic acid.

d) Adsorption Mechanism of Hydrocarbons. The adsorption of hydrocarbons depends largely on their structures [42]. Petrii and Marvet [100] carried out a detailed investigation into the chemisorption of methane, which is distinguished by lowest adsorbability on platinum compared with other hydrocarbons. The curves for the potential shift after bringing a Pt /Pt electrode into contact with methane in acidic solutions are similar to the corresponding curves for methanol, at low concentrations of the latter, [18, 19, 56] both in form (Fig. 2) and in the nature of their dependence on pH, partial pressure of methane, and temperature.

On charging curves after contact of electrode with methane there are arrests corresponding to oxidation of H_{ads} and the chemisorbed substance (Fig. 3) [46, 100]. The results lead to the assumption that in the open circuit the shift of φ_r to the cathodic side arises from dehydrogenation of methane. In order to determine the nature of the chemisorption product from methane the ratio between the amount of hydrogen Q_H adsorbed in the solution of methane at the final potential to the amount of electricity Q_M consumed in the oxidation of the chemisorbed substance was determined [100]. In the investigated temperature range (20–95°C) the $Q_H:Q_M$ ratio amounts to ~1.2–2.0. Since the methane molecule gives up eight electrons on full oxidation to carbon dioxide, to explain the obtained relationship it is necessary to postulate adsorption of oxygen-containing particles, e.g., particles of the HCO type formed during adsorption of methanol. Consequently, reaction between the organic particles and water or its oxidation products must also occur in addition to dehydrogenation. This is confirmed by agreement between the oxidation potentials of adsorbed methanol and methane.

Adsorption of ethane, propane, and n-butane on platinum at an open circuit is accompanied by a potential shift to the more negative potentials (Fig. 2) and evolution of gaseous products (methane in the case of ethane and methane and ethane in the case of butane) [41, 42]. Ethane, methane, and small amounts of propane and butane are evolved during cathodic reduction of chemisorbed ethane; in the reduction of chemisorbed propane, C_3H_8 > C_2H_6 > CH_4 ≫ C_4H_{10} [42]; in the reduction of chemisorbed butane, C_4H_{10} ≫ C_3H_8, C_2H_5, and CH_4 [103]. These results seem to be due to reactions involving dehydrogenation, cracking, and self-polymerization of the organic compounds during adsorption. After cathodic polarization a certain amount of chemisorption product, which is oxidized at positive potentials, remains at the electrode surface [42].

On charging curves recorded after contact of ethane, propane, and isobutane [42, 104, 105] with a platinum electrode there are arrests corresponding to ionization of adsorbed hydrogen and oxidation of the organic compound. The length of the second arrest increases in the order C_2H_6 < C_3H_8 < C_4H_{10}, i.e., in the order of increased-molecular weight of the adsorbate (Fig. 3).

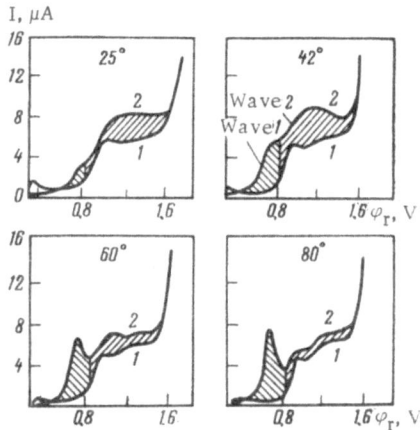

Fig. 4. Potentiostatic curves on platinum black (v = 0.1 V/sec) in 4.3 N $HClO_4$ (1) and in the presence of chemisorbed ethane at φ_r = 0.4 V (2).

By means of a multipulse potentiodynamic method, with certain additional assumptions, Gilman [107–109] obtained the formula C_2H_2 for the average stoichiometric composition of products of ethane adsorbed at $\varphi_r \leqslant 0.5$ V in 1 N perchloric acid on smooth platinum. The particles formed at $\varphi_r > 0.5$ V had lower hydrogen content. An average stoichiometric composition of C_2H_2 is characteristic of a mixture of two different types of adsorbed particles. In fact the part of the adsorbed ethane which corresponds to the second wave on the potentiostatic curve (Fig. 4) is desorbed during cathodic polarization. The easily oxidized particles which correspond to Wave 1, are not desorbed even at − 0.2 V. Similar results were obtained by Niedrach and others [101, 102] on platinum black. Here it was concluded (on the basis of the approximate coincidence of Wave 1 for oxidation of adsorbed hydrocarbons with the oxidation waves of chemisorbed carbon monoxide, carbon dioxide, and formic acid) that Wave 1 corresponds to oxidation of oxygen-containing particles with one carbon atom, and Wave 2 corresponds to dehydrogenation of "ethylene-like" particles. The overall scheme for chemisorption of ethane on platinum can therefore be represented as follows:

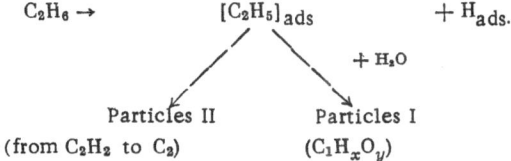

Burshtein, Pshenichnikov, and co-workers [105] found that the stoichiometric composition of ethane adsorbed on a Pt /Pt electrode at 90°C in 1 N sulfuric acid solution varied with time from C:H = 1:2 to C:H = 1:1. The change in the C:H ratio was attributed to slow and more intense dehydrogenation of the adsorbed particles, with simultaneous formation of methane:

$$C_2H_6 \rightarrow aCH_4 + b\,(CH_x)_{ads} + zH_{ads}.$$

Slow changes in the nature of the adsorbed particles in hydrocarbon solutions are also mentioned in [42, 106].

The behavior of propane during adsorption on platinum is in general similar to the behavior of ethane [101]. Grubb and Lazarus [110], combining electrochemical measurements with chromatographic determination of carbon dioxide, showed that rupture of C − C bonds occurs during adsorption of propane and particles with one carbon atom are converted preferentially into oxygen-containing particles which are not reduced during cathodic polarization of the electrode. On smooth platinum in phosphoric acid solution Brummer and co-workers [74, 111-113] found three types of particles in the layer of adsorbed propane, and one of these particles evidently contained oxygen. Similar conclusions about the nature of adsorbed propane were also reached by others [114].

Adsorption of unsaturated hydrocarbons (ethylene, acetylene, propylene) and of cyclopropane was also investigated [10, 40, 42, 104, 115-121]. Contact between these hydrocarbons and platinum leads to evolution of gaseous products [42, 117]; C_2H_6 and CH_4 ($C_2H_6 > CH_4$) are formed in the case of ethylene, C_2H_4 and C_2H_6 in the case of acetylene, and $C_3H_8 > C_2H_6 > CH_4$ in the case of propylene. During cathodic reduction $C_2H_6 \gg C_4H_{10} > CH_4 > C_3H_8$ are obtained from adsorbed ethylene and $C_3H_8 \gg CH_4 > C_2H_6$, and also traces of hexane, from propylene. Thus, a considerable degree of cracking and polymerization takes place during adsorption. The relative amounts of hydrogenation products from the adsorbed substances depend on the potential at which the adsorption took place.

In the oxidation of adsorbed ethylene and acetylene [109, 118] on smooth platinum at 30-60°C in 1 N perchloric acid solution there is only one wave, which occurs at potentials of the oxygen region and corresponds to the second oxidation wave of adsorbed ethane. The potentiostatic oxidation curves of adsorbed ethylene and

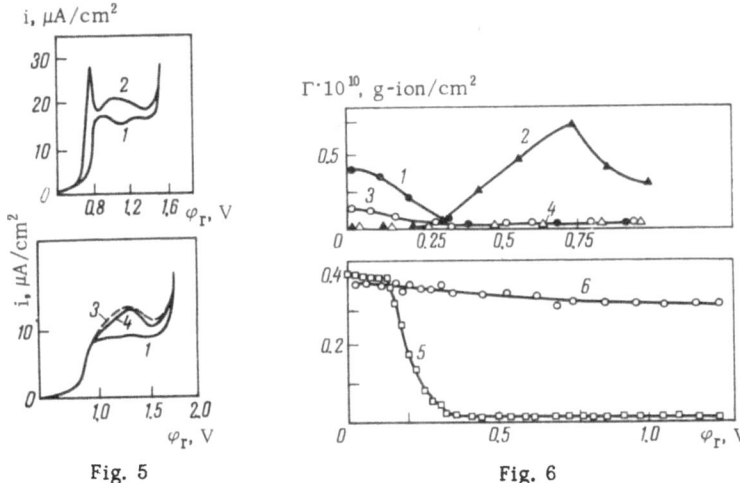

Fig. 5   Fig. 6

Fig. 5. Potentiostatic curves (v = 10 V/sec) for smooth platinum electrode in 1 N perchloric acid (1) and in presence of ethane (2), ethylene
(3), and acetylene (4) adsorbed at φ_r = 0.4 V [109].

Fig. 6. Dependence of adsorption of Na^+ (1, 3, 5, 6) and SO_4^{2-} (2, 4) ions
on potential [124] in solutions of: 1, 2) $5 \cdot 10^{-3}$ N Na_2SO_4 + $5 \cdot 10^{-3}$ N
H_2SO_4; 3, 4) $5 \cdot 10^{-3}$ N Na_2SO_4 + $5 \cdot 10^{-3}$ N H_2SO_4 + $2 \cdot 10^{-4}$ M $C_{10}H_8$; 5)
10^{-3} N Na_2SO_4 + 10^{-3} N H_2SO_4; 6) 10^{-3} N Na_2SO_4 + 10^{-3} N H_2SO_4 + 10^{-3}
M $C_6H_{13}OH$.

acetylene practically coincide (Fig. 5). Gilman proposed that the
composition of the substance adsorbed at $\varphi_r \lesssim$ 0.5 V corresponds
to the formula C_2H_2, i. e. , that an associative adsorption mechanism occurs with acetylene and a dissociative mechanism with
ethylene. However, data obtained by Sokol'skii and Beloslyudova
[116] indicate that dehydrogenation of acetylene takes place during
adsorption. The composition of adsorbed ethylene and acetylene
depends on temperature [109, 115, 118].

Thus, one of the main features of adsorption phenomena on
platinum metals is the development of chemisorption reactions between the hydrocarbons and the electrode, and this leads to considerable changes in the nature of the particles on adsorption. The
presence of water makes it possible (particularly at elevated temperatures) for reactions forming oxygen-containing particles to
occur.

Bockris and co-workers [110, 119-123] attribute primary importance to the competition between the organic compound and water molecules for sites on the surface. This conclusion is based on determination of heats and entropies of adsorption by analysis of adsorption isotherms obtained by a tracer method. However, the theories put forward in these papers do not take adequate account of all the experimental results reported for adsorption phenomena at platinum electrodes.

Electrochemical investigations discussed above have led to the development of a range of methods, by means of which information can be obtained on the nature of the chemisorbed state of substances on catalysts:

a) Detection of H_{ads}, appearing during dehydrogenation in the course of adsorption, by means of charging curves and the potentiostatic method;

b) Establishment of stoichiometric proportions between this hydrogen and the amount of adsorbed carbon-containing product;

c) Determination of the number of sites on the electrode surface occupied by the adsorbed substance;

d) Determination of the amount of adsorbed carbon-containing substance and the amount of carbon dioxide formed during its oxidation;

e) Analysis of products which appear during adsorption of organic substances;

f) Analysis of cathodic reduction products resulting from adsorbed substances;

g) Comparison of potentiostatic curves and charging curves in the presence of various adsorbed substances.

2. Effect of Adsorption of Organic Substances on Structure of Double Electric Layer

By a radioactive tracer method Kazarinov and Mansurov [124] found a large change in the structure of the double electric

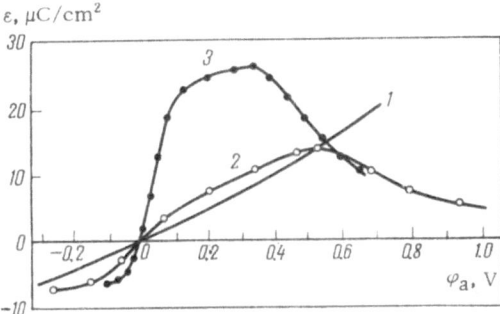

Fig. 7. Dependence of electrode surface charge on potential expressed, relative to the zero-charge point, on mercury (1) in 0.1 N sodium fluoride solution, on platinum (2) in 0.01 N sulfuric acid + 0.01 N cesium sulfate solution, and on rhodium (3) in 0.01 N sulfuric acid + 1 N sodium sulfate.

layer during adsorption of methanol, hexanol, and nephthalene on platinum (Fig. 6). A similar conclusion was reached by Reshet'ko and Shlygin [125], who determined the change in composition of solutions during formation of the double layer [126] on platinum in solutions of ethanol.

Specific adsorption of anions and cations on platinum has a considerable effect on the chemisorption processes [18, 91, 99, 100, 127-132). The effect can even be observed in the presence of phosphate anions, the specific adsorption of which is comparatively low at small positive φ_r values [133, 134]. Halide ions (Cl⁻ and Br⁻) make chemisorption of organic compounds more difficult, but they are nevertheless displaced from the surface by the adsorbed substance in the course of time [91, 129]. If the organic substance is adsorbed first on the electrode and specifically adsorbed ions are then introduced to the solution, adsorption of the ions only begins after some of the adsorption sites have been freed during anodic oxidation of the organic substance [9].

3. Effect of Adsorption of Hydrogen and Oxygen on Adsorption of Organic Compounds

Frumkin [135, 136] has used a thermodynamic approach to the phenomena of adsorption of organic substances on the surface of platinum metals. In the theory which he developed it was shown that with full reversibility of the adsorption process the ability to adsorb hydrogen and oxygen, inherent in platinum metals and also to some extent in many other solid electrodes, should lead to a number of special features in the relationships governing the adsorption of organic substances. Thus, in this case the standard free energy of the adsorption process for an organic substance ΔG_{org} is given by the following equation:

$$\left(\frac{\partial \Delta G_{org}}{\partial \varphi}\right)_{\Gamma_{org}} = -\left(\frac{\partial \varepsilon}{\partial \Gamma_{org}}\right)_{\varphi} + \left(\frac{\partial A_H}{\partial \Gamma_{org}}\right)_{\varphi}, \tag{3}$$

where Γ_{org} is the surface density of the organic substance; ε is the surface charge; A_H is the amount of hydrogen adsorbed on 1 cm^2 surface, expressed in electrical units.

Equation (3) differs from the analogous equation for the mercury electrode by the presence of the second term on the righthand side. As shown by analysis, the magnitude of this term is substantial even if the coverage of the surface by the adsorbed hydrogen is comparatively small. In a similar way the presence of adsorbed oxygen should reduce the adsorption of organic compounds. As a result, maximum adsorption of organic substances should be observed at potentials corresponding to minimum adsorption of hydrogen and oxygen.

In work by Bockris and his co-workers [1, 10, 119-123, 137, 138] the dependence of reversible adsorption of organic substances on the potential at platinum metals was interpreted on the basis of the assumption that there is competition between the organic molecules and the water molecules for sites on the surface and that the standard free energy of adsorption of water depends on the electric field at the electrode – solution interface. This approach has been criticized [9, 139] from the point of view of the analysis made

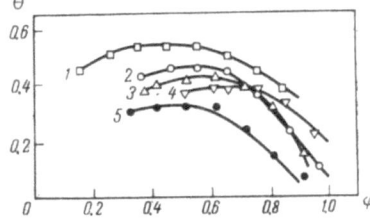

Fig. 8. Dependence of benzene adsorption on potential at platinum electrode in solutions of sulfuric acid + sodium sulfate and sodium hydroxide + sodium sulfate with various pH values [120]: 1) 0.8; 2) 3.0; 3) 3.6; 4) 5.8; 5) 11.0.

by Bockris, of the effect of adsorption of water on the adsorption of organic substances.

The effect of hydrogen and oxygen adsorption on adsorption phenomena has been detected directly in the investigation of the double electric layer structure on platinum metals [126, 136, 139, 140]. The intensity of the double layer electric field due to positive charges of the surface and negative charges of the adsorbed anions decreases at the oxygen evolution potentials and does not increase with increase in potential (Fig. 7). The same relationship, although less well defined, can be observed for the adsorption of cations with low specific adsorbability at potentials in the hydrogen region. Recently these phenomena have been investigated by the method of the isoelectric potential shifts during change in the pH value of the solution [136, 141–145].

4. Dependence of Adsorption of Organic Substances on Potential and pH Value of Solution

The theories developed by Frumkin [135, 136] can be evidently be applied to the interpretation of some of the results obtained by Bockris and his co-workers [120, 123], who studied the dependence of adsorption of naphthalene and benzene on platinum on the φ_r value. For naphthalene the adsorption maximum lies at 0.1–0.4 V in acidic solutions and at -0.4 V (normal hydrogen electrode) in alkaline solutions. The dependence of benzene adsorption on potential is also described by a bell–shaped curve, the peak of which occurs at potentials corresponding to minimum adsorption of hydrogen and oxygen in solutions of sulfuric acid + sodium sulfate and sodium hydroxide + sodium sulfate with pH values between 0.8 and 11 (Fig. 8).

With other organic compounds interpretation of the relationship between adsorption and the potential of the electrode is more complicated. The dependence of θ on φ for methanol has been determined experimentally [12, 17, 33, 146]. Thus, if during measurement of coverage the potential of the electrode is changed abruptly from 1.2 V (where methanol is not adsorbed on the electrode) to a selected value of φ_r, a dome-shaped relationship is obtained between θ and φ_r with a maximum at $\varphi_r = 0.35-0.55$ V, irrespective of the methanol concentration and the pH value of the solution [12, 146].* Under similar conditions on iridium [148] and rhodium electrodes [149] the relationship between θ and φ_r gave curves with maxima at $\varphi_r = 0.3-0.4$ V in the first case and $\varphi_r = 0.2-0.3$ V in the second, i.e., the maximum corresponds to the region of minimum hydrogen and oxygen adsorption on the electrodes.

However, if methanol was adsorbed at φ_r corresponding to maximum adsorption and the potential of the electrode was then decreased [17, 33], no appreciable decrease in θ was observed over a prolonged period [18, 39]. These results demonstrate the irreversible nature of methanol adsorption. In such a case the decrease in adsorption with decreased φ_r, found in the experiments described above, can be related to a decrease in the adsorption rate of the alcohol at a surface occupied by hydrogen, as a result of which stationary surface coverage is not reached during the measurements. Moreover, the decrease in the adsorption of methanol at anodic φ_r values is evidently due to oxidation of the adsorbed particles to form carbon dioxide. This suggestion is confirmed by experiments on a platinum — ruthenium electrode [150]. In this case the descending branch of the θ, φ_r curve is displaced by ~150 mV to the cathodic side compared with the descending branch of the analogous curve obtained on a platinum electrode. The overpotential of stationary electrochemical oxidation of

---

* The adsorption of methanol on platinum can be [147] determined by a radiochemical method, and in this case an adsorption maximum was obtained at ~0.2 V. The reasons for the difference from the results obtained in other work are not clear.

methanol on a platinum – ruthenium electrode is also ~150 mV lower than that on a platinum electrode.

The dependence of formic acid adsorption on φ_r has been investigated in some detail [75]. Considerable adsorption of formic acid is retained in the hydrogen region of potentials, and this is probably due to the possibility of reaction between formic acid and adsorbed hydrogen with simultaneous formation of an adsorption product:

$$H_{ads} + HCOOH \rightarrow H_2 + COOH_{ads}.$$

Decreased adsorption of formic acid at $\varphi_r \geqslant 0.35$ V is due to desorption resulting from oxidation of the adsorbed particles.

Adsorption of carbon monoxide, which is maximum at φ_r = 0.4–0.9 V, decreases rapidly to zero at $\varphi_r \geqslant 0.91$ V owing to oxidation of the adsorbed layer.

Maximum adsorption of methane, ethane [101, 102, 106, 151], and butane is observed at $\varphi_r \simeq 0.3$ V. Below 0.1 V for all the compounds and above 0.5 V for methane and 0.9 V for ethane the adsorption becomes close to zero. Maximum adsorption of propane [111, 112] occurs between 0.1 and 0.7 V, and that of n–octane between 0.2 and 0.3 V [152].

According to measurements by a radioactive tracer method [119, 153], maximum adsorption of ethylene on platinum in 1 N sulfuric acid solution occurs at φ_r = 0.4–0.45 V. Gilman [109, 118] made a detailed investigation of the adsorption of ethylene and acetylene on smooth platinum in 1 N perchloric acid solution by pulse methods. The amount of electricity Q_{org} consumed in the oxidation of the adsorbed ethylene and acetylene was maximum if the adsorption took place at φ_r = 0.2–0.4 V. The decrease of Q_{org} at low potentials was explained by electrochemical hydrogenation of ethylene and acetylene [154]. The decrease of Q_{org} at high φ_r values is due to a number of causes: variations in the structure of the adsorbed layer due to more intense dehydrogenation of the hydrocarbons; appearance of adsorbed oxygen, competing with the organic substance for sites on the surface; decrease in the adsorption rates of ethylene and acetylene; oxidation of the adsorbed particles.

In the literature mention is made of the irreversible nature of the adsorption of organic compounds due to irreversibility of the adsorption and ionization of oxygen on platinum when the electrode is polarized to $\varphi_r \geq 1.5$ V [33, 88, 155, 156].

Dahms and Green [157, 158] found an adsorption maximum for benzene, naphthalene, and phenanthrene on gold at $\varphi_r \sim 0.5$ V. The decreased adsorption with departure from $\varphi_r = 0.5$ V was explained by the desorbing action of the double layer. This conclusion requires further investigation [9].

5. Dependence of Adsorption on Concentration of Organic Substance. Adsorption Isotherms

The dependence of the adsorption for various substances on concentration c has been investigated in detail by Bagotskii, Vasilev, and co-workers using pulse methods on smooth platinum, iridium, and rhodium [12, 60, 75, 146, 148, 149, 159]. On platinum at moderate coverages θ (between 1.0 and 0.85) with methanol, ethylene glycol, and formic acid a linear relationship is observed between θ and log c when c is changed by four orders of magnitude, i.e., the adsorption is described by the logarithmic Temkin isotherm:

$$\theta = a + \frac{1}{f} \ln c, \tag{4}$$

where f is an inhomogeneity factor for the platinum surface.

The slopes of the isotherms for the various substances are approximately identical in the central part and close to the slope of the hydrogen adsorption isotherm. On iridium the methanol adsorption isotherms are satisfied by the Freundlich equation, where the power in the equation of the isotherm is close to that in the isotherm for hydrogen adsorption on iridium. The adsorption isotherms for methanol and hydrogen on rhodium are the same in nature. These results lead to the conclusion that the dependence of θ and c is determined not by the nature of the adsorbed substance but by the nature and character of inhomogeneity in the catalyst surface.

Chizmadzhev and Markin [160] investigated theoretically the adsorption isotherms of molecules which occupy two elementary areas on an inhomogeneous surface. Equations for the isotherms were derived for "domain" inhomogeneity, where areas with identical adsorption energies form continuous sections of fairly large size, and for "microscopic" inhomogeneity, where the energies of adjacent sections are not interrelated. In both cases the dependence of θ on c is satisfied by a logarithmic adsorption isotherm. However, in domain inhomogeneity $1/f$ is determined by a range of energies related to the whole adsorbed molecule. In the region of moderate coverages θ with microscopic inhomogeneity the slope of the isotherm is determined by a range of adsorption energies related to one area (of bond). Comparison of theory with the data given above shows that the idea of microscopic inhomogeneity corresponds more closely to the adsorption characteristics of the platinum electrode surface. A conclusion about microscopic inhomogeneity in the platinum surface can also be reached when a different approach is used [161].

For compounds of the methanol type, which are subject to intense destructive changes during adsorption, some importance attaches to the question as to whether the dependence of θ on c can be regarded as resulting from establishment of thermodynamic equilibrium between the adsorption and desorption processes, since it is possible that the isotherms obtained only express the balance between the rates of adsorption and hydrogenation at low φ_r values and the rates of adsorption and oxidation to carbon dioxide and water at high φ_r values [50, 74, 108, 150]. Further research is required to solve this problem.

Heiland, Gileadi, and Bockris [120, 121] used the Frumkin isotherm to describe the adsorption of benzene on platinum in solutions of sulfuric and phosphoric acids.

Experimental results on the adsorption of naphthalene on gold satisfy the following equation:

$$Bc = \frac{\theta}{(1 - \theta)^5},$$ (5)

which was derived for adsorption on a homogeneous surface on the assumption that the naphthalene molecule lies flat and displaces five water molecules [158].

Measurements of the relationship between θ and φ_r for ethylene by a tracer method [119] and pulse methods [108] led to greatly divergent results. According to [119], adsorption of ethylene on platinum obeys the Langmuir isotherm. In the maximum adsorption region (0.2-0.4 V), according to [108], the Q_{org} the values for ethylene and acetylene do not depend on their partial pressures. These differences are evidently due to different conditions for the preliminary treatment of the electrodes and assessment of their true surface areas.

6. Adsorption Kinetics of Organic Compounds

Adsorption of organic substances on a platinum electrode proceeds at comparatively low rates, which in a number of cases can be determined directly by experiment. Systematic investigation of the adsorption kinetics on smooth platinum, iridium, and rhodium has been carried out by Bagotskii, Vasil'ev and co-workers [12, 60, 148, 149, 162, 163]. Curves for the variation of the surface coverage with time were obtained for a solution with a constant concentration of organic substance. Such relationships are called kinetic adsorption isotherms. On platinum and rhodium in the region of moderate coverges:

$$\theta = \text{const} + \frac{1}{\alpha f}\ln t, \tag{6}$$

where $0 < \alpha < 1$. On iridium the kinetic isotherms are linear against the coordinates $\log \theta$ and $\log t$. The nature of the kinetic isotherms is thus determined by the nature of the inhomogeneity of the electrode surface.

The adsorption rates v_{ads} of CH_3OH, CH_2OHCH_2OH, and $HCOOH$ on platinum at constant potential are closely described by the Roginskii – Zel'dovich equation [164-166]:

$$v_{ads} \equiv \frac{d\theta}{dt} = K_{ads} c \exp(-\alpha f \theta), \tag{7}$$

where K_{ads} is a constant.

The overall effect of the concentration of organic substance and the coverage of the surface with adsorbed hydrogen θ_H and oxygen θ_O is determined by the following equation:

$$v_{ads} = K_{ads}\, c \exp\left[-\alpha/\theta - \alpha_H f_H \theta_H - \alpha_O f_O \theta_O\right], \tag{8}$$

where $\alpha_H f_H$ and $\alpha_O f_O$ relate to adsorption of hydrogen and oxygen at the electrode.

The differential activation energies E_a of the adsorption process for methanol, ethylene glycol, and formic acid on platinum are satisfied by the equation for the kinetics of activated adsorption on a uniformly inhomogeneous surface:

$$E_a = E_a^0 + \alpha f RT\theta. \tag{9}$$

An attempt was made [167] to interpret the curves for the potential displacement of a platinum electrode with an open circuit observed after addition of methanol to the solution on the basis of established relationships for adsorption kinetics.

Measurements on the adsorption kinetics of ethane and methane on a porous platinum–black electrode [101, 102] gave results which satisfied the Roginskii – Zel'dovich equation.

Gilman [107] found that the adsorption rate of ethane was approximately half the diffusion rate in the solution. In his opinion, kinetic measurements at low and medium coverages are described best of all by the question:

$$\frac{d\theta}{dt} = K_{ads}\, c\, (1 - \theta)^2, \tag{10}$$

This means that adsorption of ethane is a second–order reaction with respect to unoccupied sites at the electrode surface and obeys Langmuir kinetics. Gilman proposed the following mechanism for the retarded adsorption stage:

$$C_2H_6 + 2Pt \rightarrow \underset{\underset{Pt}{|}}{C_2H_5} + \underset{\underset{Pt}{|}}{H}.$$

In the presence of Cl^- the initial adsorption rate of ethane decreases according to the following equation:

$$v_{ads} = K_{ads}\, c\, (1 - \theta - \theta_{Cl^-})^2, \tag{11}$$

where θ_{Cl^-} is the degree of coverage of the surface with Cl^- anions.

The value of θ_{Cl^-} decreases with time, however, and the adsorption rate of ethane approaches the value which is obtained in the absence of Cl^- anions.

Cairns and co-workers [114] expressed the adsorption rate of propane in hydrogen fluoride solutions by an equation for a third-order reaction with respect to free sites on the platinum surface:

$$v_{ads} = K_{ads} c \, (1 - \theta)^3. \tag{12}$$

The adsorption rate of ethylene [118, 119], acetylene [118], and propane in phosphoric acid solutions [112], n-butane [106], n-octane [152], benzene [120], carbon monoxide [88, 91], and 1-hexene [168] under conditions used in these papers (stationary electrodes) during the initial period is diffusion controlled.

II. KINETICS AND MECHANISM FOR ANODIC OXIDATION OF ORGANIC COMPOUNDS

Investigation of the anodic oxidation of organic substances involves certain experimental difficulties due, on the one hand, to the complex multistage nature of the reactions and, on the other, to the special characteristics of the surface state in platinum metals. This makes it necessary to employ a range of experimental methods to determine the mechanism of the anodic processes.

1. Methods for Investigation of Kinetics and Mechanism of Electrochemical Oxidation in Organic Substances

The general course of the anodic oxidation reaction is established by analysis of the reaction products and coulometric measurements. Comparison of the amount of carbon dioxide, which is the final oxidation product from organic compounds, with the number of coulombs passing through the system allows the coulombic efficiency of the electrochemical oxidation process to be evaluated. This quantity may be a function of potential [169], and coulometric measurements are therefore best carried out under potentiostatic conditions.

To determine the kinetic parameters of electrochemical oxidation reactions stationary polarization curves are obtained by galvanostatic and potentiostatic methods. As shown by experience, the establishment of constant potential in galvanostatic measurements and constant current in potentiostatic measurements in most cases requires large intervals of time. Here not only the value of the potentials and currents but also the slopes of the polarization curves before the establishment of a stationary state are functions of the time of measurement.

On account of the need for prolonged polarization of the electrode the need arises (particularly when working with smooth electrodes) for thorough purification of the solutions from various types of contaminants capable of poisoning the catalyst. In order to obtain reliable results, in addition to purification of the solutions, it is useful to reduce the ratio between the volume of the solution and the surface area of the electrode [170, 171].

The above-mentioned circumstances have the result that the potentiostatic method, which is used with success for investigating simple electrode reactions (evolution and ionization of hydrogen or oxygen), makes it possible to obtain information on the stationary oxidation process only with appropriate selection of potential scanning rate (e.g., for investigation of the oxidation of ethylene the rate should be less than 10^{-4} V/sec) [172].

Palladium membrane electrodes have been used to investigate the mechanism of electrochemical oxidation [12, 77]. The mechanism of formic acid decomposition of palladium has been investigated by means of partly deuterated compounds [79, 80].

Daniél'-Bek and co-workers [173, 174] have developed a method for combining gasometric and polarization measurements during anodic and cathodic polarization of an electrode in solutions of organic compounds.

2. Some Criteria for Establishment of Anodic Oxidation Mechanisms

As possible oxidation mechanisms we will examine: a) direct transfer of electrons from the reactants to the surface [175] (electrochemical mechanism) and b) dehydrogenation and reaction of adsorbed substances with OH_{ads} or other forms of adsorbed oxygen (chemical mechanism) [12, 15, 19, 26, 176]. Choice of mechanism can be based on determination of the dependence of the oxidation currents i on the pH value of the solution.

For the electrochemical mechanism, in accordance with the laws of electrochemical kinetics, it should be expected (regardless of pH) that it will be constant with constant electrode potential φ, measured with reference to the same reference electrode; $(\partial \log i / \partial pH)_{\varphi} = 0$. For the chemical mechanism the oxidation rate should remain constant with identical φ_r values, i.e. $(\partial \log i / \partial pH)_{\varphi_r} = 0$.

In fact, when an organic compound RH is slowly dehydrogenated according to the following scheme:

$$RH \xrightarrow{\text{slow}} R_{ads} + H_{ads} \tag{13}$$

the following equation can, as shown by Frumkin and Podlovchenko [176], be written for the reaction rate:

$$i = K_1 c \exp\left(-\frac{\beta \mu_H}{RT}\right), \tag{14}$$

where μ_H is the chemical potential of the adsorbed hydrogen; K_1 and β are constants $(0 < \beta < 1)$. Since,

$$\mu_H = -F\varphi_r + \text{const}, \tag{15}$$

then

$$i = Kc \exp\left(\frac{\beta F \varphi_r}{RT}\right) \tag{16}$$

and

$$\left(\frac{d \log i}{d pH}\right)_{\varphi_r, c} = 0. \tag{17}$$

The derivation of Eq. (16) was based on the assumption that the energy of the bond between the hydrogen and the surface varies with the degree of coverage and that the dehydrogenation kinetics are determined by the maximum energy of the bond at sections which remain uncovered by adsorbed hydrogen. If there is no adsorbed hydrogen on the electrode surface, the dehydrogenation rate should not depend on potential.

Oxidation of an organic substance involving adsorbed oxygen can be represented as follows:

$$H_2O - \bar{e} \rightleftarrows OH_{ads} + H^+ \text{ (acidic solutions)} \tag{18a}$$

or

$$OH^- - \bar{e} \rightleftarrows OH_{ads} \text{ (alkaline solutions).} \tag{18b}$$

$$OH_{ads} + RH_{ads} \xrightarrow{\text{slow}} R_{ads} + H_2O. \tag{19}$$

With constant surface coverage with organic substance θ the kinetic equations corresponding to such a scheme take the following form [12, 15, 19, 26, 176]:

$$i = K \exp\left(\frac{F\varphi_r}{RT}\right), \tag{20}$$

$$\left(\frac{d\lg i}{d\text{pH}}\right)_{\varphi_r} = 0, \tag{21}$$

$$\left(\frac{d\varphi_r}{d\lg i}\right)_{\text{pH}} = 2.3 \frac{RT}{F}. \tag{22}$$

According to Eq. (22), the slope of the polarization curve against the coordinates φ_r and $\lg i$ should amount to 2.3 RT/F and can thus serve as an additional criterion for the establishment of an anodic reaction mechanism.

In order to realize the mechanism described by Eq. (19) it is necessary that adsorbed oxygen should be present on the platinum metals even at $\varphi_r \simeq 0.4$ V. As shown by Petrii and Marvet [177] by the method of isoelectric potential shifts and adsorption curves, in acidic solutions which do not contain specifically adsorbed anions and organic substances the regions of hydrogen and oxygen adsorption on platinum overlap, and the first portion of adsorbed oxygen may appear even at 0.35-0.4 V. In the literature, however, there are no direct data to show that oxygen can be adsorbed in this φ_r range if adsorbed organic substance is present at the surface.

Bogdanovskii, Vovchenko, and co-workers [178] employed poisoning of the electrode with mercury and arsenic, which act as poisons for the dehydrogenation process, to determine the oxidation mechanism of organic substances.

The possibility of oxidation by a mixed mechanism, including both direct electrochemical oxidation and dehydrogenation, is not excluded. Daniél'-Bek and co-workers [173, 174] proposed a method for combining gasometric and polarization determination in order to determine the proportions of each mechanism in the overall process. Such a division is based on the assumption that the dehydrogenation rate does not depend on potential, whereas the electrochemical oxidation rate bears an exponential relationship to φ [see, however, Eqs. (16) and (20) and the experimental results cited below].

2. Modern Theories on the Oxidation Mechanism for Various Classes of Organic Compounds

It is convenient to start a discussion of the theories on the oxidation mechanism of organic compounds with the electrochemical oxidation of methanol, which at present has been investigated in greatest detail.

a) <u>Electrochemical Oxidation Mechanism of Methanol</u>. The relationships governing electrochemical oxidation of methanol on a free surface or a surface slightly covered with adsorbed substance and under stationary conditions are different [17-26]. The process on a clean surface can be studied by determining the nonstationary anodic current which develops during introduction of methanol into contact with a platinum electrode kept at constant potential. The data given below show that this current is due to ionization of hydrogen formed during dehydrogenation of the methanol, i.e., it is determined by the adsorption kinetics of methanol. In fact the maximum nonstationary current observed initially depends little on the pH of the solution in acidic solutions when φ_r = constant. At potentials in the double-layer region the reaction rate increases by only an order of magnitude when the potential changes by 350-400 mV (Fig. 9). The oxidation rate varies linearly

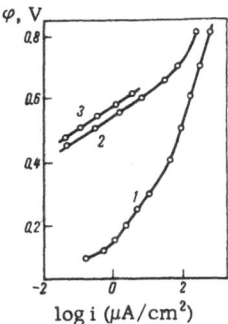

Fig. 9. Dependence of methanol oxidation rate (1, 2) at initial moment of time (1) and under stationary conditions (2) and oxidation rate of strongly chemisorbed substance (3) on potential of Pt/Pt electrode in 1 N sulfuric acid solution [19].

with the methanol concentration. The nonstationary current [22] decreases with time, according to the following equation:

$$i = \frac{\text{const}}{t}. \tag{23}$$

The rate of the stationary methanol oxidation process at low potentials is considerably lower than the adsorption rate. Thus, the difference amounts to more than four orders of magnitude at $\varphi_r = 0.4$ V. In acidic solutions $(\partial \log i/\partial pH)_{\varphi_r} \simeq 0$. In the $\varphi_r = 0.45-0.55$ V range the polarization curve is satisfied by a Tafel equation with a slope of ~ 0.06 V. A large amount of data indicates participation of the methanol adsorption products in the stationary process and parallelism between the relationships governing oxidation of strongly chemisorbed methanol and oxidation of methanol in solution. In the region of medium coverages with methanol at $\varphi_r = \text{const}$ [26]:

$$i = K \exp(\beta/\theta), \tag{24}$$

where $\beta \simeq 0.45$. Equation (24) for the oxidation current of methanol is analogous to the expression for the desorption rate with a uniformly inhomogeneous surface [164-166]. With $\varphi_r = \text{const}$ and pH = const,

$$i = Kc^\beta, \tag{25}$$

where β is close to 0.5. The power relationship can be explained on the basis of Eqs. (4) and (24). In the region of potentials between 0.55 and 0.75 V the polarization curve for oxidation of methanol, obtained on the condition that $\theta = \text{const}$ [26], is described by the following equation:

$$i = K \exp\left(\frac{\beta' F \varphi_r}{RT}\right), \tag{26}$$

where the experimental β' value is close to unity. On the basis of these results the most likely suggestion is that in acidic solutions the slow stage is reaction between the adsorbed particles and OH_{ads} according to Eq. (19).

With $\varphi_r > 0.8$ V the methanol oxidation rate decreases on account of adsorption of oxygen on the electrode. Participation of OH_{ads} particles in electrochemical oxidation of adsorbed methanol makes the reaction rate dependent on the proportion of free surface. This evidently explains the nature of the charging curves in the presence of adsorbed organic substances, namely, the fact that φ_r is independent of the amount of electricity passed in the initial section of the arrest for oxidation of the adsorbed substance and even the appearance of a minimum on the curves corresponding to fairly high coverages [20, 21].

In addition to carbon dioxide, the products from electrochemical oxidation of methanol with $\varphi_r = 450-500$ mV under stationary conditions in the initial period of electrolysis (10–12 h) were found to contain formaldehyde and formic acid, the current yields of which amounted to ~3 and 10%, respectively, in 1 N sulfuric acid solution and ~1 and 70% in 1 N potassium hydroxide solution [51]. On the basis of these data Podlovchenko, Frumkin, and Steinin [51] proposed that the process involves several parallel reactions:

$$CH_3OH_{sol} \rightarrow CH_3OH_{ads} - \begin{cases} \rightarrow SCS \xrightarrow{i_1} CO_2 \\ \rightarrow A_1 \xrightarrow{i_2} CO_2 \\ \rightarrow A_2 \xrightarrow{i_3} HCOOH\ (HCOO^-) \\ \rightarrow A_3 \xrightarrow{i_4} H_2CO \end{cases}$$

where SCS is a strongly chemisorbed substance; A_1, A_2, and A_3 are relatively weakly adsorbed particles. A similar scheme was recently proposed by Breiter [179]. For acidic solutions $i_2 > i_3, i_4$, and i_1; for alkaline solutions $i_3 > i_2 > i_1, i_4$.

Petrii and Éntina [150, 180] investigated the electrochemical oxidation of methanol on a platinum − ruthenium electrode in detail. On a clean surface the process is determined by dehydrogenation of methanol, and under stationary conditions by the oxidation of methanol adsorption products with participation of OH_{ads}. The kinetic relationships governing the process on the platinum − ruthenium electrode are found to be simpler than on the Pt /Pt electrode. In particular, the equations

$$(d\log i/d\mathrm{pH})_{\varphi_r} = 0 \quad \text{и} \quad (d\varphi_r/d\log i)_{\mathrm{pH,\ }c} = 2.3 \frac{RT}{F}$$

are valid over a wider range of pH values than on platinum. The results from analysis of the solution for methanol and formaldehyde in the course of the stationary process are found to be identical in 1 N sulfuric acid and 1 N potassium hydroxide solutions.

b) <u>Anodic Oxidation of Carbon Monoxide.</u> Although in composition carbon monoxide is the simplest substance capable of being oxidized, its anodic oxidation has been investigated comparatively little. On the basis of galvanostatic measurements Warner and Schuldiner [83] proposed that oxidation of carbon monoxide involved adsorbed oxygen:

$$O_{ads} + CO_{ads} \rightarrow CO_2. \tag{27}$$

Gilman [89] investigated the anodic oxidation of carbon monoxide on smooth platinum by the method of triangular sweeps. He observed a rapid increase in the rate of the process as the adsorbed substance became oxidized. To explain this result a mechanism of "reacting pairs" was postulated, according to which the CO_{ads} reacted with molecules of water adsorbed at adjoining sites on the surface. The reaction rate thus depends both on the coverage of the surface with adsorbed substance and on the number of free sites. Here it was necessary to assume that the adsorbed carbon monoxide aggregates into islets, and that oxidation only occurs at the boundaries of islets. With φ = const it is then possible to write (with certain simplifying assumptions) the following equation for the reaction rate:

$$i = k\,(\theta_{CO} - a)^p\,(\theta_F - b)^q, \tag{28}$$

where θ_{CO} is the proportion of the surface covered with carbon monoxide; θ_F is the proportion of free surface; p and q are constants between 0.5 and 1; a and b are the proportions of inactive sections on the surface, where oxidation does not occur.

The possibility of oxidation of carbon monoxide on platinum with participation of surface oxides was proposed by Stonehart [92]. He related the decrease in oxidation rate with time needed for accumulation (occurring in parallel with the main reaction) of the one-electron oxidation product CO^+ at the surface; this is then slowly removed by an electrochemical mechanism [181].

Miller, Tyurikova, and Veselovskii [182] carried out an investigation into the electrochemical oxidation of carbon monoxide on palladium in 5 N potassium hydroxide solution and on platinum in 0.1 N sulfuric acid. Two sections were found on the polarization curves for palladium. An anodic process evidently takes place on the first through electrochemical conversion of carbon monoxide followed by oxidation of the hydrogen formed. On the second section oxidation of carbon monoxide takes place in the region of oxygen adsorption on the electrode and may be realized either through discharge of OH^- with subsequent reaction between OH_{ads} and CO_{ads} or through discharge of an adsorbed $Pt - COOH^-$ complex. On platinum in acidic solution the rate of the process is determined by the reaction between carbon monoxide and adsorbed oxygen (OH_{ads}).

c) Electrochemical Oxidation of Carboxylic Acids at Low Anodic Potentials. Experimental data which allow conclusions to be reached about the oxidation mechanism of carboxylic acids at low anodic potentials are at present only available for formic acid, oxalic acid, and maleic acid, although many other organic acids [1] are also subject to anodic oxidation at low φ_r values. According to data from [12, 77-80], oxidation of formic acid and the formate ion at a palladium electrode takes place by a dehydrogenation mechanism. This conclusion is directly confirmed by experiments made by Bagotskii, Vasil'ev, and co-workers [12, 77] on a palladium membrane electrode.

If formic acid is brought into contact with a platinum electrode kept at constant potential, a nonstationary ionization current is found for the hydrogen formed during dehydrogenation of the acid, as in the case of methanol. With φ_r = 0.4-0.7 V the dehydrogenation rate is practically independent of φ_r. With φ_r < 0.4 V the slope of the polarization curve for dehydrogenation amounts to ~140-150 mV.

Conclusions regarding the oxidation mechanism of formic acid and formate ion on platinum under stationary conditions are considerably less definite.*   The stationary oxidation rate of formic acid at φ_r < 0.4 V is much lower than the rate of adsorption with dehydrogenation [75]. The slopes of the Tafel sections of the electrochemical oxidation curves give a value of ~50-60 mV in the 300-450 mV range for formic acid in 1 N sulfuric acid and ~60-70 mV in the 280-360 mV range for sodium formate in 0.1 N potassium hydroxide solution. The overpotential for electrochemical oxidation of substances which are strongly chemisorbed in solutions of formic acid and sodium formate is approximately 200 mV higher than the overpotential for electrochemical oxidation of these substances in solution. It is most likely that stationary electrochemical oxidation of formic acid and formate ion takes place by at least two parallel mechanisms [10, 21, 74] — through strongly adsorbed particles and through less strongly bound and more readily oxidized particles. The contribution of the first reaction to the total oxidation current is small, and the strongly chemisorbed substance can then be regarded as an unreactive side product, which blocks the surface. By taking the low values of the oxidation potentials for formic acid and sodium formate into account, it is possible to regard the dehydrogenation of weakly adsorbed particles as the slow stage in a process at a surface occupied by a strongly adsorbed substance and hydrogen. The complex nature of electrochemical oxidation of formic acid is demonstrated by the slow establishment of the stationary potentials when polarization curves were recorded by the galvanostatic method [20].

Daniél'-Bek and Glazatova [174] investigated the oxidation of formic acid on porous carbon platinized electrodes at low positive potentials (up to 200 mV) in 1 M sulfuric acid solutions, by

---

* In some papers [183, 184] it is stated that the formate ion is not oxidized on platinum. However, the possible oxidation of sodium formate is strongly alkaline solution at low φ_r values was frequently indicated [20, 76, 81, 185, 186].

combining polarization and gasometric measurements. They put forward the suggestion that a mixed mechanism of electrochemical oxidation including both dehydrogenation and direct electrochemical oxidation of formic acid was possible.

During galvanostatic oxidation of formic acid on a rhodium electrode periodic changes in potential were observed under certain conditions [6, 55]. Periodic effects have also been described in electrochemical oxidation processes [62, 187, 188]. It has been suggested that the potential fluctuations may be due to periodic formation and oxidation of an adsorbed layer of organic substances on the electrode [55].

Oxidation of oxalic acid in 1 N sulfuric acid solution takes place quantitatively to give carbon dioxide. Oxalic acid is not oxidized in alkaline solutions at low φ_r values. Only undissociated species are thus subject to oxidation. Experimental results on the oxidation kinetics at 80°C were represented by the following equation [189]:

$$i = kc_{H_2C_2O_4}^{0.35}\, c_{H^+}^{-0.55} \exp\left(\frac{F\varphi}{RT}\right). \tag{29}$$

A number of possible oxidation mechanisms were examined on the assumption that the electrode surface was uniform (Langmuir conditions) or uniformly inhomogeneous (Temkin conditions). The following were put forward as most likely schemes:

$$H_2C_2O_4 \rightleftarrows HC_2O_4{}_{ads} + H^+ + e^-,$$
$$HC_2O_4{}_{ads} \xrightarrow{\text{slow}} 2CO_2 + H^+ + e.$$

and

$$H_2C_2O_4 \rightarrow HC_2O_4{}_{ads} + H^+ + e,$$
$$HC_2O_4{}_{ads} \xrightarrow{\text{slow}} HCO_2{}_{ads} + CO_2,$$
$$HCO_2{}_{ads} \rightarrow CO_2 + H^+ + e.$$

The two schemes under "Temkin conditions" lead to relationships which are consistent with the experimental data.

According to data from Johnson and Gilmartin [190] maleic acid is oxidized to carbon dioxide on platinum in acidic solution at 80°C. At $\varphi_r = 0.35\text{–}0.5$ V the slope of the polarization curve amounts to ~140–165 mV. The following kinetic parameters were

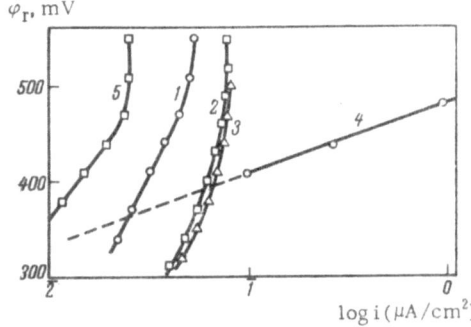

φ_r, mV

$\log i\,(\mu A/cm^2)$

Fig. 10. Polarization curves for electrochemical oxidation of methane
(1, 2, 3, 5) and its chemisorption
product (4) on Pt /Pt electrode in 1
N sulfuric acid solution [100]. 1) and
4) 60°C; 2) 80°C; 3) 95°C, p_{CH_4} = 1
atm; 5) 60°C, p_{CH_4} = 0.33 atm.

also obtained for the reaction:

$$\left(\frac{d\log i}{d\mathrm{pH}}\right)_\varphi = 0; \quad \left(\frac{d\log i}{dc_{C_4H_4O_4}}\right)_\varphi < 0;$$

only $C_4H_4O_4$ molecules take part in the reaction. The proposed oxidation scheme includes the slow stage of H_2O or OH^- discharge to
form OH_{ads} and is similar to the scheme which is examined below
for oxidation of unsaturated hydrocarbons with double bonds.

d) Electrochemical Oxidation of Hydrocarbons. Petrii and
Marvet [100] made a detailed investigation of the electrochemical
oxidation of methane at a platinum electrode. The stationary oxidation current of methane increases little with increase in φ_r (Fig.
10). Thus, when the potential is increased from 300 to 500 mV at
60°C, the reaction rate increased by only about three times. In
the region where $\varphi_r \geqslant 550$ mV at 60°C, and at somewhat lower
φ_r values with higher temperatures, passivation evidently begins
through adsorption of oxygen. The methane oxidation rate decreases linearly with decrease in the partial pressure of methane
p_{CH_4} and changes little when the pH is changed from 0.37 to 2.34.
Calculation of the activation energy for the methane oxidation in 1
N sulfuric acid solution at $\varphi_r = 0.4$ V gave ~17 kcal /mole, which
is close to the value quoted by Niedrach [101, 102]. In the presence of 0.5 N hydrochloric acid the methane oxidation current becomes close to that of the supporting electrolyte.

Comparison of the oxidation curve for methane with the oxidation curve for its chemisorption products, plotted from the potentials corresponding to the beginning of oxidation of chemisorbed

methane at various current densities, showed that with $\varphi_r > 0.35$ V the oxidation rate of the chemisorption products considerably exceeds the rate of the stationary methane oxidation process. In accordance with this the chemisorbed substance does not accumulate during polarization of the electrode under stationary conditions in the region where $\varphi_r > 0.35$ V, and the reaction proceeds on a platinum surface which is practically free from chemisorption products.

The results obtained for the electrochemical oxidation of methane on platinum can be explained on the supposition that dehydrogenation of methane at the electrode surface is retarded. As shown above, similar relationships are found in the dehydrogenation of methanol at potentials of the double-layer region. The dehydrogenation rate of methanol at 20°C, calculated for the same volume concentration as methane, is found to be of the same order as the dehydrogenation rate of methane at 60°C. The assumption about dehydrogenation as the rate-determining step in the stationary methane oxidation process is consistent with the effect of pH in the acidic region, the effect of specific adsorption of Cl⁻ ions, and the effect of the methane partial pressure. In contrast oxidation of the methane chemisorption products is evidently determined by interaction with OH_{ads} particles.

The electrochemical oxidation mechanism of a large group of unsaturated hydrocarbons (ethylene, acetylene, propylene, 1-butene, 2-butene, allene, butadiene, cyclohexadine, benzene) was investigated in detail by Bockris and co-workers [1, 15, 16, 170, 191-196, 200]. The experimental results obtained in these papers can be formulated in the following way:

1) At $\varphi_r \simeq 0.4$–0.75 V in 1 N sulfuric acid solution (80°C) on smooth and platinized platinum electrodes, $(\partial \varphi / \partial \log i)_{pH} \simeq 0.14$–0.19 V for unsaturated carbons with double bonds and ~0.07 V for acetylene;

2) In the Tafel region of potentials $(\partial \log i / \partial p_{RH})_{\varphi, pH}$ amounts to between -0.11 and -0.2 (p_{RH} is the partial pressure of the hydrocarbons) (Fig. 11);

3) $(\partial \log i / \partial pH)_{\varphi} \simeq 0.8$ for acetylene and 0.39–0.47 for other hydrocarbons;

4) The apparent activation energies for the oxidation of various hydrocarbons lie between ~20 and ~24 kcal/mole;

5) The oxidation rate of the hydrocarbons increases in the following order: benzene < butadiene < butenes < propylene < ethylene < allene. However, the differences in the rate constants lie within an order of magnitude;

6) At $\varphi_r > 0.85$–1.05 V (depending on the nature of the hydrocarbon) the oxidation process is retarded through adsorption of oxygen at the surface;

7) The oxidation rate of hydrocarbons at the electrode increases if the potential of the electrode is temporarily increased to 0.9 V. The decrease in current with time after activation of the electrode depends on the nature of the hydrocarbon and its partial pressure and is due to adsorption of the hydrocarbon on the surface;

8) Carbon dioxide is the principal product from oxidation of unsaturated hydrocarbons. Some new important data are given in [251].

The scheme proposed on the basis of these data for the oxidation of hydrocarbons with double bonds involves the slow step of water molecule discharge to form adsorbed OH particles and rapid interaction of these particles with hydrocarbon adsorbed at the electrode surface:

$$H_2O \xrightarrow{\text{slow}} OH_{ads} + H^+ + e, \tag{30}$$

$$RH_{ads} + OH_{ads} \rightarrow \text{products.} \tag{31}$$

This scheme corresponds to the following equation:

$$i = k\,(1 - \theta_t)\exp\left(\frac{\alpha F \varphi}{2RT}\right), \tag{32}$$

where θ_t is the total coverage of the electrode surface with OH_{ads} and RH_{ads}.

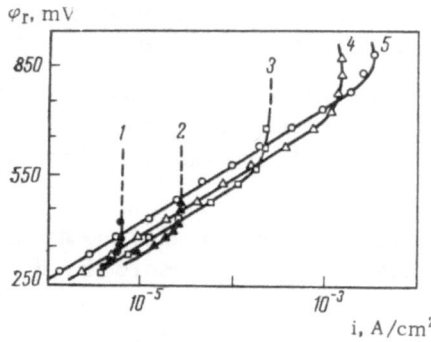

Fig. 11. Polarization curves for electrochemical oxidation of ethylene (pH 0.5) at various $p_{C_2H_4}$ values[170]: 1) 10^{-4} atm; 2) 10^{-3} atm; 3) 10^{-2} atm; 4) 10^{-1} atm; 5) 1 atm.

This equation makes it possible to explain the decrease in oxidation rate with increase in the partial pressure of hydrocarbon in the system, since the proportion of the surface on which the reaction (30) proceeds decreases. However, according to Eq. (32) $(\partial \log i / \partial pH)_\varphi = 0$, whereas in practice $(\partial \log i / \partial pH)_\varphi \simeq 0.5$. An attempt was made in [170] to explain the effect of pH on the assumption that the potential taken with reference to the zero-charge (zc) point of platinum should enter into Eq. (32) and that the zero-charge point of platinum changes with pH according to the following equation [197, 198]:

$$\varphi_{zc} = \varphi^0_{zc} + \frac{RT}{F} \ln [H^+].$$ (33)

It is, however, not possible to agree with these suggestions [136, 199].

In the electrochemical oxidation of acetylene [200] reaction (31) evidently becomes slow, since the heat of adsorption of acetylene on platinum is higher than the heat of adsorption of unsaturated hydrocarbons with a double bond. Moreover, the heat of dissociation of the CH bonds is also greater for acetylene than for ethylene. On the basis of these suppositions Bockris and co-workers [200] wrote the equation for the oxidation rate of acetylene in the following form:

$$i = k_1 \theta_{C_2H_2} \theta_{OH}.$$ (34)

From the preceding stage, which occurs under quasiequilibrium,

$$\theta_{OH} = k_2 (1 - \theta_t) \exp \left(\frac{F\varphi_r}{RT} \right). \qquad (35)$$

Substitution of Eq. (35) in Eq. (34) gives:

$$i = k\theta_{C_2H_2} (1 - \theta_t) \exp \left(\frac{F\varphi_r}{RT} \right). \qquad (36)$$

Since θ_{OH} is small in the Tafel region, $\theta_t \simeq \theta_{C_2H_2}$. Consequently, a negative reaction order should be observed with $\theta_{C_2H_2} > 0.5$.

Johnson and others [201] investigated the anodic oxidation of acetylene on gold at 80°C. Unlike platinum, where the main product is carbon dioxide, on gold the process proceeds with partial formation of polymeric products. Two linear Tafel regions, separated by a transitional region, were obtained on the polarization curves. Below this region $\partial\varphi / \partial\log i \approx 140$ mV, $(\partial i/\partial p_{C_2H_2})_\varphi < 0$, and $(\partial\log i /\partial pH)_\varphi \simeq 1$ in weakly acidic and alkaline solutions. Above the transitional region $\partial\varphi / \partial\log i \simeq 140$ mV, $(\partial i/\partial pH)_\varphi = 0$, and $(\partial\log i /\partial pH)_\varphi = 0$ in moderately and strongly acidic solutions, and $(\partial\log i /\partial pH)_\varphi = 1$ in weakly acidic and basic solutions. They proposed that below the transitional region discharge of water or OH⁻ is the slow stage and that then a rapid reaction takes place between adsorbed oxygen or OH_{ads} and adsorbed hydrocarbon. Electron transfer according to the following scheme may be slow above the transitional region:

$$C_2H_2{}_{ads} + H_2O \rightarrow C_2H_2OH_{ads} + H^+ + e,$$

$$C_2H_2{}_{ads} + OH^- \rightarrow C_2H_2OH_{ads} + e.$$

III. CATALYSTS FOR ANODIC OXIDATION OF ORGANIC SUBSTANCES

A large number of catalytically active materials have now been tested in electrochemical oxidation of various organic substances. It has been found empirically that the most active are metals of the platinum group (particularly platinum), their alloys, nickel, and silver.

For practical use the electrode must possess simultaneously high and stable catalytic activity, a well developed surface, high electrical conductivity, sufficient chemical resistance, and must be made of comparatively inexpensive materials.

1. Types of Platinum Catalyst

The platinum catalyst is usually obtained from solutions of H_2PtCl_6 and its salts by chemical or electrochemical deposition. Hydrogen, formic acid (or formates), formaldehyde, hydrazine, sodium and lithium borohydrides, and lithium aluminum hydride are used as reducing agents for chemical deposition. The reduction is carried out directly from the solution, with subsequent sedimentation of the platinum black, or on a carrier previously impregnated with platinum salts. The electrodes are prepared from the powdered platinum black by compression with Teflon [202, 203] or by compression into a dense gauze of platinum, nickel or other material. In some cases softer metals (e.g., gold [204]) are used as binding agent when compressing electrodes from platinum black. The "impact method" is also used for studying powdered catalysts, where catalyst particles in the suspended condition impose a potential on an indicator electrode [35].

The nature of the reducing agent and the deposition conditions primarily affect the specific surface area of the platinum catalyst. Shropshire and co-workers [205] found differences in the specific surface area (of two to three times) and physical properties of platinum blacks obtained with various reducing agents. According to [206], the surface area obtained in deposition of platinum black from dilute solutions of K_2PtCl_6 (0.1%) by sodium borohydride in alkaline solution can be twice as large as that obtained with formaldehyde.

Recently data have also been obtained to show that the activity of the platinum changes with change in its degree of dispersion.* Thus, it was found in [171] that the specific activity of smooth platinum in stationary electrochemical oxidation of methanol is two to three times greater than that of platinized platinum. The activity of a platinized platinum electrode increases with aging. According to data in [207], the adsorption and electrochemical

---

* Obrucheva [211] found large differences in the properties of oxide layers on smooth and platinized platinum.

oxidation rates of methanol decrease by an order of magnitude on switching from a smooth to a highly platinized electrode. During an investigation into electrochemical oxidation of formic acid [64] and of hydrocarbons [100, 170], however, it was concluded that the platinizing of electrodes only increases the true surface area, without substantially changing the catalytic activity.

Adams platinum, obtained by reduction of platinum dioxide [203, 204], has been used in a number of papers [203, 204, 210] on electrochemical oxidation. In [203] a platinum catalyst, prepared by deposition and subsequent reduction of oxidized forms of platinum on a carbon substrate, was used. Adams platinum is an active form of the catalyst, which is stable towards heat treatment and the action of contaminants [208].

Krupp and others [204] proposed that skeletal platinum, prepared by decomposition of an alloy between platinum and aluminum, should be used as catalyst. As shown by Petrii and Marvet [209], hydrogen adsorption isotherms for disperse platinum catalysts (platinized platinum, platinum black, Adams platinum, and various forms of skeletal platinum) in aqueous solutions of electrolytes exhibit relatively small differences.

The optimum conditions for electrochemical deposition of platinum have been determined [109, 212, 216]. In deposition from dilute solutions of H_2PtCl_6 (1-2%) [209] the specific surface area is found to be ~20-25 m^2/g with current densities between 10^{-3} A/cm^2. Prigent [213] found an increase of approximately seven times in the activity of electrodes (calculated for identical weights of catalysts) when the deposition potential was changed from −50 mV (normal hydrogen electrode) to 250 mV. In accordance with this [214], an increase was observed in the true surface area and the particle size decreased from 0.024 to 0.015 μ when the deposition potential was increased from 0.0 to 0.05 V. Prigent and Koch [213, 214] suggested that high coverage of the platinum surface with hydrogen at $\varphi < 0$ prevents the formation of nuclei during electrodeposition and leads to the formation of large crystals.

Bianchi [215] investigated the effect of heavy metal ion additions to the solution from which the platinum is electrodeposited on the nature of the deposits. Lead, mercury, cadmium, and thallium increase the lattice constants; chromium, manganese, iron, cobalt, and nickel reduce them; bismuth, tin, arsenic, and gold do not affect the lattice constants, although they do change the outward appearance of the deposits.

The effect of lead additions on electrolytic platinum deposits is mentioned in the literature [217–219]. Thacker [216, 220] and Prigent showed that during oxidation of propane, propylene, and n-butane in phosphoric acid good results are obtained by means of an electrode prepared in a solution containing 0.003% lead acetate, where maximum development of the platinum surface occurs. Besides increasing their specific surface area, addition of lead (and also of cadmium) to platinum deposits increases the resistance of the electrodes to recrystallization.

Binder and others [221, 222] proposed that the platinum surface should be modified with sulfur or selenium. The overpotential for oxidation of carbon monoxide and formic acid in sulfuric acid is reduced by ~200 mV on the modified electrodes. Maximum activity is given by electrodes where the coverage with sulfur amounts to ~0.3.

During investigation of a platinum catalyst on a carrier it was found that the surface area and possibly the specific activity of platinum depends on the nature of the substrate. Platinum catalysts, deposited on a carbon [203], graphite, titanium and tungsten carbides [223], granulated activated charcoal [224, 225], carbon deposited on a nickel gauze [224, 225], boron carbide [226], polyvinyl chloride [227], and tantalum [228] have been tested in anodic oxidation reactions. The authors of these papers noticed that platinum catalysts were more effective when deposited on substrates with fully developed surfaces. Thus, Cairns and McInerney [203], who tested a platinum catalyst deposited on a carbon substrate, came to the conclusion that the consumption of platinum could be reduced by an order of magnitude compared with platinum black electrodes without the carrier but with the same activity. It was also noticed that the surface areas of catalysts on carriers were not reduced so much when used in electrolyte solutions at elevated temperatures.

An improvement in the catalytic qualities of platinum in the oxidation of carbon monoxide was observed when materials, known to be effective catalysts for various heterogeneous processes (cobalt molybdate, tungsten oxides), were added [229]. The reaction rate increased by two orders of magnitude when 10-50% of the oxide was added to the catalyst.

2. Comparison of Catalytic Activity of Various Platinum Metals

In comparing the activities of catalysts in electrochemical oxidation reactions difficulties arise from the fact that the slopes of the Tafel relationships differ for different electrodes. The order of activity thus depends on the potential range in which the comparison is made.

Binder and others [230] found that Raney platinum was the best single catalyst for electrochemical oxidation of hydrocarbons. Data obtained by Bianchi [231], Grubb [232], and Cairns and McInerney [203], who compared the catalytic activities of various platinum blacks in the anodic oxidation of propane, are consistent with this conclusion. As found by Petrii and Marvet [100], the oxidation rate of methane decreases sharply on changing from platinum to other platinum catalysts. Bockris and Dahms [192] showed that the oxidation rate of ethylene decreases in the series Pt > Rh > Ir and Pd > Au, where platinum is considerably more active than palladium. It was found in [233] that the oxidation rate of propane in alkali decreases in the order Pd > Pt > Ag, and those of ethane ethylene, propylene, and n-butane decrease in the order Pt > Pd > Ag.

The catalytic activities of various noble metals in the oxidation of methanol, formaldehyde, and formic acid were first compared by Müller and his co-workers [2-8]. It was found that silver was the best catalyst for oxidation of formaldehyde in alkaline solution. According to data in [8], the metals can be placed in the following order: Pd > Rh > Au > Pt according to their activity towards the oxidation of methanol in alkaline solution. This order differs from that obtained recently [234]: Pt > Pd > Ru = Rh > Ir > Os > Au. The difference may be due to the fact that in [8] the

properties of the smooth electrodes were compared at relatively
negative potentials, whereas in [234] it was possible owing to the
use of Raney catalysts to determine the oxidation rates at more
positive potentials. For oxidation of methanol in alkaline solution
preference was given to platinum, palladium, and silver [235]. In
the oxidation of methanol vapor in sulfuric acid (65°C) Longhi [236]
found that platinum is a better catalyst than palladium, ruthenium, or
osmium.* The inertness of ruthenium at room temperature was
pointed out by Binder and co-workers [234] and by Petrii [237]. It
was shown that ruthenium, which is passive at 20°C, surpassed
platinum in activity when the temperature was increased to 80°C
[234]. Petrii and Éntin [238] explained this feature of the behavior
of ruthenium by a change in the controlling step of the process with
increase in temperature. Chemisorption of methanol on ruthenium
was found to be retarded at 20°C owing to previous and strong adsorption of oxygen. With increase in temperature the adsorption
rate increased, the appearance of adsorption products led to a
change in the surface characteristics, and at slightly positive potentials the oxidation of chemisorbed methanol became the rate-
determining step.

According to data in [21], at 20°C in the region of slightly
positive potentials and on disperse electrodes, the activity of the
platinum metals in the oxidation of methanol varies as follows: in
acidic solutions, $Pt \simeq Rh > Os > Ru > Pd$; in alkaline solutions
$Pt > Rh > Os > Pd > Ru$. With smooth platinum metals Breiter
[239] obtained the following order of activity in acidic solutions:
$Pt > Pd > Rh > Ir > Au$. The criterion for activity in this case
was the height of the peak of the first methanol oxidation wave on
the cyclic potentiostatic curve, i.e., the data were obtained in a
nonstationary process and with comparatively positive potentials.

Palladium is the best catalyst for electrochemical oxidation
of formic acid and formate ion [78, 224].

---

* In [203] it was shown that osmium electrodes are unstable, due possibly to formation
of volatile OsO_4.

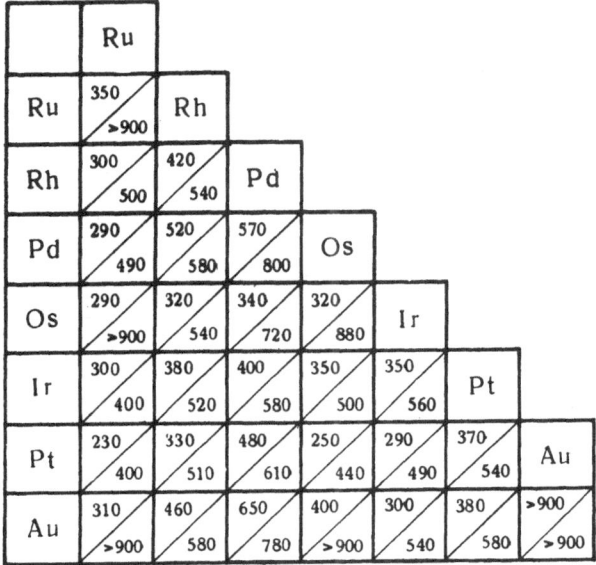

Fig. 12

3. Use of Mixed Catalysts for Anodic Oxidation Reactions

Numerous results show that in anodic oxidation reactions the activity of catalysts consisting of several metals is higher than that of the individual components [192, 237, 240, 241]. Bockris and co-workers [193] investigated a wide range of binary and ternary catalysts on the oxidation of methanol. The results (Fig. 12) show that alloys of platinum containing ruthenium and molybdenum are the most active: $Ru_2Rh_{10} < Pt < Pt_{67}Re_{33} = RuPt < Pt_{70}Ru_{15}Mo_{15}$. (The subscripts indicate the proportions of the components by weight.) It was noticed [240] that the catalytic properties of platinum are improved by addition of molybdenum.

Petrii [237] showed that the electrochemical oxidation rate of methanol and its chemisorption products in acidic solutions increases by more than two orders of magnitude when 10-20% ruthenium is added to platinum. Binder and others [234] also consider that catalysts containing ruthenium are the best. Figure 12 shows the potentials (mV) established in a solution of 2 mole/liter methanol + 4.5 N sulfuric acid at 25°C (lower figures) and 80°C (upper figures) with current (i) equal to 50 mA/cm^2 and with the alloy components in the atomic ratio 1:1 [234].

Accelerated oxidation of alcohols was found when even 5% ruthenium was added to platinum [241]. It was noticed that alloys containing ruthenium have greater activity than pure platinum in the oxidation of propane, and alloys of platinum with rhodium and iridium were found to be less active. However, the oxidation rate of methane [100] decreased when even small amounts of ruthenium were added to the platinum. Accelerated oxidation of formic acid [224] and of hydrogen – carbon monoxide mixture [242] was observed in the presence of platinum and ruthenium alloys. An improvement in the characteristics of platinum electrodes with the addition of ruthenium was observed by Vielstich [224].

Petrii and Éntin [54, 57, 78, 150, 180, 243] investigated the adsorption and electrochemical characteristics of platinum and ruthenium alloys in detail. The optimum alloy compositions for oxidation of methanol at various temperatures were determined, the anodic oxidation reactions of various compounds of the alloys were investigated, the stability of the alloys after prolonged use was investigated, and the characteristics of platinum – ruthenium alloys prepared by various methods (skeletal electrodes, electrolytically mixed deposits on a platinum and titanium carbide bases, powders deposited by sodium borohydride, smooth alloys) were compared. It was found that heat treatment of platinum – ruthenium alloys at 800°C in an atmosphere of inert gas led to loss of their high catalytic properties and formation of catalysts which behave similarly to platinum. This phenomenon is explained by diffusion of ruthenium atoms from the surface layer into the volume. The same explanation for the effect of heat treatment on the characteristics of alloys was given by Dam'yanovich and Bruzik [224], who investigated alloys of platinum and palladium with gold in the electrochemical reduction of oxygen.

Vielstich [224] proposed a platinum – palladium catalyst for oxidation of methanol – formate mixture in alkaline solutions, and this has found practical application.

TABLE 1. Comparison of Catalytic Activity of Platinum Black and Alloys in Oxidation of Methanol [163]

Metal or alloy	Overpotential in 2 N H_2SO_4, V (at 100°C and 20 mA/cm^2)	Remarks
Pt black	0.44	
Binary Pt alloys		
with main metals of Groups		Best alloys
IV and V	0.37—0.39	$Pt_{67}Mo_{33}$ — 0.20; $Pt_{80}Re_{20}$ —0.29; $Pt_{67}Re_{33}$ — 0.31
Groups VI and VII	0.20—0.38	
Fe, Ni, Co, Be, Pb, Cu, Bi, Sb	$\geqslant$0.44	
with noble metals		
Ru	0.24—0.32	Optimum 20% Ru (Pt_2Ru); 33% Ru (RuPt)
Os	0.30—0.35	
Rh	0.44	
Binary alloys without Pt		
Ru — Rh	0.29—0.38	Optimum $Ru_{90}Rh_{10}$
Ru with Pd, Ir, Rh; Os with Pd, Ir, Rh	0.38—0.46	
Ternary alloys of Pt + Ru with		
W	0.25—0.37	
Ta	0.24—0.27	
Zr	0.30	
Mo	0.23	Optimum $Pt_{70}Ru_{15}Mo_{15}$

4. Theories of Effect of Electrode Nature on Anodic Oxidation Processes

The differing activities of electrodes in anodic oxidation reactions arise from the surface characteristics, i.e., from the conditions for adsorption of the reagents on the electrode. Change in the adsorption conditions can lead both to change in the rate of the determining stage of the process and to change in the course of the reaction.

Thus, Niedrach and Weinstock [229] related the accelerated oxidation of carbon monoxide on a mixed platinum—vanadium oxide catalyst to the fact that, in this case, the reaction which takes place is not direct electrochemical oxidation of carbon monoxide but reaction of carbon monoxide with water followed by ionization of the hydrogen formed. The high reaction rates on platinum modified by sulfur [221] are also explained by optimum conditions for formation of a binary $H_2O_{ads} - CO_{ads}$ complex, where the water is adsorbed on the sulfur atoms and carbon monoxide on the platinum atoms.

The increased activity of platinum — molybdenum alloys is evidently due to a change in the general course of the reaction [193, 240]. Addition of molybdenum (on account of its ready dissolution in acidic solutions and the possibility of change in the valence state of the ions) can lead to formation of a redox system ($Mo^{5+} - Mo^{6+}$) near the electrode, and this leads to a change in the course of the reaction. This point of view is confirmed by the fact that oxidation of methanol and formaldehyde is accelerated by molybdate ions [245]. Shropshire [245] proposed the following scheme to explain this effect:

$$n\,[Mo^{6+}]_{ads} + \begin{Bmatrix} HCHO \\ CH_3OH \end{Bmatrix} \rightarrow n\,[Mo^{5+}]_{ads} + CO_2 + (n+1)\,H_2O,$$
$$n\,[Mo^{5+}]_{ads} + H_2O \rightleftarrows n\,[Mo^{6+}]_{ads} + H^+ + e.$$

Like the molybdates, the perrhenates [246] also accelerate the electrochemical oxidation of formaldehyde and methanol.

In a number of papers attempts have been made to find a correlation between the catalytic activity and physicochemical characteristics of metals and alloys. It has been suggested that the strong catalytic properties of platinum — ruthenium alloys can be related to their electronic structure [247, 248]. The number of unpaired d electrons on the catalyst atom can serve as a quantitative characteristic of the electronic structure. According to [247], for platinum and palladium the number of d electrons on the atoms is 0.6; for rhodium 1.4; for iridium 1.7; and for ruthenium 2.2. It is suggested that in homogeneous alloys there is a linear relationship between the number of d electrons and the alloy composition. Increased activity is due to an optimum number of unpaired d electrons. Binder and others [234] have compared the activities of catalysts possessing paramagnetic susceptibility, which bears a direct relationship to the number of unpaired d electrons. In

[247, 248], however, there is large amount of experimental material which is difficult to explain on the basis of electronic theory.

Bockris [1, 196] obtained a hump-shaped relationship (i.e., a curve with a maximum) between the activity of certain metals and alloys in the oxidation of ethylene. In [1] mention is made of the possible role of surface microstructure (orientation of crystallographic faces, concentration, and nature of defects) in anodic oxidation reactions.*

With electrochemical processes, as mentioned above, it is necessary to take account of the effect of the double-layer electric field and adsorption of ions, solvent molecules, hydrogen, and oxygen on the electrode surface. Thus, the absence of electrochemical oxidation of methane at appreciable rates on rhodium, iridium, ruthenium, and osmium is explained by previous adsorption of oxygen and stronger adsorption of anions on the surface of these metals compared with platinum, and this retards the process [100]. Further research into adsorption and electrochemical characteristics of metals and alloys is therefore required for determining the reasons for differing catalytic activities.

LITERATURE CITED

1.   B. J. Piersma and E. Gileadi, Modern Aspects of Electrochemistry, Vol. 4, ed. J. O'M. Bockris, Plenum Press, New York (1966), p. 47.
2.   E. Müller, Z. Elektrochem., 28:101 (1922); 29:264 (1923).
3.   E. Müller and K. Sponsel, Z. Elektrochem., 28:307 (1922).
4.   E. Müller and H. Hindemith, Z. Electrochem., 33:561 (1927).
5.   E. Müller and K. Schwabe, Z. Elektrochem., 33:568 (1927).
6.   E. Müller and S Tanaka, Z. Elektrochem., 34:256 (1928); 35:38 (1929).
7.   E. Müller and K. Schwabe, Z. Elektrochem., 34:170 (1928).
8.   E. Müller and S. Takegami, Z Elektrochem., 34:704 (1928).
9.   B. B. Damaskin, O. A. Petrii, and V. V. Batrakov, Adsorption of Organic Compounds on Electrodes [in Russian], Nauka, Moscow (1968).
10.   E. Gileadi, J. Electroanalyt. Chem., 11:137 (1966).
11.   S. Gilman, Electroanalytical Chemistry, Vol. 2, ed. A. J. Bard, Dekker, New York (1967), p. 111.

---

* The dependence of the adsorption and electrochemical behavior of iron on surface microstructure was recently discovered by Iofa, Batrakov, and Nikiforova [249].

12.   V. S. Bagotskii and Yu. B. Vasil'ev, in: Advances in Electrochemistry of Or-
      ganic Compounds [in Russian], Nauka, Moscow (1966); p. 38; Electrochim. Acta,
      11:1439 (1966).
13.   S. Gilman and M. W. Breiter, J. Electrochem. Soc., 109:1099 (1962).
14.   S. Gilman, J. Phys. Chem., 66:2657 (1962); 67:1898 (1963).
15.   J. O'M. Bockris, Proceedings of the First Australian Conference on Electrochem-
      istry, 1963, Pergamon Press, New York (1965); p. 691.
16.   H. Wroblowa, B. J. Piersma, and J. O'M. Bockris, J. Electrochem. Soc., 111:
      863 (1964).
17.   Khira Lal, O. A. Petrii, and B. I. Podlovchenko, Élektrokhimiya, 1:316 (1955).
18.   B. I. Podlovchenko, O. A. Petrii, and E. P. Gorgonova, Élektrokhimiya, 1:182
      (1965).
19.   O. A. Petry (Petrii), B. I. Podlovchenko, A. N. Frumkin, and Hira (Khira) Lal,
      J. Electroanalyt. Chem., 10:253 (1965).
20.   B. I. Podlovchenko, O. A. Petry (Petrii), A. N. Frumkin, and Hira (Khira) Lal,
      J. Electroanalyt. Chem., 11:12 (1966).
21.   O. A. Petrii, B. I. Podlovchenko, and A. N. Frumkin, Modern Problems in
      Physical Chemistry [in Russian], Vol. 2, Izd. MGU, Moscow (1967).
22.   S. S. Beskorovainaya, Yu. B. Vasil'ev, and V. S. Bagotskii, Élektrokhimiya,
      2:167 (1966).
23.   O. A. Khazova, Yu. B. Vasil'ev, and V. S. Bogotskii, Izv. Akad. Nauk SSSR,
      Ser. Khim., 1531:1778 (1965).
24.   O. A. Khazova, Yu. B. Vasil'ev, and V. S. Bogotskii, Élektrokhimiya, 2:267
      (1966).
25.   S. S. Beskorovainaya, O. A. Khazova, Yu. B. Vasil'ev, and V. S. Bagotskii,
      Élektrokhimiya, 2:932 (1966).
26.   V. S. Bagotzky (Bagotskii) and Yu. B. Vassilyev (Vasil'ev), Electrochim. Acta,
      12:1323 (1967).
27.   A. I. Shlygin and M. E. Manzhelei, Uch. Zap. Kishinevskogo Univ., 8:13
      (1953).
28.   T. O. Pavela, Ann. Acad. Sci. Fennica, Ser. A., 59 (1954).
29.   A. I. Shlygin, Proceedings of Conference of Electrochemistry [in Russian], Izd.
      Akad. Nauk SSSR, Moscow (1953), p. 322.
30.   G. P. Khomchenko and G. D. Vovchenko, Vestn. MGU, Ser. Estestv. Nauk,
      No. 8, 91 (1955).
31.   Yu. A. Podvyazkin and A. I. Shlygin, Zh. Fiz. Khim., 30:1521 (1956).
32.   T. I. Pochekaeva, Zh. Fiz. Khim., 35:1606 (1961).
33.   M. W. Breiter and S. Gilman, J. Electrochem. Soc., 109:622 (1962).
34.   M. W. Breiter, Electrochim. Acta, 8:477 (1963).
35.   D. V. Sokol'skii, Hydrogenation in Solutions [in Russian], Izd. Akad. Nauk
      KazSSR, Alma-Ata (1962), Ch. 2.
36.   J. Giner, Electrochim. Acta, 8:857 (1963); 9:63 (1964); Thesis at 15th CITCE
      Meeting, London (1963).
37.   A. N. Frumkin and B. I. Podlovchenko, Dokl. Akad. Nauk SSSR, 150:349 (1963).

38. B. I. Podlovchenko, O. A. Petrii, and A. N. Frumkin, Dokl. Akad. Nauk SSSR, 153:379 (1963).
39. B. I. Podlovchenko and E. P. Gorgonova, Dokl. Akad. Nauk SSSR, 156:673 (1964).
40. M. J. Schlatter, Fuel Cells, ed. G. J. Young, Reinhold, New York (1963), p. 190; 145th National Meeting American Chem. Soc., Div. of Fuel Chemistry, Vol. 7, No. 4, New York, (1963).
41. W. T. Grubb, Nature, 198:883 (1963); J. Electrochem. Soc., 111:1086 (1964).
42. L. W. Niedrach, J. Electrochem. Soc., 111:1309 (1964).
43. G. C. Bond, Catalysis of Metals, Academic Press, New York (1962).
44. D. O. Hayword and B. M. T. Trapnell, Chemisorption, Butterworths, London (1964).
45. R. W. Roberts, Trans. Faraday Soc., 58:1159 (1962); Ann. N. Y. Acad. Sci., 101:766 (1963); J. Phys. Chem., 67:2035 (1963); 68:2718 (1964).
46. D. W. McKee, J. Am. Chem. Soc., 84:1109,4427 (1962); J. Phys. Chem., 67:841 (1963).
47. G. Blyholder and L. D. Neff, J. Catalysis, 2:138 (1963); J. Phys. Chem., 70:1738 (1966); G. Blyholder and M. W. Wyatt, J. Phys. Chem., 70:1745 (1966).
48. B. A. Morrow and N. Sheppard, J. Phys. Chem., 70:2406 (1966).
49. R. S. Hansen, J. R. Arthur Jr., N. Y. Mimeault, and R. R. Rye, J. Phys. Chem., 70:2787 (1966).
50. T. Biegler and D. F. A. Koch, J. Electrochem. Soc., 114:904 (1967).
51. B. I. Podlovchenko, A. N. Frumkin, and V. F. Stenin, in: Advances in Electrochemistry of Organic Compounds [in Russian], Nauka, Moscow (1968), p. 4; Élektrokhimiya, 4 (1968).
52. M. W. Breiter, J. Electroanalyt. Chem., 14:407 (1967).
53. M. W. Breiter, J. Electroanalyt. Chem., 15:221 (1967).
54. V. S. Éntina, O. A. Petrii, and V. T. Rysikova, Élektrokhimiya, 3:758 (1967).
55. O. A. Petrii and N. Lokhanyai, Élektrokhimiya, 4:514, 656 (1967).
56. V. F. Stenin and B. I. Podlovchenko, Élektrokhimiya, 3:481 (1967).
57. V. S. Éntina, O. A. Petrii, and Yu. N. Zhitnev, Élektrokhimiya, 3:344 (1967).
58. B. I. Podlovchenko, Élektrokhimiya, 1:101 (1965).
59. B. I. Podlovchenko and V. F. Stenin, Élektrokhimiya, 3:649 (1966).
60. J. Weber, Yu. B. Vasil'ev, and V. S Bogotskii, Élektrokhimiya, 2:515, 522 (1966).
61. P. R. Johnson and A. T. Kuhn, J. Electrochem. Soc., 112:599 (1965).
62. D. F. A. Koch, Proceedings of the First Australian Conference on Electrochemistry, 1963, Pergamon Press, New York (1965), p. 657.
63. R. A. Munson, J. Electrochem. Soc., 111:572 (1964).
64. M. H. Gottlieb, J. Electrochem. Soc., 111:465 (1965).
65. C. W. Fleischmann, G. K. Johnson, and A. T. Kuhn, J. Electrochem. Soc., 111:602 (1964).
66. W. Vielstich and U. Vogel, Ber. Bunsen Ges. Physikal. Chem., 68:688 (1964).
67. M. W. Breiter, J. Electrochem. Soc., 111:1298 (1964).

68. S. B. Brummer and A. C. Makrides, J. Phys. Chem., 68:1448 (1964).
69. D. R. Rhodes and E. F. Steigelman, J. Electrochem. Soc., 112:16 (1965).
70. M. W. Breiter, Electrochim. Acta, 10:503 (1965).
71. S. B. Brummer, J. Phys. Chem., 69:562 (1965).
72. S. B. Brummer, J. Phys. Chem., 69:1363 (1965).
73. M. W. Breiter, J. Electrochem. Soc., 112:1244 (1965).
74. S. B. Brummer, J. Electrochem. Soc., 113:1041 (1966).
75. N. Minakshisundaram, Yu. B. Vasil'ev, and V. S. Bagotskii, Élektrokhimiya, 3:193, 283 (1967).
76. J. Eckert, Electrochim. Acta, 12:307 (1967).
77. A. G. Polyak, Yu. B. Vasil'ev, and V. S. Bagotskii, Élektrokhimiya, 1:968 (1965).
78. V. S. Éntina, O. A. Petrii, and I. V. Shelepin, Élektrokhimiya, 2:457 (1966).
79. A. G. Polyak, Yu. B. Vasil'ev, V. S. Bagotskii, and R. M. Smirnova, Élektrokhimiya, 3:1076 (1967).
80. A. G. Polyak, Yu. B. Vasil'ev, and V. S. Bagotskii, in: Advances in Electrochemistry of Organic Compounds [in Russian], Nauka, Moscow (1968), p. 13.
81. R. A. Munson, J. Electroanalyt. Chem., 5:292 (1963).
82. A. B. Fasman, G. N. Padyukova, and D. V. Sokol'skii, Dokl. Akad. Nauk SSSR, 150:856 (1963).
83. T. B. Warner and S. Schuldiner, J. Electrochem. Soc., 111:992 (1964).
84. S. B. Brummer and J. I. Ford, J. Phys. Chem., 69:1355 (1965).
85. G. N. Padyukova, A. B. Fasman, and D. V. Sokol'skii, Élektrokhimiya, 2:885 (1966).
86. A. B. Fasman, Z. N. Novikova, and D. V. Sokol'skii, Zh. Fiz. Khim., 40:556 (1966).
87. Z. N. Novikova, A. B. Fasman, and D. V. Sokol'skii, Élektrokhimiya, 2:1015 (1966).
88. S. Gilman, J. Phys. Chem. 66:2657 (1962); 67:78, 1898 (1963).
89. S Gilman, J. Phys. Chem., 68:70 (1964).
90. D. R. Rhodes and E. F. Steigelman, J. Electrochem. Soc., 112:16 (1965).
91. S. Gilman, J. Phys. Chem., 70:2880 (1966).
92. P. Stonehart, J. Electroanalyt. Chem., 15:239 (1967).
93. H. Pichler, Adv. Catalysis, Vol. 4, Academic Press, New York—London (1952), p. 271.
94. T. B. Warner and S. Schuldiner, J. Phys. Chem., 69:4048 (1965).
95. S. B. Brummer, J. Phys. Chem., 69:4049 (1965).
96. W. Vielstich and U. Vogel, Z. Élektrochem., 68:688 (1964).
97. B. J. Piersma, T. B. Warner, and S. Schuldiner, J. Electrochem. Soc., 113:841 (1966).
98. M. W. Breiter, Electrochim. Acta, 12:1213 (1967).
99. B. I. Podlovchenko, A. A. Ekibaeva, and V. F. Stenin, in: Advances in Electrochemistry of Organic Compounds [in Russian], Nauka, Moscow (1968), p. 3.
100. R. V. Marvet and O. A. Petrii, Élektrokhimiya, 3:153 (1967).

101. L. W. Niedrach, S. Gilman, and J. Weinstock, J. Electrochem. Soc., 112:1161 (1965).
102. L. W. Niedrach, J. Electrochem. Soc., 113:645 (1966).
103. J. A. Shropshire and H. H. Horowitz, J. Electrochem. Soc., 113:490 (1966).
104. R. Kh. Burshtein, A. G. Pshenichnikov, and V. S. Tyurin, Dokl. Akad. Nauk SSSR, 160:629 (1965).
105. R. Kh. Burshtein, A. G. Pshenichnikov, V. S. Tyurin, and L. L. Knots, Élektrokhimiya, 1:1268 (1965).
106. R. J. Flannery and D. C. Walter, Hydrocarbon Fuel Cell Technology, ed. S. Baker, Academic Press, New York—London (1965), p. 335.
107. S. Gilman, Trans. Faraday Soc., 61:2546, 2561 (1965).
108. S. Gilman, J. Electrochem. Soc., 113:1036 (1966).
109. S. Gilman, Hydrocarbon Fuel Cell Technology, ed. S. Baker, Academic Press, New York—London (1965), p. 349.
110. W. T. Grubb and M. E. Lazarus, J. Electrochem. Soc., 114:360 (1967).
111. S. B. Brummer, J. I. Ford, and M. J. Turner, J. Phys. Chem., 69:3424 (1965).
112. S. B. Brummer and M. J. Turner, Hydrocarbon Fuel Cell Technology, ed. S. Baker, Academic Press, New York—London (1965), p. 409.
113. S. B. Brummer and M. J. Turner, Ext. Abstr. Battery Division, Vol. 11, Electrochem. Soc. Meeting, Philadelphia (1966), p. 13.
114. E. J. Cairns and A. M. Breitenstein, J. Electrochem. Soc., 114:764 (1967).
115. V. S. Tyurin, A. G. Pshenichnikov, and R. Kh. Burshtein, Élektrokhimiya, 2: 948 (1966).
116. T. M. Beloslyudova and D. V. Sokol'skii, Élektrokhimiya, 1:1182 (1965).
117. G. Bianchi, L. de Carlo, and G. Faita, Chim. e Ind., 47:830 (1965); 49:16 (1967).
118. S. Gilman, Trans. Faraday Soc., 62:466, 481 (1966).
119. E. Gileadi, B. T. Rubin, and J. O'M. Bockris, J. Phys. Chem., 69:3335 (1965).
120. W. Heiland, E. Gileadi, and J. O'M. Bockris, J. Phys, Chem., 70:1207 (1966).
121. E. Gileadi, Compt. Rend. 2-me Symposium European sur Les Inhibiteurs de Corrosion, Vol. II, Ferrara (Italia) (1966), p. 543.
122. J. O'M. Bockris and D. A. J. Swinkels, J. Electrochem. Soc., 111:736 (1964).
123. J. O'M. Bockris, M. Green, and D. A. J. Swinkels, J. Electrochem. Soc., 111: 742 (1964).
124. V. E. Kazarinov and G. N. Mansurov, Élektrokhimiya, 2:1338 (1966).
125. P. K. Reshet'ko and A. I. Shlygin, Uch. Zap. Dal'nevostochnogo Gos. Univ., 8:79 (1966).
126. A. Shlygin, A. Frumkin, and V. Medvedovskii, Acta Physicochim URSS, 4:911 (1936).
127. B. I. Podlovchenko and Z. A. Iofa, Zh. Fiz. Khim. 38:211 (1964).
128. O. A. Petrii, R. V. Marvet, and Zh. N. Malysheva, Élektrokhimiya, 3:1141 (1967).
129. S. Gilman, Ext. Abstr. Battery Division, Vol. 11, Electrochem. Soc. Meeting, Philadelphia (1966), p. 9.

130. E. J. Cairns, Hydrocarbon Fuel Cell Technology, ed. S. Baker, Academic Press, New York—London (1965), p. 465.
131. L. W. Niedrach and M. Tochner, J. Electrochem. Soc., 114:17 (1967).
132. L. W. Niedrach and M. Tochner, J. Electrochem. Soc., 114:233 (1967).
133. S. Gilman, J. Phys. Chem. 68:2098, 2112 (1964).
134. V. E. Kazarinov and N. A. Balashova, Coll. Czechosl. Chem. Communs., 30: 4184 (1965).
135. A. N. Frumkin, Dokl. Akad. Nauk SSSR, 154:1432 (1964).
136. A. N. Frumkin, N. A. Balashova, and V. E. Kasarinov, J. Electrochem. Soc., 113:1011 (1966).
137. H. Dahms and M. Green, J. Electrochem. Soc., 110:1075 (1963).
138. H. Wroblowa and M. Green, Electrochim. Acta, 8:679 (1963).
139. A. N. Frumkin and B. B. Damaskin, Modern Aspects of Electrochemistry [in Russian], Mir, Moscow (1967); p. 170.
140. N. A. Balashova and V. E. Kazarinov, Usp. Khim., 34:1721 (1965).
141. A. N. Frumkin, Élektrokhimiya, 2:387 (1966).
142. A. Frumkin, O. Petry (Petrii), and R. Marvet, J. Electroanalyt. Chem., 12:504 (1966); Élektrokhimiya, 3:1311 (1967).
143. O. A. Petrii, R. V. Harvet, and A. N. Frumkin, Élektrokhimiya, 3:116 (1967).
144. A. Frumkin, O. Petry (Petrii), A. Kossaja, V. Entina, and V. Topolev, J. Electroanalyt. Chem., 16:175 (1968).
145. O. A. Petrii, A. M. Kossaya-Tsybulevskaya, Yu. M. Tyurin, Élektrokhimiya, 3:617 (1967).
146. O. A. Khazova, Yu. B. Vasil'ev, and V. S. Bagotskii, Élektrokhimiya, 1:82 (1965).
147. R. E. Smith, H. B. Urbach, J. H. Harrison, and N. L. Hatfield, J. Phys. Chem., 71:1250 (1967).
148. S. S. Sedova, Yu. B. Vasil'ev, and V. S. Bagotskii, in: Advances in Electrochemistry of Organic Compounds [in Russian], Nauka, Moscow (1968), p. 76.
149. O. A. Khazova, Yu. B. Vasil'ev, and V. S. Bogotskii, in: Advances in Electrochemistry of Organic Compounds [in Russian], Nauka, Moscow (1968), p. 19.
150. V. S. Éntina and O. A. Petrii, Élektrokhimiya, 3:1237 (1967).
151. L. W. Niedrach, Hydrocarbon Fuel Cell Technology, ed. S. Baker, Academic Press, New York—London (1965), p. 377.
152. M. L. Savitz, K. L. Januszeski, and G. R. Frysinger, Hydrocarbon Fuel Cell Technology, ed. S. Baker, Academic Press, New York—London (1965), p. 443.
153. H. Dahms, M. Green, and J. Weber, Nature, 196:1310 (1962).
154. L. D. Burke, F. A. Lewis, and C. Kemball, Trans. Faraday Soc., 60:913, 919 (1964).
155. M. W. Breiter, Electrochim. Acta, 8:447, 457 (1963).
156. V. E. Kazarinov, Élektrokhimiya, 2:1170 (1966).
157. M. Green and H. Dahms, J. Electrochem. Soc., 110:465 (1963).
158. H. Dahms and M. Green, J. Electrochem. Soc., 110:1075 (1963).
159. V. S. Bogotzky (Bagotskii) and Yu. B. Vassilyev (Vasil'ev), Electrochim. Acta, 9:869 (1964).

160. Yu. A. Chizmadzhev and V. S. Markin, Élektrokhimiya, 3:127 (1967).
161. B. I. Podlovchenko, Khira Lal, and O. A. Petrii, Élektrokhimiya, 1:744 (1965).
162. S. S. Beskorovainaya, Yu. B. Vasil'ev, and V. S. Bagotskii, Élektrokhimiya, 1:691 (1965).
163. S. S. Beskorovainaya, Yu. B. Vasil'ev, and V. S. Bagotskii, Élektrokhimiya, 1:1020 (1965).
164. M. I. Temkin, Zh. Fiz. Khim., 15:296 (1941).
165. S. Z. Roginskii, Adsorption and Catalysis on Inhomogeneous Surfaces [in Russian], Izd. Akad. Nauk SSSR, Moscow (1948).
166. S. L. Kiperman, Introduction to Kinetics of Heterogeneous Catalytic Reactions [in Russian], Izd. Akad. Nauk SSSR, Moscow (1964).
167. S. S. Beskorovainaya, Yu. B. Vasil'ev, and V. S. Bagotskii, Élektrokhimiya, 2:44 (1966).
168. M. L. Savitz, R. L. Carreras, and G. R. Frysinger, Ext. Abstr. Battery Division, Vol. 11, Electrochem. Soc. Meeting, Philadelphia (1966), p. 7.
169. E. Gileadi and S. Srinivasan, J. Electroanalyt. Chem., 7:452 (1964).
170. H. Wroblowa, B. J. Piersma, and J. O'M. Bockris, J. Electroanalyt. Chem., 6:401 (1963).
171. V. F. Stenin and B. I. Podlovchenko, Vestn. MGU, Ser. Khim.', No. 4, 21 (1967).
172. E. Gileadi, G. Stoner, and J. O'M. Bockris, J. Electrochem. Soc., 113:585 (1966).
173. V. S. Daniél'-Bek and G. V. Vitvitskaya, Zh. Prikl. Khim., 37:1724 (1964); Élektrokhimiya, 1:354, 494, 759 (1965).
174. T. N. Glazatova and V. S. Daniél'-Bek, Élektrokhimiya, 2:1042 (1966).
175. A. I. Shlygin and G. A. Bogdanovskii, Proceedings of the Fourth Conference on Electrochemistry, 1956 [in Russian], Izd. Akad. Nauk SSSR, Moscow (1959); p. 282; Zh. Fiz. Khim., 31:2428 (1957); 33:1769 (1959); 34:57 (1960); G. A. Martinyuk and A. I. Shlygin, Zh. Fiz. Khim., 32:16 (1958); A. K. Korolev and A. I. Shlygin, Zh. Fiz. Khim., 36:314 (1962).
176. A. N. Frumkin and B. I. Podlovchenko, Dokl. Akad. Nauk SSSR, 156:673 (1963).
177. R. V. Marvet and O. A. Petrii, Élektrokhimiya, 3:901 (1967).
178. G. A. Bogdanovskii, Zh. I. Bobanova, G. D. Vovchenko, and Yu. N. Bogulavskii, Heterogeneous Catalytic Processes and Properties of Catalysts, Sofia (1967); p. 7; in: Advances in Electrochemistry of Organic Compounds [in Russian], Nauka, Moscow (1968), p. 9.
179. M. W. Breiter, Z. Phys. Chem., in press; Faraday Disc., Apr. (1967).
180. V. S. Éntina and O. A. Petrii, Élektrokhimiya, 4:678 (1968).
181. P. Stondhart, Electrochim. Acta, 12:1185 (1967).
182. N. B. Miller, O. G. Tyurikova, and V. I. Veselovskii, in: Advances in Electrochemistry of Organic Compounds [in Russian], Nauka, Moscow (1968); p. 17.
183. R. Buck and L. Griffith, J. Electrochem. Soc., 109:1005 (1962).
184. C. Liang and T. C. Franklin, Electrochim. Acta, 9:517 (1964).

185. K. Schwabe, Z. Electrochem., 61:743 (1957).

186. Yu. B. Vasil'ev and V. S. Bagotskii, Dokl. Akad. Nauk SSSR, 148:132 (1963).

187. J. A. Shropshire, Electrochim. Acta, 12:253 (1967).

188. H. F. Hunger, Ext. Abstr., 1-4, Electroorganic Division, Am. Electrochem. Soc., 3:3 (1967).

189. J. W. Johnson, H. Wroblowa, and J. O'M. Bockris, Electrochem. Acta, 9: 639 (1964).

190. J. W. Johnson and L. D. Gilmartin, J. Electroanalyt. Chem., 15:231 (1967).

191. M. Green, J. Weber, and V. Dražic, J. Electrochem. Soc., 111:721 (1964).

192. H. Dahms and J. O'M. Bockris, J. Electrochem. Soc., 111:728 (1964).

193. J. O'M. Bockris and H. Wroblowa, J. Electroanalyt. Chem., 7:328 (1964).

194. J. O'M. Bockris, H. Wroblowa, E. Gileadi, and B. J. Piersma, Trans. Faraday Soc., 61:2531 (1965).

195. J. O'M. Bockris, B. J. Piersma, E. Gileadi, and B. D. Cahan, J. Electroanalyt. Chem., 7:487 (1964).

196. J. O'M. Bockris and S. Srinivasan, J. Electroanalyt. Chem., 11:350 (1966).

197. E. Gileadi, S. D. Argade, and J. O'M. Bockris, J. Phys. Chem., 70:2044 (1966).

198. S. D. Argade and E. Gileadi, Electrosorption, ed. E. Gileadi, Plenum Press, New York (1967), p. 87.

199. A. N. Frumkin, Élektrokhimiya, 1:394 (1965).

200. J. W. Johnson, H. Wroblowa, and J. O'M. Bockris, J. Electrochem. Soc., 111: 863 (1964).

201. J. W. Johnson, J. L. Reed, and W. J. James, J. Electrochem. Soc., 114:572 (1967).

202. L. W. Niedrach and H. R. Alford, J. Electrochem. Soc., 112:117 (1965).

203. E. J. Cairns and E. J. McInerney, J. Electrochem. Soc., 114:980 (1967).

204. H. Krupp, H. Rabenhorst, G. Sandstede, G. Walter, and R. McJones, J. Electrochem. Soc., 109:553 (1952).

205. J. A. Shropshire, E. H. Okrent, and H. H. Horowitz, Hydrocarbon Fuel Cell Technology, ed. S. Baker, Academic Press, New York—London (1965), p. 539.

206. M. R. Tarasevich, K. A. Radyushkina, and R. Kh. Burshtein, Élektrokhimiya, 1:1391 (1965).

207. O. A. Khazova, Yu. B. Vasil'ev, and V. S. Bagotskii, Élektrokhimiya, 3:1020 (1967).

208. R. V. Marvet and O. A. Petrii, Élektrokhimiya, 3:518 (1967).

209. R. V. Marvet and O. A. Petrii, Élektrokhimiya, 3:591 (1967).

210. M. Fukuda, C. L. Rulfs, and P. J. Elving, Electrochim. Acta, 9:1563 (1964).

211. A. D. Obrucheva, Zh. Fiz. Khim., 26:1448 (1952).

212. M. J. Joncich and N. Hackerman, J. Electrochem. Soc., 111:1286 (1964).

213. M. Prigent, Revue Générale d'Électricité, 74:69 (1965).

214. D. F. A. Koch, Ext. Abstr. Theoret. Electrochem., Vol. 5, Electrochem. Soc. Meeting, Dallas (1967), p. 67.

215. G. Bianchi, Ann. Chim. (Rome), 40:222 (1950).

216. R. Thacker, Hydrocarbon Fuel Cell Technology, ed. S. Baker, Academic Press, New York—London (1965), p. 525.

217. H. T. Beans and L. D. Hamett, J. Am. Chem. Soc., 47:1215 (1925).

218. P. Popov, A. H. Kunz, and R. D. Snow, J. Phys. Chem., 32:1056 (1928).

219. A. I. Shlygin, Uch. Zap. Kishinevskogo Univ., 7:3 (1958).

220. R. Thacker, Nature, 212:182 (1966).

221. H. Binder, A. Köhling, and G. Sandstede, Advanced Energy Conversion, 7:77 (1967).

222. H. Binder, A. Köhling, and G. Sandstede, Nature, 114:262 (1967).

223. J. D. Voorhies, J. S. Mayell, and H. P. Landi, Hydrocarbon Fuel Cell Technology, ed. S. Baker, Academic Press, New York—London (1965), p. 455.

224. W. Vielstich, Hydrocarbon Fuel Cell Technology, ed. S. Baker, Academic Press, New York—London (1965), p. 79.

225. H. Schmidt and W. Vielstich, Z. Analyt. Chem., 224:84 (1967).

226. W. T. Grubb and D. W. McKee, Nature, 210:192 (1966).

227. K. R. Williams, M. R. Andrew, and F. Jones, Hydrocarbon Fuel Cell Technology, ed. S. Baker, Academic Press, New York—London (1965), p. 143.

228. R. Thacker, Electrochem. Technol., 3:312 (1965).

229. L. W Niedrach and J. B. Weinstock, Electrochem. Technol., 3:270 (1965).

230. H. Binder, A. Köhling, H. Krupp, K. Richter, and G. Sandstede, J. Electrochem. Soc., 112:355 (1965).

231. G. Bianchi, J. Electrochem. Soc., 112:233 (1965).

232. W. T. Grubb, J. Electrochem. Soc., 113:191 (1966).

233. R. Thacker and D. D. Bump, Electrochem. Technol., 3:9 (1965).

234. H. Binder, A. Köhling, and G. Sandstede, Hydrocarbon Fuel Cell Technology, ed. S. Baker, Academic Press, New York—London (1965), p. 91.

235. M. Jamano and H. Ikeda, Hydrocarbon Fuel Cell Technology, ed. S. Baker, Academic Press, New York—London (1965), p. 169.

236. P. Longhi, La Chim. e l'Ind., 47:606 (1965).

237. O. A. Petrii, Dokl. Akad. Nauk SSSR, 160:871 (1965).

238. V. S. Éntina and O. A. Petrii, Élektrokhimiya, 4:678 (1968).

239. M. W. Breiter, Electrochim. Acta, 8:973 (1963).

240. Plat. Met. Review, 9:119 (1965).

241. O. J. Aldhart, H. Shilds, and S. Pudick, Chem. Abs., 64:3011h (1966).

242. L. W. Niedrach, D. W. McKee, J. Paynter, and J. F. Danzig, Ext. Abstr. Electrochem. Soc. Meeting, Vol. 11, Philadelphia (1965), p. 32.

243. V. S. Éntina and O. A Petrii, Élektrokhimiya, 4:111 (1968).

244. A. Damjanovic and V. Brusic, J. Electroanalyt. Chem., 15:29 (1967).

245. J. A. Shropshire, J. Electrochem. Soc., 112:465 (1965).

246. J. A. Shropshire, J. Electrochem. Soc., 114:773 (1967).

247. D. W. McKee and F. J. Norton, J. Catalysis, 3:252 (1964).

248. G. C. Bond and D. E. Webster, Plat. Met. Revs., 10:10 (1964).

249. Z. A. Iofa, V. V. Batrakov, and Yu. A. Nikiforova, Vestn. MGU, Ser. Khim., No. 6, 11 (1967).

250. B. I. Podlovchenko, V. F. Stenin, and V. E. Kazarinov, in: Double Layer and Adsorption on Solid Electrodes [in Russian], Izd. Tartuskogo Gos. Univ., Tartu (1968), p. 121.

251. A. T. Kuhn, H. Wroblowa, and J. O'M. Bockris, Trans. Faraday Soc., 63:1458 (1967).

Electrochemical Methods for Investigation of Catalytic Hydrogenation Mechanism in Solutions

D. V. Sokol'skii

The use of electrochemical methods for investigation of catalysts and the mechanism of heterogeneous and homogeneous catalytic reactions in solutions has acquired ever-increasing significance in recent years. Systematic investigations have been carried out in this field by Frumkin and his school, using large electrode-catalysts [1].

Through the use of electrochemical methods it has become possible to determine the concentration of the reactants on the catalyst surface during the reaction, the proportions of various forms of hydrogen in the catalyst, the mean energy of the bond between the hydrogen and the surface, the presence of charged forms of the organic compounds, and the surface area of the catalyst itself. Of special significance in the practical use of catalytic hydrogenation was the application of these methods to powdered catalysts [2]. In a number of cases, e. g., by measuring the potential of the catalyst during hydrogenation of acetylene derivatives or fats, it has been possible to determine the selectivity of the process and the extent of adsorption of the reaction products on the surface. Recently it has been possible to use electrochemical methods for investigating the reaction mechanisms of processes in dielectric media and even in the gas phase. The results were treated systematically [3]. The present article gives the major results obtained in recent years.

Very interesting results have been obtained for the hydrogenation of dimethylethynylcarbinol (DMEC) on mixed ruthenium– platinum catalysts with various compositions. These results confirmed the relationships which had been established earlier in the hydrogenation of 1-heptene, cyclohexene, benzaldehyde, and o-nitrophenol [4].

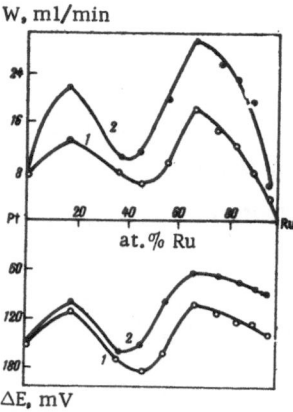

W, ml/min

at. % Ru

ΔE, mV

Fig. 1. Dependence of hydrogenation rate and potential displacement of catalyst on composition of ruthenium — platinum
alloys during hydrogenation of
dimethylethynylcarbinol in ethanol at 20°C. 1) Hydrogenation of
triple bond; 2) hydrogenation of
double bond.

Curves for the relationship between activity and composition of the catalyst are complex in nature and contain two maxima and a minimum (Fig. 1), the positions of which depend on the nature of the compound and the solvent. During hydrogenation of DMEC in various solvents it was noticed that the first maxima for hydrogenation of the triple and double bonds coincide and correspond to a catalyst containing 17.6 at. % ruthenium. The position of the second maximum depen on the medium; maximum activity occurs at 65.9 at. % ruthenium in ethanol and at 81.8 at.% ruthenium in water and alkali. The position of the minimum is the same in all solvents and corresponds to 45.2 at. % ruthenium for the triple and 32.5 at. % ruthenium for the double bonds.

Curves for the dependence of the potential shift on the composition of the catalyst are also complex in nature and pass through maxima and a minimum. Increase in the ruthenium content of the catalys changes the reaction mechanism, which is consistent with potentiometric data, where the variations in activity and potential shift are opposite in form.

The highest hydrogenation rate corresponds to minimum potential shift in the catalyst (and vice versa), and this is explained by the different generation rates of active hydrogen on specimens of catalys with different compositions. Investigation of the temperature dependence of the reaction showed that ΔE is smaller on specimens with large

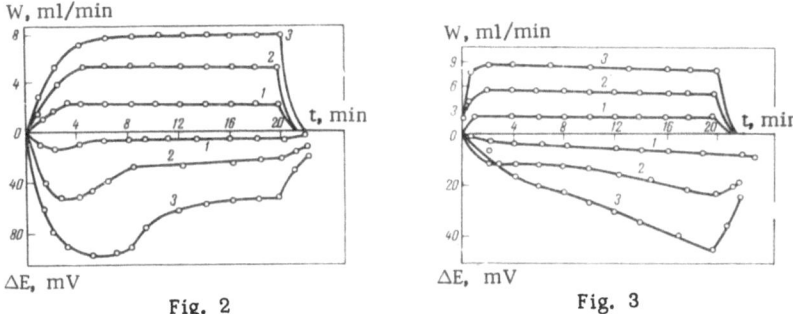

Fig. 2   Fig. 3

Fig. 2. Hydrogenation of dimethylethynylcarbinol under conditions of constant transport to a ruthenium — platinum catalyst (17.6 at. % ruthenium) in water at 20°C. Transport rate of DMEC (as hydrogen equivalent): 1) 2.8; 2) 5.3; 3) 7.8 ml H_2 per min.

Fig. 3. Hydrogenation of dimethylethynylcarbinol on ruthenium — platinum catalyst (74.3 at. % ruthenium) in water at 20°C. Transport rate of DMEC (as hydrogen equivalent): 1) 2.3; 2) 5.3; 3) 8.3 ml H_2 per min.

ruthenium contents, which indicates that the hydrogen generation rate increases with temperature.

Investigations into hydrogenation of the carbinol with constant concentration of unsaturated compound [3] have shown that two groups of catalyst can be distinguished according to the nature of the potential curves:

1. Platinum-rich (up to 56.3 at.% ruthenium);

2. Ruthenium-rich (74.3, 81.8, 88.5, 93.9 at.% ruthenium) (Figs. 2-4).

On platinum black, kinetic equilibrium is established after 12-14 min and adsorption equilibrium after 6-10 min (Fig. 4). On catalysts belonging to the first (platinum-rich) group constant rates are established 5-10 min after the beginning of the experiment (Fig. 2). When the ruthenium content is increased a steady state is established more quickly. The potential of the catalyst at the beginning of the experiment is shifted towards positive potentials and during the reaction has a tendency to move progressively towards negative values.

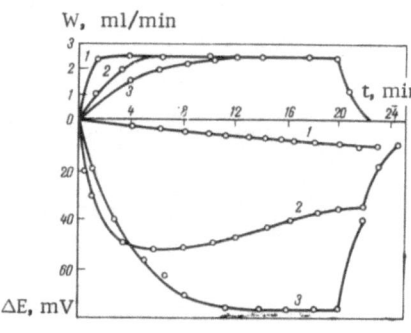

Fig. 4. Hydrogenation of dimethyl-ethynylcarbinol on ruthenium−platinum catalysts of various compositions with compound delivered at constant rate in water at 20°C. 1) 74.5 at. % Ru; 2) 56.3 at.% Ru; 3) Pt black.

Such a shift of potential to the negative potentials during the reaction can probably be explained by the formation of the catalyst surface.

With catalysts of the second group (ruthenium-rich) steady state is established more quickly − after 2-3 min (Fig. 3). The catalysts can be placed in the following order according to the time taken for establishment of steady state: Pt > Pt − Ru > Ru − Pt.

It is characteristic of these specimens that their potential is not established during the experiment. The positive shift of the potential continues to increase throughout the whole experiment. On the basis of results obtained during hydrogenation of DMEC in water with single-shot addition of the compound and with proportioned delivery it can be concluded that strong adsorption of the reaction product takes place in water, since the potential of the catalyst does not return to the reversible hydrogen potential at the end of the experiment. This is not observed on platinum or on catalysts which are rich in platinum; adsorption of the reaction product is evidently more clearly expressed on ruthenium than on platinum.

The synergic effect of platinum on ruthenium helps to decrease the active hydrogen generation rate at the catalyst surface. Results obtained from hydrogenation of DMEC by dripping onto ruthenium−platinum catalysts of various compositions show that the generation rate of active hydrogen is much greater for active specimens than for specimens with low activity. The hydrogen generation rate is 21 ml/min on specimens with highest activity (81.8 at. % ruthenium) and 2.8 ml/min on specimens with lowest activity (56.3 at. % ruthenium). On highly active catalysts hydrogenation takes place through weakly bound hydrogen, and its generation rate is not less than the hydrogenation rate. These results are also consistent with results obtained in normal hydrogenation, where the reaction proceeds with small potential shifts on specimens with maximum activity and is distinguished by low activation energy.

A straight-line relationship is obtained between the potential shift and the reaction rate, and in some cases between the potential shift and the logarithm of the reaction rate. This is consistent with data obtained by Sokol'skaya and co-workers for other catalysts [5].

The method of hydrogenation with the unsaturated compound delivered at constant rate thus creates new possibilities for investigating the catalytic hydrogenation mechanism in solutions and selecting optimum conditions and catalysts [6]. Particularly interesting is the case where all the compound is completely hydrogenated. By adding various unsaturated compounds to the solution it is possible to observe different potential shifts with the same hydrogen adsorption rate, and by performing the reaction at various temperatures it can be seen that the higher the temperature the smaller the potential shift at a given rate. These results make it possible to calculate the adsorption isotherm and activation energy of the reaction. If addition of the unsaturated compound is interrupted, absorption of hydrogen very quickly ceases. This demonstrates the absence of unreacted substance in the solution. However, the hydrogen adsorption rate only increases to a certain limit and does not increase any more with further increase in the delivery rate of the compound. Unreacted compound then begins to accumulate in the solution. The potential of the catalyst continuously moves to positive potentials during the reaction owing to accumulation of hydrogenation product. This displacement becomes particularly large under conditions where the original compound accumulates in the solution.

It was found that during hydrogenation of, e.g., cinnamic alcohol the highest hydrogenation rate is obtained under conditions where there is no appreciable excess of unsaturated compound in the solution. This is characteristic of compounds which show zero reaction order with respect to the hydrogenated substance. Consequently the increase in hydrogenation rate is determined by activation of the hydrogen. The considerable potential shift during hydrogenation, particularly on nickel, leads to rapid deactivation of the catalyst, and it is therefore most convenient to work at the maximum rate and without excess unsaturated compound. Increase in the potential shift during accumulation of the reaction product makes it possible to calculate the adsorption isotherm of β-phenylpropiolic alcohol. By carrying out the reaction under the same conditions but at a different temperature it is possible to calculate temperature coefficients for the optimum rate and for the adsorption of the reaction product. Thus, for example, during hydrogenation of dimethylethynylcarbinol on nickel at the same given rate of 4.8 ml H_2 in 1 min the potential shift amounts to 120 mV at 15°C and only 60 mV at 30°C. Consequently a lower surface ratio is required at higher temperature to give the same reaction rate. Within the hydrogen region this makes it possible to calculate the activation energy from the reciprocal values of the potential shift and to determine the reproducibility of the working surface.

Rather different relationships are obtained in the hydrogenation of relatively poorly adsorbed compounds, such as 3-hexene. The absorption rate increases with accumulation of 3-hexene, and the reaction retains first order with respect to this compound up to considerable concentrations. 3-Hexene accumulates in the solution even with very low addition rates. Such compounds can be hydrogenated at fairly high concentrations in the solution.

If the hydrogenation product does not poison the catalyst surface (e. g., sodium maleate, o-nitrophenol), it is possible to compare the surface formation characteristics of the catalyst during the reaction. Thus, on nickel and palladium, constant hydrogen absorption rate is established significantly earlier than constant potential; on platinum, on the other hand, constant potential is established quickly, and the rate increases for a long time (the platinum "develops" gradually). This feature of platinum becomes understandable, if account is taken of the fact that the energy of the cohesive bond amounts to 20.3 kcal/mole for platinum, 16.8 kcal/mole for nickel, and only 15.5 kcal/mole at each bond for palladium. Hydrogen breaks the cohesive bonds with nickel and palladium relatively easily, and the catalyst surface (by absorbing the heat of reaction) quickly develops through the reaction itself. In the general case the relationship between the reaction rate and the potential shift can be expressed by the following equation [5]:

$$\Delta\varphi = a + KW, \tag{1}$$

where a is the intercept on the potential axis; K is a proportionality constant.

In accordance with this the potential shift can arise partly from adsorption of the unsaturated compound or hydrogenation product, with displacement of hydrogen from the surface. In some cases the curve passes through the origin of the coordinates, and the equation becomes simpler:

$$\Delta\varphi = KW. \tag{2}$$

According to other data, a relationship of the following type is more likely [7]:

$$\Delta\varphi = a + b \log W. \tag{3}$$

This equation is similar to the Tafel equation and the equation expressing the relationship between $\Delta\varphi$ and W in electrochemical hydrogenation. After the optimum conditions for hydrogenation of unsaturated compounds on various catalysts in ethanol at 30°C had been determined it was possible to calculate the optimum hydrogenation rates under these conditions (Table 1).

As seen from Table 1, the optimum reaction rates for various compounds differ by a factor of four on skeletal nickel, a factor of 30 on platinum black, and a factor of 25 on palladium – BaSO$_4$.

Thus, the specific hydrogen generation rate is determined not only by the nature of the catalyst but also by the nature of the unsaturated compound (the heat of reaction and other factors). The catalysts in their turn are arranged in the following order: palladium on barium sulfate is on average an order of magnitude more active than platinum black, and platinum black is an order of magnitude more active than skeletal nickel.

The possibilities of the potentiometric method are clearly illustrated in the hydrogenation of p-nitrochlorobenzene. Figure 5 shows the kinetic and potentiometric data obtained during hydrogenation of

TABLE 1. Optimum Hydrogenation Rates in Ethanol*

Hydrogenated compound	Optimum reaction rate, ml /min·m^2		
	skeletal Ni	Pt black	Pd on BaSO$_4$
1-Hexene	0.4	64.0	62.2
Cyclohexene	0.2	3.8	53.3
3-Heptene	0.1	0.6	2.6
Allylbenzene	0.2	4.0	11.1
Phenylacetylene	0.2	3.6	12.2
Dimethylethynylcarbinol	0.4	5.7	-
Acrolein	0.2	0.4	13.3
Benzalacetone	0.1	0.2	4.4
Allyl alcohol	0.4	0.7	38.9
Cinnamic alcohol	0.2	0.9	13.3
Eugenol	0.2	0.4	28.9
Diallyl ether	0.3	-	-
Allyl ether	0.4	-	-

* Data from Sokol'skaya and Ryabinina.

various quantities of p-nitrochlorobenzene in isopropyl alcohol (10% rhodium on aluminum oxide). As seen from Fig. 5, after absorption of approximately 3 mole of hydrogen a slight inflection is seen on the kinetic curves; this can be attributed to termination of reduction of the nitro-group and the beginning of chlorine removal or deamination. The potential curves were found to be much more expressive. After an initial shift of potential by 200 mV or more to positive values a shift to the negative potentials, which is due to formation of p-amino-chlorobenzene, immediately commences. The potential of the catalyst reaches a maximum with the absorption of 3 moles of hydrogen and then decreases sharply. It is clear that this is due to acidification of the solution by the hydrochloric acid formed. Chromatographic analysis, performed at the point corresponding to maximum potential, demonstrated the presence of p-aminochlorobenzene in the solution and the absence of chloride ions. The selectivity of the process can thus be controlled directly by the potential of the catalyst. Similar possibilities were observed earlier in the hydrogenation of fats [3],

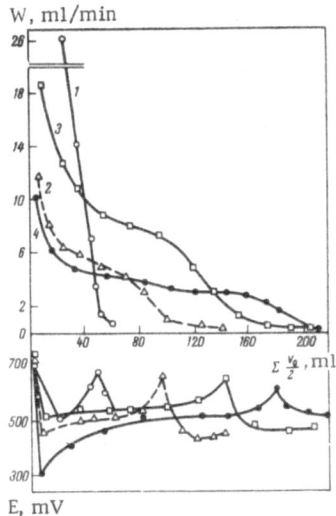

Fig. 5. Hydrogenation of various weights of p-nitrochlorobenzene in isopropanol (30 ml) on 10% Rh /Al$_2$O$_3$ catalyst (0.3 g) at 30°C. Amount of compound: 1) 0.118 g; 2) 0.236 g; 3) 0.354 g; 4) 0.472 g.

acetylene derivatives, and other unsaturated compounds. Under the appropriate conditions the change in potential of the catalyst during hydrogenation is similar to polarography on a solid cathode.

Recently the very interesting question as to the effect of the nature of the carrier on the rate and mechanism of hydrogenation has again been raised in a number of papers. Deposition of palladium on various carriers gave the following general relationship [8]. Up to a certain limit the hydrogenation rate of unsaturated compounds in the region of small coverages of the carrier surface with catalyst increases in proportion to the amount of active phase on the carrier surface. Then, at first sight quite unexpectedly, the hydrogenation rate sharply increases and finally begins to approach a limiting value (Fig. 6). The length of the first section is directly proportional to the surface of the carrier. The inflection occurs at a palladium content of 0.6% on carbon with a surface area of 308 m^2/g, at 0.45% on silica gel with a surface area of 180 m^2/g, and at 0.1% on calcium carbonate with a surface area of 14 m^2/g. Account must be taken of the fact that the palladium catalysts on calcium carbonate and on barium sulfate were prepared by coprecipitation, for which purpose it is necessary to increase the amount of palladium. The inflection on the relationship

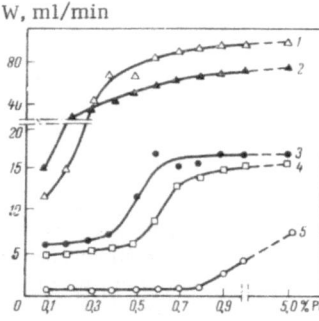

Fig. 6. Dependence of hydrogenation rate for dimethylethynylcarbinol in water at 40°C on lead content with various carriers: 1) Barium sulfate; 2) calcium carbonate; 3) silica (II); 4) charcoal; 5) silica (I).

between the rate and the amount of active phase arises from the appearance of considerable amounts of dissolved hydrogen and from a change in the mechanism of the process. This is clear from the form of the kinetic and potential curves. In the region of low coverages the hydrogenation rate of the triple bond is less than that of the double bond, and the curve has a "platinum-like" form. In the region corresponding to sharp increase in rate the triple bond, as on pure palladium, is hydrogenated at a higher rate than the double bond.

X-ray analysis and electrical conductivity measurements showed that palladium crystallizes on carriers long before the formation of a monolayer. Investigations into the behavior of the reactant on the catalyst and above all investigation of the form of the kinetic and potential curves perhaps characterize the structure of this catalyst more closely than any other physicochemical method. Palladium catalysts on carriers with various degrees of coverage differ greatly in a number of characteristics. Increase in the degree of coverage leads to increase of the stability of the catalysts on storage, decrease in the relative adsorption of the reaction products, decrease in the activation energy of the reaction (on account of a change in the slow stage), increase in the degree of reversibility of the process, and increase in the isomerizing ability of the catalyst.

It would not be an exaggeration to say that we are dealing with two types of catalyst: the first hydrogenates through adsorbed hydrogen, i.e., behaves like platinum, and the other, which contains a relatively high percentage of palladium, retains the character of the deposited active metal. Hydrogenation of the triple bond on the second catalyst occurs mainly through dissolved hydrogen.

It has been shown before [9] that platinum and palladium on aluminum oxide with small degrees of coverage (0.1-0.5%) adsorb ten and five hydrogen atoms respectively for each atom. This demonstrates the special condition of the carrier surface at a considerable distance from the center of hydrogen molecule atomization (Pt_n and Pd_n), the transfer of hydrogen atoms by the adsorption center, and the diffusion of molecular hydrogen along the surface from the adsorption centers to the centers where the hydrogen molecules dissociate.

It is particularly interesting that addition of an "inert" carrier (aluminum oxide, silica gel) to a system containing a catalyst (platinum, palladium, nickel), during hydrogenation in solutions, can effect a considerable increase in the hydrogenation rate [10]. Figure 7 shows the dependence of the hydrogenation rate for various unsaturated compounds with platinum black on the amount of aluminum oxide powder added to the solution. As seen from Fig. 7, there is a considerable increase in the hydrogenation rate of nitrobenzene and dimethylethynylcarbinol while the hydrogenation rate of 1-heptene remains unchanged. The hydrogenation rate on aluminum oxide thus increases with compounds which displace the potential of the catalyst significantly to positive potentials during hydrogenation, i.e., compounds where the hydrogenation rate is determined by activation of hydrogen. The potential curve for hydrogenation of dimethylethynylcarbinol (Fig. 7) shows that the potential shift decreases from 300 to 60 mV when the amount of aluminum oxide is increased. Such an effect may be due to adsorption of the unsaturated compounds on the aluminum oxide (redistribution of the molecules of unsaturated compounds between the solid phases) or to improved conditions for activation of hydrogen and increase in the amount of hydrogen in the system. The first suggestion was dropped since adsorption of nitrobenzene on aluminum oxide is small. The second suggestion was confirmed by the fact that during hydrogenation on palladium black, where the reaction rate is not limited by activation of hydrogen, the same aluminum oxide has no effect on the reaction rate. When they meet particles of platinum the aluminum oxide particles clearly give up their reserves of molecular hydrogen to the platinum and receive hydrogen atoms, which are sufficiently active to hydrogenate the nitrobenzene and dimethylethynylcarbinol.

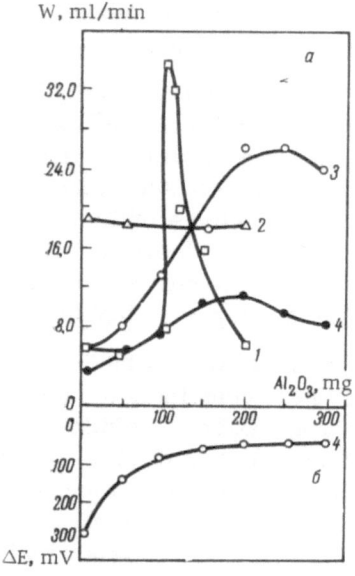

Fig. 7. Variation of hydrogenation rate for various compounds on platinum in the presence of aluminum oxide (a) and dependence of potential shift of platinum black during hydrogenation of DMEC on amount of aluminum oxide at 20°C (b). 1) Nitrobenzene; 2) 1-heptene; 3) double bond of dimethylvinylcarbinol; 4) triple bond of dimethylethynylcarbinol.

In connection with the above it is of interest to determine the amounts of hydrogen sorbed by various catalysts. Among the various methods, electrical conductivity measurement in powdered catalysts is of special importance [11].

It was shown that when powdered nickel, platinum, or palladium saturated with hydrogen are deposited in hexane, benzene, or absolute ethanol, their resistance decreases and rapidly attains a constant value. When the hydrogen is removed from the surface (e.g., by cyclohexene or phenylacetylene) the resistance of the powder increases and with full removal of hydrogen reaches several hundred thousand ohms. It is clear that the hydrogen atoms and molecules are gradually replaced by solvent molecules or hydrogenation products and the conduction mechanism changes. If the rate of establishment of constant resistance in the powder when part of the hydrogen is rapidly removed is determined, it is possible to calculate the diffusion rate of hydrogen along the catalyst surface. When catalyst which has been freed from hydrogen is saturated with hydrogen the resistance returns to its original value. If the hydrogen is replaced by organic molecules, the

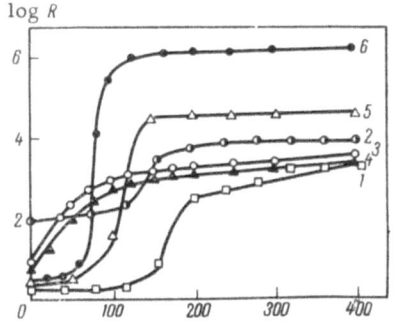

Fig. 8. Effect of degree of dehydrogenation by phenylacetylene on the resistance of platinum and palladium black and skeletal nickel in ethanol. 1) Palladium black (2.4 g); 2) palladium black while shaking; 3) platinum black (3.4 g); 4) platinum black, repeated experiment; 5) skeletal nickel (3.0 g); 6) skeletal nickel in ethylbenzene (in place of ethanol).

resistance depends on their nature. The final resistance is lower in benzene than in toluene and lower in toluene than in xylenes.

The curves for the dependence of the logarithm of the resistance for various metals on the amount of hydrogen removed show features similar to those of the charging curves for the same metals. Figure 8 shows such curves for the resistance of platinum, palladium, and nickel powders. Platinum does not contain dissolved hydrogen, since the resistance of the powder increases continuously as the hydrogen is removed (Curves 3 and 4). The resistance of palladium powder remains constant for a long time (Curve 1) owing to removal of dissolved hydrogen; nickel occupies an intermediate position (Curves 5 and 6). As seen in the case of nickel, the total amount of hydrogen removed depends on the nature of the solvent. From the very beginning ethylbenzene displaces more hydrogen from the catalyst surface than does ethanol. From the resistance curves it is possible to calculate the total amount of hydrogen sorbed by the catalyst and the amounts of adsorbed and dissolved hydrogen separately. The proportions of the various forms of hydrogen for palladium and nickel depend on the degree of dispersion of the catalyst and the nature of the solvent. When the solvent is completely removed the resistance of the catalyst powder saturated with hydrogen remains the same as in the presence of an inert solvent (hexane, heptane, etc.). This confirms the supposition that surface conduction takes place through exchange of electrons by relay. The method makes it possible to compare the adsorption of organic compounds on metals and the degree of mobility of the electrons in them.

For the promoted catalyst it is possible to determine the effect of the nature of the promoter on the total sorption of hydrogen, the surface mobility of hydrogen, and the energy of its bond with the surface. Determination of the amount of hydrogen sorbed by skeletal nickel in organic solvents gives values in the order of 30-60 ml. This agrees well with data obtained by physical methods. Here the nickel surface is not oxidized at all. Measurement of electrical conductivity could become a standard method for testing hydrogenation catalysts. Change in the composition of the solvent can alter the proportions of the various forms of active hydrogen in the catalyst, and consequently the activity and selectivity of the catalyst, within wide limits. This can be seen particularly clearly in the effect of cations in the solution on the adsorption of hydrogen and on the hydrogenation mechanism.

In investigations by Frumkin and co-workers [12, 13] the specific adsorption of cations on metals of the platinum group has been demonstrated unequivocally and measured quantitatively by a tracer method. Adsorption of cations near the reversible hydrogen potential is accompanied by ionization of the adsorbed hydrogen and leads to a decrease in its amount. However, the energy characteristics of hydrogen adsorbed on platinum are different and obey the Frumkin – Temkin equation, which was derived for a uniformly inhomogeneous surface [14]. In connection with the inhomogeneity of the active centers on the platinum surface the question arises as to the adsorption mechanism of the cations and their effect on the nature of the adsorbed hydrogen, since the activity and selectivity of the catalysts in hydrogenation reactions depend on the energy state of the hydrogen [3].

A nonstationary potentiostatic method was used to determine the mutual effects of adsorbed cations and hydrogen on platinum. Figure 9 shows the potentiostatic I, φ curves for a platinized platinum electrode in solutions of sulfuric acid and with additions of zinc sulfate at 20°C. The curve for 1 N sulfuric acid has characteristic form with two maxima in the hydrogen region and clearly defined double-layer and oxygen regions [15].

In 10^{-2} N zinc sulfate solution the heights of the peaks in the hydrogen and oxygen region decrease, and with further increase in the zinc sulfate concentration (0.1-1 N) the form of the I, φ curves also changes considerably. In the hydrogen region at a potential of 0.12 V

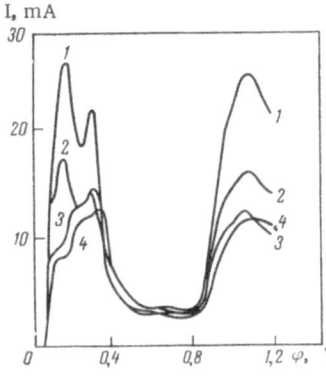

Fig. 9. Potentiostatic anodic I, φ curves for a platinized platinum electrode in solutions of sulfuric acid and with additions of zinc sulfate at 20°C. 1) N H_2SO_4; 2) 10^{-4} N $ZnSO_4$ + 1 N H_2SO_4; 3) 10^{-1} N $ZnSO_4$ + 1 N H_2SO_4; 4) 1 N $ZnSO_4$ + 1 N H_2SO_4.

there is a new maximum, which demonstrates the appearance of weakly bound hydrogen at the modified platinum surface. Under these conditions the current maximum at 0.18 V decreases sharply and is displaced by 0.02 V to the positive side, while the height of the peak of the strongly bound oxygen at 0.3 V changes little compared with the curve obtained in 10^{-3} N zinc sulfate solution. The transition region from the hydrogen to the double-layer region in zinc sulfate solutions is poorly defined owing to desorption of zinc from the platinum surface at these potentials.

The weakly bound hydrogen, the maximum of which appears at 0.12 V on the I, φ curve, is evidently molecular hydrogen adsorbed at mixed platinum – zinc centers on the surface, where atomization is difficult owing to the low heat of adsorption. We have shown by calculation that the heat of adsorption of hydrogen decreases in the presence of zinc and cadmium sulfates.

By integration of the hydrogen region of I, φ curves [15], obtained in sulfuric acid and zinc sulfate and cadmium sulfate solutions at 28, 40, and 60°C, hydrogen adsorption isotherms were plotted with θ against log p (p = pressure), from which it was seen that they obey a logarithmic isotherm equation:

$$\theta = a + \frac{1}{f} \ln p . \tag{4}$$

The inhomogeneity factor f varies between 11.8 and 12.5.

On the basis of these data the heats of adsorption of hydrogen were calculated from curves for log p against $1/T$, obtained at various degrees of coverage θ. The dependence of the heat of adsorption on the degree of coverage $(0.2 < \theta < 0.8)$ is linear in nature. In sulfuric acid solution with moderate degrees of coverage the heat of adsorption of hydrogen increases from 6.2 to 11.2 kcal/mole, while in solutions of surface-active cations it increases from 4.8 to 8.0 and from 6.4 to 8.4 kcal/mole for zinc and cadmium cations respectively.

Thus, specific adsorption of cations takes place on the most active centers and leads to a considerable change in the bond energy of hydrogen adsorbed at the surface of modified platinum. There is no doubt that increase in the proportion of weakly bound molecular and atomic hydrogen affects the activity and selectivity of platinum. As an example the hydrogenation of dimethylethynylcarbinol on platinum black in zinc sulfate solutions can be cited (Fig. 10).

The curve for hydrogenation of the carbinol in water shows a complex form; up to absorption of 90 ml hydrogen from the gas phase hydrogenation proceeds with an average rate of 9.2 ml/min, and the rate then increases to 28 ml/min and passes through a maximum. As shown by chromatographic analysis, the triple bond is mostly hydrogenated in the first section. The selectivity of the process during absorption of 1 mole of hydrogen amounts to 87.3%.

The form of the kinetic curves changes when zinc sulfate is added; the maximum, observed in water, is very poorly defined and practically disappears with increase in the zinc salt concentration to 1 N. The reason for this change in the nature of the kinetic curve becomes clear on chromatographic analysis of the reaction products. It was found that with increase in the concentration of zinc salt the selectivity to the period of time corresponding to absorption of 1 mole of hydrogen decreases from 87.3 to 70.2 in 10^{-3} N zinc sulfate solution and to 61.9% in 1 N zinc sulfate solution. The reduction in the selectivity of the process indicates that the triple and double bonds are being hydrogenated simultaneously. The change found in the hydrogenation mechanism on modified platinum surface arises from a decrease in the bond energy of the absorbed hydrogen and the appearance of molecular hydrogen.

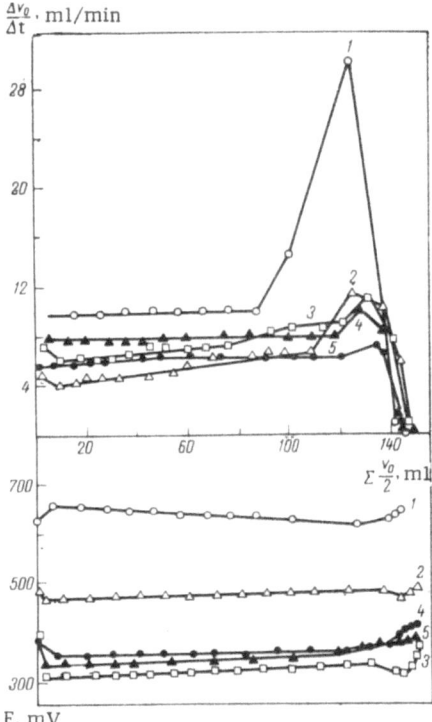

Fig. 10. Effect of zinc sulfate on hydrogenation kinetics of dimethylethynylcarbinol on
platinum black at 20°C. 1) Water; 2) 10^{-4}
N zinc sulfate; 3) 10^{-3} N zinc sulfate; 4) 10^{-2}
N zinc sulfate; 5) 10^{-1} N zinc sulfate.

On palladium catalysts the main hydrogenating agent is hydrogen
dissolved in the bulk of the metal. Specific adsorption of cations, the
cadmium ion in particular, greatly increases the overpotential of the
$\beta \rightarrow \alpha$ phase transition [16]. The heat of the $\beta \rightarrow \alpha$ transition in sulfuric acid and cadmium sulfate solutions (Table 2) was calculated from
charging curves obtained at 20, 40, and 60°C in the region of the phase
transition for the palladium – hydrogen system.

TABLE 2. Effect of Specific Adsorption of Cadmium
on Heat of $\alpha \rightarrow \beta$ and $\beta \rightarrow \alpha$ Transitions in Palladium
– Hydrogen System

Electrolyte	Equilibrium potential of $\beta \rightarrow \alpha$ transition, mV	Heat of $\alpha \rightarrow \beta$ transition, kcal/mole	Heat of $\beta \rightarrow \alpha$ transition, kcal/mole
$1N$ H$_2$SO$_4$	73	10.0	10.0
$10^{-2}N$ CdSO$_4$	85	10.0	15.5
$10^{-1}N$ CdSO$_4$	105	10.0	21.0
$1N$ CdSO$_4$	115	—	24.9

In sulfuric acid the heat of sorption and desorption of hydrogen
in the region of the phase transition amounts to 10.0 kcal/mole, which
is in agreement with the published data [17]. Specific adsorption of
cadmium cations on the palladium surface increases the heat of the
$\beta \rightarrow \alpha$ phase transition to 25 kcal/mole. The difference found in the
heats of the direct and reverse transition is thermodynamically impossible, if the processes occur at the same centers, uniform in chemical composition.

Adsorption of cations is only selective at those centers where
there is adsorbed hydrogen, since adsorption of cations is accompanied by ionization of hydrogen. It follows from published data on the diffusion of hydrogen through palladium membranes that the processes of
sorption and desorption of hydrogen take place at different active centers. Moreover, atomization and diffusion of hydrogen into the depth
of the crystal lattice can only take place at palladium centers, since
mixed centers of the palladium – cadmium type, as was shown on platinum, are not in a state to split off molecular hydrogen owing to the insufficient energy of the bond between the active center and the hydrogen. The $\beta \rightarrow \alpha$ transition can take place at the mixed centers but
involves absorption of a considerable amount of heat. Of course change
in the energy state of the dissolved hydrogen will substantially affect
the activity and selectivity of palladium.

Figure 11 shows curves for hydrogenation of propargyl alcohol
on palladium in water and in solutions of cadmium sulfate at 20°C.
The kinetics of hydrogenation of propargyl alcohol on palladium black

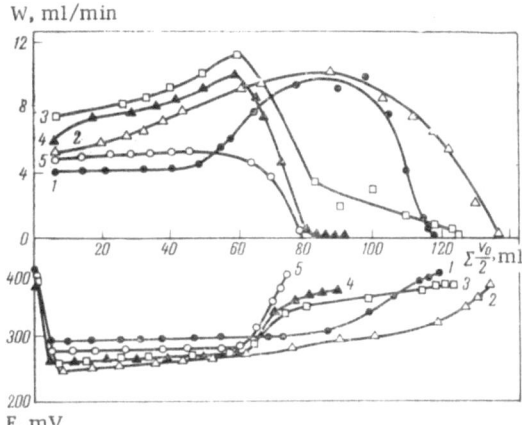

Fig. 11. Effect of cadmium sulfate on hydrogenation kinetics of propargyl alcohol on palladium black at 20°C. 1) Water; 2) 10^{-2} N cadmium sulfate; 3) 10^{-1} N cadmium sulfate; 4) $5 \cdot 10^{-1}$ cadmium sulfate; 5) 3 N cadmium sulfate.

are complex; the hydrogenation rate of the triple bond (4.2 ml/min) is less than half the hydrogenation rate of the double bond. The hydrogenation reaction is complete when 118 ml hydrogen has been absorbed from the gas phase instead of the theoretical 150 ml.

Chromatographic analysis has shown that this inconsistency arises from the presence of various side reactions. Contact of propargyl and allyl alcohols with the palladium – hydrogen system leads to the formation not only of propanol but also propionaldehyde and gaseous products (propane, propylene, and others), although the absolute percentage of the latter is small. At the end of the reaction the products were found to contain 68.8% propanol and 31.7% propionaldehyde, i.e., one third of the allyl alcohol formed isomerizes.

In solutions of cadmium sulfate (10^{-4} to 0.1 N) the amount of hydrogen consumed in the hydrogenation of propargyl alcohol increases and approaches the theoretical value. This is due to lower isomerizing ability of the catalyst modified with cadmium. In 0.1-1 N solutions of cadmium sulfate the kinetics of the process change as well. The

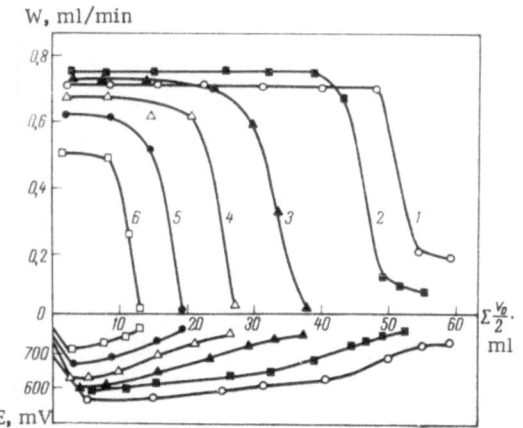

Fig. 12. Electrolytic hydrogenation of nitrobenzene
in 1 N solution of sulfuric acid in 50% ethanol with
cathodic polarization of electrode. Current density:
1) 0; 2) $4 \cdot 10^{-4}$; 3) $8 \cdot 10^{-4}$; 4) $12 \cdot 10^{-4}$; 5) $16 \cdot 10^{-4}$; 6)
$20 \cdot 10^{-4}$ A/cm^2.

hydrogenation rate of the triple bond increases to 11 ml/min, while
the double bond is hardly hydrogenated at all; the reaction ceases
when approximately 1 mole hydrogen has been absorbed.

An analysis, made in 1 and 2 N solution of cadmium sulfate,
showed the absence of propionaldehyde and a practically quantitative
yield of allyl alcohol. The change in the behavior of organic compounds when they are adsorbed on the modified palladium surface is
undoubtedly due to an energetic hindrance to the $\beta \rightarrow \alpha$ phase transition. The preferential hydrogenation of acetylene derivatives before
ethylene derivatives is determined by the greater exothermic nature
of adsorption and hydrogenation in acetylene derivatives [3].

The use of electrochemical methods in catalysis makes it possible substantially to influence the intensity and direction of the reactions which take place and in a number of cases also makes it possible
to investigate the mechanism of catalytic processes in more detail.
Thus, for example, if the catalyst is cathodically polarized during
catalytic hydrogenation, it is possible to a considerable degree to vary

TABLE 3. Conditions for Reduction of Nitrobenzene to Aniline

$D \cdot 10^4$, A/cm^2	Duration of experiment, min	Amount of hydrogen, ml			
		evolved during electrolysis	absorbed from gas phase	absorbed during hydrogenation	total
0	109	—	54.5	0	54.5
4	90	12.6	43.0	0	55.6
8	80	22.5	33.5	0	56.0
12	75	31.0	23.0	0	54.0
16	75	42.0	10.6	0	52.6

the hydrogenation rate of unsaturated organic compounds, and the nature and magnitude of the change in rate depends on the nature of the hydrogenated compound [18].

Figure 12 shows the kinetic and potentiometric curves for electrochemical catalytic reduction of nitrobenzene in 1 N solution of sulfuric acid in 50% ethanol. Curve 1 represents the kinetics of the process without the application of polarization current. In this case the reaction follows a zero-order equation. The amount of hydrogen absorbed correspond to the amount required theoretically for reduction of the nitrobenzene to aniline. When current is applied to the electrode-catalyst the average hydrogen absorption rate from the gas phase at first hardly changes and then decreases slightly (Curves 2, 3, and 4), whereas the potential drop decreases systematically; the volume of hydrogen absorbed from the gas phase also decreases. Analysis of the reaction mixture after absorption of hydrogen from the gas phase had ceased showed quantitative reduction of the nitrobenzene to aniline. This shows that electrochemical reduction compensates for the theoretical deficiency of hydrogen absorbed from the gas phase. The total amount of hydrogen released during electrolysis and the amount of hydrogen absorbed from the gas phase are practically equal to the volume of hydrogen required theoretically for reduction of the nitrobenzene to aniline (Table 3). The current yield thus amounts to about 100%. With application of polarization current the duration of the experiment decreases from 105 to 75 min.

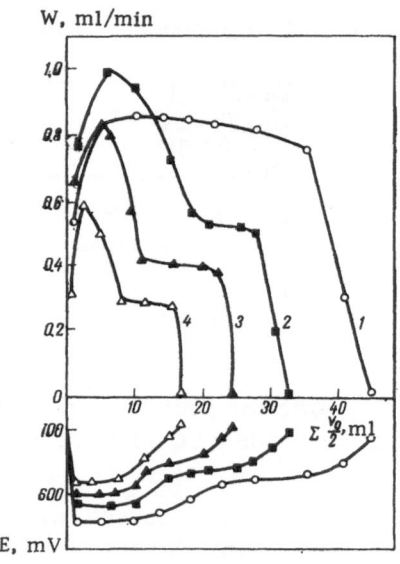

Fig. 13. Electrochemical hydrogenation of maleic acid with cathodic polarization of electrode. Current density: 1) 0; 2) $4 \cdot 10^{-4}$; 3) $8 \cdot 10^{-4}$; 4) $12 \cdot 10^{-4}$; 5) $16 \cdot 10^{-4}$ A/cm^2.

A similar effect was obtained by the application of polarization current in the hydrogenation of benzoquinone. The only difference lay in the fact that with low applied current densities some increase in the absorption rate of hydrogen from the gas phase was observed in the case of benzoquinone, which evidently results from activation of the quinone by the polarizing current.

Hydrogenation of maleic acid under similar conditions takes place somewhat differently (Fig. 13). As with electrochemical catalytic reduction of nitrobenzene and quinone, the application of a current and increase of the current density from $4 \cdot 10^{-4}$ to $16 \cdot 10^{-4}$ A/cm^2 in the electrochemical catalytic hydrogenation of maleic acid lead to a decrease in the volume of hydrogen absorbed from the gas phase. This evidently also occurs through an increase in the contribution of the electrochemical hydrogenation reaction to the overall process. However, the absorption rate of hydrogen from the gas phase decreases sharply even at the lowest current densities, and the overall rate of the process remains constant throughout the range of current densities investigated. These two factors provide evidence for the fact that

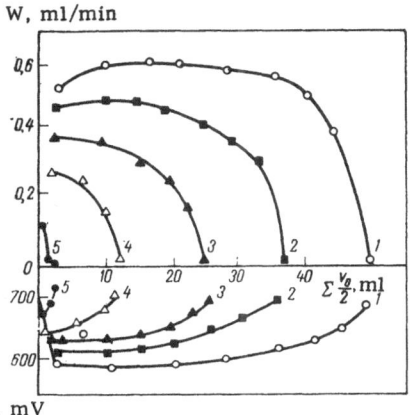

Fig. 14. Electrochemical hydrogenation of dimethylethynylcarbinol with cathodic polarization of electrode. Current density: 1) 0; 2) $4 \cdot 10^{-4}$; 3) $8 \cdot 10^{-4}$; 4) $12 \cdot 10^{-4}$ A/cm².

polarization does not promote activation of maleic acid at the catalyst surface and also that the active forms of hydrogen which are essential for catalytic hydrogenation and electrochemical hydrogenation of maleic acid are equally effective.

Hydrogenation of nitrobenzene and quinone thus take place by the following mechanism:

1) $R + e \rightarrow R^-$,

2) $H - e \rightarrow H^+$,

3) $R^- + H^+ \rightarrow RH$.

The application of polarization current accelerates process (1), slightly retards process (2) (which in some cases impedes the absorption of hydrogen from the gas phase), and supplies the proton (H^+) required for process (3). In the case of maleic acid hydrogenation proceeds by the following radical mechanism:

4) $R + H \rightarrow RH$,

but an additional source of atomic hydrogen arises when polarization current is applied; it forms through discharge of the solvent protons:

   5) $H^+ + e \rightarrow H$.

   An interesting effect of the application of polarization current on the catalyst is observed in the case of multistage hydrogenation processes, particularly with hydrogenation of dimethylethynylcarbinol (Fig. 14). Without polarization of the electrode-catalyst (Curve 1) the process proceeds at a constant rate to the end of the reaction. Application of cathodic current to the electrode (Curves 2, 3, and 4) is reflected differently in the hydrogenation kinetics of the double and triple bonds; a clearly defined maximum appears on the kinetic curve, and the absorption rate of hydrogen from the gas phase then decreases sharply, after which it remains constant to the end of the experiment. Analysis of the reaction mixture shows that electrochemical catalytic hydrogenation of dimethylethynylcarbinol takes place selectively, where hydrogenation of dimethylethlnylcarbinol to dimethylvinylcarbinol takes place at the highest rate; as this product forms it is displaced from the cathode surface into the solution by the original compound. The double bond is hydrogenated at the section of the kinetic curve which is distinguished by constant hydrogen absorption rate. It is characteristic that, as in the hydrogenation of quinone, during electrochemical catalytic hydrogenation of dimethylethynylcarbinol (with a current density of $4 \cdot 10^{-4}$ A/cm^2) the absorption rate of the first mole of hydrogen from the gas phase is higher than the hydrogenation rate of the acetylene bond without polarization. The hydrogenation kinetics of the double bond of dimethylvinylcarbinol are reminiscent of the electrochemical catalytic hydrogenation of maleic acid, i.e., the absorption rate of hydrogen from the gas phase decreases sharply with application of the polarization current.

   Investigation of the rates of individual steps and of the overall process, with and without application of polarization current, showed that polarization of the electrode-catalyst considerably reduced the duration of the experiment by shortening the time for adsorption of the first mole of hydrogen. The length of the second step does not depend on the presence and magnitude of polarization current. The results obtained thus show that the mechanism of the process changes on passing from one step to another, and if hydrogenation of the triple bond to a double bond proceeds by an electronic mechanism further hydrogenation of the double bond proceeds by an atom-radical mechanism.

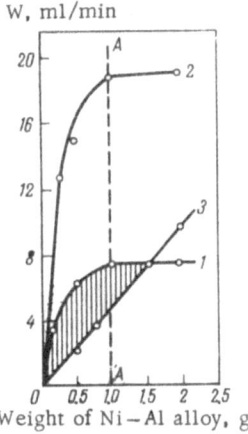

W, ml/min

Weight of Ni – Al alloy, g

Fig. 15. Dependence of electrochemical hydrogenation rate of
dimethylethynylcarbinol on
amount of catalyst. 1) At D = 20
A/dm²; 2) at D = 78 A/dm²; 3)
catalytic hydrogenation.

The rate, and sometimes the selectivity, of hydrogenation of unsaturated organic compounds can be increased at cathodes activated
with powdered catalysts. The catalysts are retained on the cathode
surface by means of a magnetic field [19]. The electrode proposed in
[19] differs from the known designs by the possibility of choice of electrode material, high catalytic activity, and high mobility of the active
layer.

Figure 15 shows the dependence of the electrochemical hydrogenation rate of dimethylethynylcarbinol at a copper cathode in alkaline
solution on the amount of skeletal nickel added. As seen from Fig. 15,
the maximum rate is obtained at current densities of 20 and 78 A/dm²
(for a given copper cathode surface) with the same amount of catalyst.
The maximum rate amounts to 7 and 18 ml hydrogen in 1 min respectively. During catalytic hydrogenation on the same amount of skeletal
nickel the rate varies between 4 and 5 ml/min. Here the current yield
is very low on the copper cathode but amounts to 95% when skeletal
nickel is added; only a slight amount of hydrogen enters the gas phase.

The use of electrochemical methods also makes it possible in
some cases to differentiate between the effects of individual factors
on the adsorption of a compound and its reactivity under the conditions
of heterogeneous – catalytic conversion. Thus, for example, an investigation into the effect of substituents on the kinetics and mechanism

of catalytic reduction of the nitro group in nitrobenzene derivatives showed [3, 20] that it is only possible to detect a relationship between the polarizing action of the substituent and its effect on the rate of the process under conditions where the reduction of all the investigated substituted derivatives is controlled by a single step. Potentiometric investigation of the catalyst during the reduction of nitrobenzene derivatives showed that the presence of a substituent in the molecule is reflected primarily in its adsorption capacity. If no special measures are adopted, introduction of electron-donor substituents (CH_3, OH, NH_2, C_2H_5) into the nitro compound as a rule leads to a decrease in the potential shift of the catalyst, i.e., a weakening of the adsorption capacity, whereas electron-acceptor substituents (COOH, CHO) increase the potential shift of the catalyst, i.e., intensify the interaction between the adsorbed molecules of the nitro compound and the catalyst surface. The way in which such a change in adsorption capacity is reflected in the overall rate of the catalytic process depends on the nature of the catalyst and the conditions under which the process operates.

On a skeletal nickel catalyst and nickel catalysts alloyed with titanium and molybdenum, under conditions where the process is controlled by activation of hydrogen, electron-acceptor substituents reduce the hydrogen concentration at the surface even more by increasing the adsorption of the nitro compound, and this in the final count leads to a decrease in the rate or complete cessation of reduction. By reducing the adsorption capacity of the nitro compound, electron-donor substituents under same conditions create a more favorable ratio of reactants at the catalyst surface and accelerate the process.

On platinum black in 50% ethanol at 25°C the reduction process is controlled by activation of the other reactant (nitrobenzene). In this case there is a linear relationship between the logarithms of the reduction rate constants for the nitrobenzene derivatives and the corresponding Hammett substituent constants, where the electron-acceptor substituents accelerate the reduction and electron-donor substituents retard it. A similar relationship was obtained between the rate of the process and the polarizing action of electron-donor substituents on skeletal nickel alloyed with titanium when the nitro compounds were added continuously to the reaction mixture at such concentrations as to ensure that the catalyst acted under optimum conditions.

These data demonstrate the electronic mechanism of catalytic reduction in aromatic nitro compounds and also that under conditions where electron-donor substituents lead to retardation of the reaction the first step of catalysis is the addition (or displacement) of the electron of the catalyst to the nitro group which is being reduced. Hydrogenation in solutions can in principle proceed by two different mechanisms [3], depending on the nature of the catalyst and the unsaturated compound:

1) During the reaction the potential of the catalyst remains close to the reversible hydrogen potential in the given solution. Hydrogenation takes place with almost complete coverage of the catalyst surface with hydrogen. This case corresponds to a first-order reaction with respect to the unsaturated compound and zero or other order with respect to hydrogen. It is realized with substances which weakly displace hydrogen from the surface (hexene, substituted olefines, and others), even at normal pressure and particularly readily in alkaline solutions. The hydrogenation rate of such substances decreases with increase in pH. With substances which remove hydrogen readily and rapidly from the catalyst surface (acetylene and its derivatives, nitro compounds, compounds with conjugated bonds) zero order with respect to hydrogen (when its generation rate at the surface is higher than its rate of removal) is established at excess pressures in the order of 40-60 atm. The activation energy depends on the structure of the unsaturated compounds and varies between 4 and 9 kcal/mole for acetylene and ethylene hydrocarbons respectively.

2) During the reaction the potential is shifted towards positive potentials so much that the appearance of hydrogen at the surface is practically excluded ($\Delta E > 300$ mV). In this case zero order reaction is observed with respect to the unsaturated compound and first order with respect to hydrogen. This mechanism is realized during hydrogenation of substances which remove hydrogen readily and completely from the catalyst surface (quinone, nitrobenzene, vinylacetylene, and others). The activation energies are practically identical for hydrogenation of all the compounds and amount to 12-14 kcal/mole, which is

characteristic for the atomization of hydrogen. In some cases the first step in the reaction is transfer of an electron from the catalyst surface to the unsaturated compound to form a negatively charged radical (hydrogenation of oxygen, quinone, nitrobenzene, acetylene on palladium). The hydrogenation rate of such compounds does not depend on the pH of the solution.

Intermediate cases are possible, where the degree of coverage of the surface with hydrogen is a function of the structure of the hydro genated compound. Here the highest hydrogenation rate is observed where the concentrations of the reactants at the surface (hydrogen and unsaturated compound) are close to stoichiometric. The hydrogenation rate of such compounds can pass through a maximum or a minimum with increase in pH. Relationships with extrema are particularly characteristic of polyfunctional compounds. In a number of cases it has been shown that in acidic solutions organic molecules at the surface of platinum, palladium, and nickel undergo intense decomposition even at low temperature (20-40°C). A linear relationship between the potential shift of the catalyst and the degree of coverage of its surface with hydrogen (to complete removal of hydrogen) corresponds to a logarithmic relationship with the hydrogen pressure. In accordance with this a linear relationship was found between the activation energy of the hydrogenation reaction and potential shift in solutions which conduct current. The activation energy of the hydrogenation reaction can be calculated from the following equation:

$$A = 23\,\Delta E \pm B, \qquad\qquad (5)$$

where A is activation energy, kcal/mole; ΔE is the potential shift during the reaction, V; B is a constant which depends on the nature of the catalyst and amounts to 4-7 kcal/mole.

It should be borne in mind that in calculating the true activation energy from hydrogenation rate constants according to the Arrhenius formula it is necessary to be sure that the potential of the catalyst (ΔE) is the same at the various temperatures. If this is not so, the difference in the potentials makes it possible to determine the preexponential term directly.

The effect of the nature of the solvent on the rate and mechanism of catalytic hydrogenation is determined by the following main factors:

1. The distribution coefficients of the hydrogenated compound and reaction products between the solvent and the catalyst surface;

2. The energy of the bond between the atoms of the reactants and the catalyst surface (these energies depend on solvent);

3. The ratio between the activation rates of hydrogen and unsaturated compound and their rates of removal from the surface;

4. The adsorption capacity of the solvent itself;

5. The presence in the solution of ions or polar substances which are capable of selective adsorption and consequently affect the reaction rate and its selectivity;

6. The solubility of hydrogen in the liquid and its diffusion rate at the gas — liquid interface, which depends on the surface tension and viscosity of the liquid.

The first two are most frequently the deciding factors.

Reactions in solutions frequently take place in a considerably more complex manner than reactions in the gas phase, since the solvent molecules (being adsorbed on the catalyst surface) exert a direct influence on the state of the surface, its charge, its free energy, and the electron distribution. This leads to a change in the kinetic and energy characteristics during adsorption and interaction of the components on the surface. In isolated cases the reaction in the liquid phase proceeds in the same way as in the gas phase, e.g., when the solvent is inert with respect to the metallic catalysts at a given temperature (heptane) or when the unsaturated compound is adsorbed so quickly and strongly that it displaces the hydrogen, solvent, and reaction products from the surface. However, in spite of the complexity of processes in solutions, it is in this region that very interesting results have been obtained from the application of electrochemical methods to the investigation of catalysts.

In analyzing all possible cases of hydrogenation it is necessary to take account of the following equilibria at the catalyst surface:

$$H_2 \text{ (gas)} \rightleftarrows H_2 \text{ (solution)} \rightleftarrows H_2 \text{ (ads)} \rightleftarrows 2H \text{ (ads)} \rightleftarrows 2H^+ \text{ (ads)} \rightleftarrows 2H^+ (H_2O)_n$$

$$\downarrow CH_2=CH_2 \qquad \downarrow \begin{matrix} OH^- \\ R^- \end{matrix}$$

$$CH_3-CH_3 \qquad H_2O; \ RH$$

Free radical   Ionic

reaction   reaction

Ionic and free-radical reactions occur at the catalyst surface in solutions; the adsorbed hydrogen most frequently has a positive charge, while the unsaturated compounds acquire positive or negative charges

The relative hydrogenation rates of unsaturated compounds on nickel [21], platinum, and palladium [22] differ greatly owing to the contents of the various forms of hydrogen and the differing rates of generation at the catalyst surface. The greatest differences are found between skeletal nickel and platinum.

In alkaline or neutral solutions the effect of the structure of the unsaturated compounds on the hydrogenation rate may differ from the effect in acidic solutions.

It has recently been shown that the promotion of skeletal nickel catalyst by additions of molybdenum, rhenium, platinum, and other transition metals has the same effect on the kinetics of anodic oxidation of hydrogen and catalytic hydrogenation. A particularly clearly defined correlation was observed in the hydrogenation of potassium maleate and o-nitrophenolate in 0.1 N potassium hydroxide solution on nickel – platinum [23] and nickel – palladium [24] alloys. The only difference lies in the fact that during hydrogenation the curve for the dependence of potential shift on reaction rate does not pass through the origin of the coordinates but makes an intercept, proportional to the adsorption of the hydrogenated compound or reaction products, on the potential axis.

Thus, the connection between catalytic and electrochemical activity of the transition metals arises from the ability of liquid-phase hydrogenation catalysts to atomize and ionize molecular hydrogen. As known, the structure of the double electric layer and the nature of electron exchange between the adsorbent and the surface have a considerable effect on the kinetics of catalytic hydrogenation in solutions. This is explained by the fact that ionized forms of the reagents take part in the formation of an activated complex.

The existence of a close analogy in the kinetics and mechanism of electrocatalytic processes controlled by activation of hydrogen substantially simplifies selection of the best catalyst. It becomes possible to model slow reactions taking place in complicated apparatus more rapidly and simply. In the development of effective gas-diffusion electrodes for low-temperature fuel cells it is feasible to study the catalytic activity of multicomponent alloys in ortho-para conversions, deutero exchange, and hydrogenation.

Electrochemical stabilization of homogeneous catalysts for hydration of acetylene and oxidation of carbon monoxide shows great possibilities. It is known that in the presence of various oxidizing agents complex ions of divalent mercury and divalent palladium are fairly active catalysts for hydration of acetylene and oxidation of carbon monoxide and ethylene in aqueous solutions [25-31]. A kinetic investigation of these processes showed that catalysis takes place through intermediate formation of a complicated ternary complex, which comprises a complex-forming cation, an oxidizing agent, and a substrate [28-31]. In this complex the central atom must undergo essentially no changes in its electron shell and remain in an active valence state throughout the whole catalytic synthesis. The active ion here is only a bridge for transporting electrons from the substrate to the oxidizing agent and accelerates the redox reaction between partners in the coordination sphere.

However, owing to the fairly large differences in the redox characteristics of the partners, the catalytic mechanisms include steps which involve formation of a reduced form of the complex-forming ion and destruction of the active particle. In spite of the relatively small proportion of these side reactions in the overall balance of the process, they lead to progressive deactivation of the catalyst.

In this connection it is of interest to investigate methods for stabilizing secondary catalytic systems. In our opinion the electrochemical method is of practical importance [31]. The contacting solution is in this case placed in an electrochemical system. An external emf secures a specific potential at the metal-complex catalyst which changes through a secondary redox route. The design of the electrochemical system must accordingly satisfy the requirements of the catalytic and electrode processes.

Figure 16 shows characteristic kinetic and potential curves describing the behavior of a catalytic solution containing mixed chlorobromoacetate complexes of divalent palladium, monovalent copper, divalent copper, divalent iron, and trivalent iron in the oxidation of

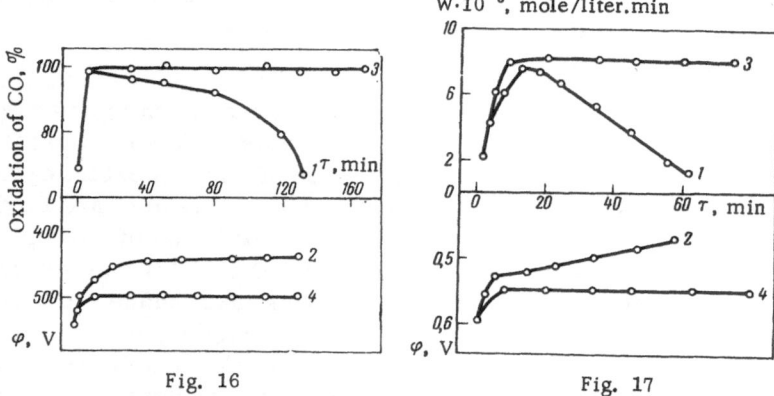

Fig. 16.

Fig. 17.

Fig. 16. Variation of activity and potential of catalyst for oxidation of carbon monoxide with time. 1) and 2) Unstabilised; 3) and 4) electrochemically
stabilised catalyst.

Fig. 17. Variation of activity and potential of catalyst for hydration of acetylene with time. 1) and 2) Hydration of acetylene without anodic polarization
of catalyst; 3) and 4) hydration of acetylene in electrochemical system.

carbon monoxide (1 vol.%) by atmospheric oxygen at 20°C in a flow reactor. In the absence of current the activity of the catalytic system
decreases continuously with time. In spite of the presence of a constant excess of the O_2 as an oxidizing agent in the system, the potentia
of the catalyst decreases in the course of the experiment. The decrease in the oxidative contact potential during the synthesis can be
explained by the fact that the regeneration of active valence states in
the palladium and copper ions is the limiting stage. For this reason
acido complexes of monovalent copper, divalent iron, and zerovalent
palladium gradully accumulate. Owing to decrease in the concentration of strong acceptors capable of quickly completing the metal complex, intraspherical conversion of carbon monoxide takes place, and
the central ion is reduced to the zero-valent state.

As seen from Fig. 16, anodic polarization of the reaction system
secures completely stable action in the catalyst. After a slight shift
at the beginning of the experiment the potential of the system remains
constant to the end. In view of the composition of the catalytic system,

the nature of the anode (graphite), and the thermodynamic conditions of the reaction, it can be supposed that a new oxidizing agent (Br_2 molecules) appears during anodic polarization. The presence of small concentrations of Br_2 in the solution stabilizes the system, since during oxidation of halogens they may add to divalent palladium by the third coordinate which forms during the formation of intermediate carbonyl complexes with divalent palladium.

The catalytic system $Fe(SO_4)_2^- - Hg(SO_4)_2^{2-} - H_3O^+$, which accelerates the hydration of acetylene, rapidly changes in activity with time on account of a secondary redox process between the components of the solution (Fig. 17). The nature of the potential curve shows that during catalysis the number of potential-forming ions continuously decreases and zero-valent mercury appears. An electrochemical system involving use of metallic mercury as anode gives stationary thermodynamic and kinetic activity in the mercury complexes during the hydration process. The constant oxidizing activity of a metal ion with variable valence in the system makes it possible to investigate the reaction kinetics under stationary conditions.

LITERATURE CITED

1. A. N. Frumkin and A. I. Shlygin, Acta Physicochimica URSS, 4:7, 911 (1936); Dokl. Akad. Nauk SSSR, 2:173 (1934).
2. D. V. Sokol'skii and V. A. Druz', Dokl. Akad. Nauk SSSR, 73:949 (1950); Yu. A. Skopin and D. V. Sokol'skii, Vestn. Akad. Nauk KazSSR, No. 6, 89 (1956).
3. D. V. Sokol'skii, Hydrogenation in Solutions [in Russian], Izd. Akad. Nauk KazSSR, Alma-Ata (1962); Vestn. Akad. Nauk KazSSR, No. 6, 3 (1965); A. M. Sokol'skaya and D. V. Sokol'skii, Kinetika i Kataliz, 4:658 (1965).
4. D. V. Sokol'skii, A. G. Sarmurzina, and K. A. Dzhardamalieva, Trudy Inst. Khim. Nauk Akad. Nauk KazSSR, 17:100 (1967).
5. A. M. Sokol'skaya, S. A. Ryabinina, and V. A. Druz', Trudy Inst. Khim. Nauk Akad. Nauk KazSSR, 14:210 (1966).
6. D. V. Sokol'skii and G. F. Tret'yakova, Dokl. Akad. Nauk SSSR, 138:399; 140:844 (1961).
7. A. M. Sokol'skaya, S. M. Reshetnikov, S. S. Ryabinina, and D. V. Sokol'skii, Dokl. Akad. Nauk SSSR, 174:884 (1967).
8. D. V. Sokol'skii and N. A. Gogol', Dokl. Akad. Nauk SSSR, 166:1140 (1966).
9. D. V. Sokol'skii and E. I. Gil'debrand, Dokl. Akad. Nauk SSSR, 133:3 (1960).
10. D. V. Sokol'skii and B. O. Zhusunbekov, Dokl. Akad. Nauk SSSR, 170:1096 (1966).

11. D. V. Sokol'skii and V. V. Malakhov, Dokl. Akad. Nauk SSSR, 117:455 (1957);
     D. V. Sokol'skii and I. K. Toibaev, Elektrokhimiya, 1(6):673 (1965); D. V.
     Sokol'skii, N. S. Samsonova, and A. M. Pak, Trudy Inst. Khim. Nauk KazSSR,
     17:61 (1967).

12. A. D. Obrucheva, Zh. Fiz. Khim., 32(9):2155 (1958); Dokl. Akad. Nauk SSSR,
     120(5):1072 (1958).

13. A. N. Frumkin, G. N. Mansurov, V. E. Kazarinov, and N. A. Balashova,
     Élektrokhimiya, 2:1358, 1443 (1966); Coll., Chem. Communs., 31:806 (1966).

14. M. I. Temkin, Zh. Fiz. Khim., 15:296 (1941).

15. F. Will, J. Electrochem. Soc., 112:451 (1965).

16. G. D. Zakumbaeva, N. A. Zakarina, and D. V. Sokol'skii, Dokl. Akad. Nauk
     SSSR, 148:630 (1963); 153:133 (1963); 162:814 (1965); G. D. Zakumbaeva and
     D. V. Sokol'skii, Trudy Inst. Khim. Nauk Akad. Nauk KazSSR, 17:22 (1967).

17. A. I. Fedorova and A. N. Frumkin, Zh. Fiz. Khim., 27:247, 517 (1953).

18. D. V. Sokol'skii, A. A. Tsoi, and V. P. Shmonina, in: Catalytic Reactions in
     Liquid Phase [in Russian], Izd. Akad. Nauk KazSSR, Alma-Ata (1967), p. 321.

19. I. V. Kirilyus, D. V. Sokol'skii, I. N. Azerbaev, M. A. Zhuk, and A. Ya.
     Matviichuk, Papers of Second All-Union Conference on Catalytic Reactions in
     Liquid Phase [in Russian], Alma-Ata (1966), p. 97.

20. V. P. Shmonina, Trudy Inst. Khim. Nauk Akad. Nauk KazSSR, 14:78 (1966).

21. A. M. Sokol'skaya, Trudy Inst. Khim. Nauk Akad. Nauk KazSSR, 17:32 (1967).

22. A. M. Sokol'skaya, S. M. Reshetnikov, K. K. Kuzembaev, S. A. Ryabinina,
     and É. N. Bakhanova, Trudy Inst. Khim. Nauk Akad. Nauk KazSSR, 14:57
     (1966).

23. A. B. Fasman, A. Isabekov, D. V. Sokol'skii, A. A. Presnyakov, and K. T.
     Chernousova, Zh. Fiz. Khim., 40:2086 (1966).

24. A. Isabekov, A. B. Fasman, and B. K. Almashev, Zh. Fiz. Khim., 41:1890
     (1967).

25. Smidt, Chem. and Ind., 2:54 (1962).

26. P. T. Henry, J. Am. Chem. Soc., 86(16):3246 (1964).

27. I. I. Moiseev, M. N. Vargaftik, and Ya. K. Syrkin, Izv. Akad. Nauk SSSR,
     Otd. Khim. Nauk, 6:1147 (1963).

28. A. B. Fasman, V. A. Golodov, and D. V. Sokol'skii, Dokl. Akad. Nauk SSSR,
     155:5 (1964); 151:98 (1963).

29. G. D. Zakumbaeva, N. F. Noskova, Zh. Konaev, and D. V. Sokol'skii, Dokl.
     Akad. Nauk SSSR, 156:6, 820 (1964).

30. K. N. Matveev, I. F. Bukhtoyarov, N. I. Shul'ts, and O. A. Emel'yanova,
     Kinetika i Kataliz, 5:649 (1964).

31. Ya. A. Dorfman and D. V. Sokol'skii, Trudy Inst. Khim. Nauk Akad. Nauk
     KazSSR, 17:80 (1967).

Electrochemical Reduction Mechanism in Organic Peroxides

É. S. Levin and A. V. Yamshchikov

An organic peroxide is represented by the general formula R_1 – O – O – R_2. If R_1 is an alkyl or acyl group and R_2 is hydrogen, it is an alkyl or acyl hydroperoxide respectively (the latter are also called peracids); if R_1 and R_2 are alkyl groups (identical or different) the compound is a dialkyl peroxide (or diacyl peroxide in the case of acyl groups); finally if R_1 is an acyl group and R_2 is an alkyl group, the compound is a perester. We are also including the inorganic compound hydrogen peroxide in the discussion because it is very similar in its electrochemical behavior to the other hydroperoxides and can serve as the simplest model.

Interest in organic peroxides (including hydrogen peroxide) arose in the first place from the role which they play in low-temperature oxidation reactions, a fact which was first demonstrated in the last century (the Bach – Engler theory). Somewhat later, attention was drawn to the outstanding ability of the peroxides to accelerate polymerization processes. In our own time the theory of chain reactions has explained the role of peroxides in oxidation and polymerization processes by establishing that they represent a source of free radicals which participate in the formation of chains. In recent years the use of peroxides in industry as initiators of polymerization and as intermediate products in synthesis has increased enormously; some applications have reached the scale of many tons. Hydrogen peroxide, the simplest hydroperoxide, has long been an important industrial material on account of its mild oxidizing properties and its cheapness. Interest in the chemical and physicochemical properties of peroxides has been intensified by the appearance within a short period of time of a series of monographs devoted to them.

Electrochemical research on peroxides has been directed almost exclusively to hydrogen peroxide. This has attracted special attention as an intermediate product in the electrochemical reduction of oxygen, and work in this direction is concerned with the problems of increasing the effectiveness of fuel cells. Organic peroxides proper have been mainly investigated with a view to developing polarographic methods of analysis. Meanwhile the electrochemical behavior of the peroxides is by no means trivial. On the i-E curves there are many outstanding features; these include clearly defined dependence of the wave form and half-wave potential ($E_{1/2}$) on the nature of the solvent and the pH value, the appearance under certain conditions of abnormally extended waves, nonturbulent current maxima, difficultly explainable division of the wave, and other features. Investigation of these phenomena should undoubtedly deepen our understanding of the electrode processes as a whole.

To a considerable extent the present review rests on the results which have been obtained for hydrogen peroxide by various methods, including the most up-to-date methods. With respect to other peroxides information on the mechanism of the electrode reactions is extremely incomplete, but the similarity in the general pattern allows one to suppose that many of the conclusions reached in the investigation of hydrogen peroxide can be extended to them.

1. RADICAL – CHAIN MECHANISM OF
CHEMICAL REACTIONS INVOLVING PEROXIDE

Oxidation – reduction changes in peroxides in solutions, without current, and also processes taking place at the metal interface, began to be investigated considerably earlier than the corresponding electrochemical changes. The principles were set out in papers by Haber, Weiss, and others [1-3]; the dissociation of hydrogen peroxide in solution in the presence of a cation with variable valency is interpreted as a combined oxidation – reduction process proceeding by a chain mechanism. The process commences with the following reactions

$$M^{z+} + H_2O_2 \rightarrow M^{(z+1)+} + OH + OH^-, \tag{1}$$
$$M^{(z+1)+} + HO_2^- \rightarrow M^{z+} + HO_2. \tag{2}$$

The radicals formed give rise to the following chain:

$$OH + H_2O_2 \rightarrow HO_2 + H_2O, \quad \Delta G^0 = -29.7 \text{ kcal/mole} \tag{3}$$

$$HO_2 + H_2O_2 \rightarrow OH + H_2O + O_2, \quad \Delta G^0 = -20.3 \text{ kcal/mole} \tag{4}$$

and so on, and this is broken by the following reaction:

$$OH + HO_2 \rightarrow H_2O + O_2, \quad \Delta G^0 = -67.4 \text{ kcal/mole}. \tag{5}$$

Reactions (3)-(5) are thermodynamically possible, as shown by the changes in standard free energy (ΔG^0) calculated from tabular data. Apart from the process examined above, the Haber – Weiss scheme also provided an explanation for a number of other changes in hydrogen peroxide. It was thus possible to interpret the photolysis of hydrogen peroxide in aqueous solutions quantitatively [4]. It was shown that, after the initial event $H_2O_2 + h\nu \rightarrow 2OH$, with low absorbed light intensity (I_{abs}) a chain is formed from Reactions (3) and (4); with higher I_{abs} values the chain is rapidly broken by Reaction (5) and (or) the following reaction:

$$2HO_2 \rightarrow H_2O_2 + O_2. \tag{6}$$

Another example is the cathodic reduction of hydrogen peroxide, initiated by oxygen; in the presence of hydrogen peroxide the limiting current caused by traces of oxygen shows a ten-fold increase [5, 6]. In this case an electron is first transferred $O_2 + e \rightarrow O_2^-$, and the O_2^- initiates the above-mentioned chain in the order (4), (3); the chain is, however, broken not by Reaction (5) but by the electron transfer: OH + e $\rightarrow$ OH⁻. The scheme proved able to explain the dependence of the current amplification factor on various parameters.

Baxendale and others [7] applied the scheme to polymerization in the presence of hydrogen peroxide and $Fe^{2+} - Fe^{3+}$ ions on the assumption that free radicals formed from the peroxide act as intitiators; the conclusions were found to be completely consistent with the experimental data. The authors assumed that a major part was played by the OH radical but could not prove it conclusively.

Important for verification of the Haber – Weiss scheme was proof of the existence of the OH radical and then the HO_2 radical in solutions by physical methods. Certain thermodynamic quantities and normal electrode potentials relating to systems involving OH, HO_2, and H_2O_2 were also determined by calculation [8-12].

2. SPONTANEOUS DECOMPOSTION OF PEROXIDES AT A METAL SURFACE AND MECHANISM FOR ESTABLISHMENT OF EQUILIBRIUM POTENTIAL

The earliest explanation for the heterogeneous decomposition of hydrogen peroxide was given by Haber [13], who considered that the process occurred through intermediately formed metal oxides and arose from two linked chemical reactions, such as:

$$Pt + H_2O_2 \rightarrow PtO + H_2O,$$
$$PtO + H_2O_2 \rightarrow Pt + H_2O + O_2.$$

This point of view was later reinforced by the work of Hickling and Wilson [15], and also of Bianchi and others [16], who established that even at small concentrations the peroxide reacts rapidly with platinum oxides.

In the ensuing years more supporters were acquired for the electrochemical theory of decomposition developed by Weiss [14], which was based on analogy with the theory for the homogeneous case. In the heterogeneous process the roles of Reactions (1) and (2) are played by the following linked electrochemical reactions:

$$H_2O_2 + e \rightarrow OH + OH^-, \tag{1a}$$
$$HO_2^- \rightarrow HO_2 + e. \tag{2a}$$

The radicals formed can initiate the chain (3)-(4), which can be broken in two ways — either by Reaction (5) or as a result of the following reactions at the phase boundary:

$$OH + e \rightarrow OH^-, \tag{7}$$
$$HO_2 \rightarrow O_2 + H^+ + e. \tag{8}$$

The Weiss theory has been refined and verified experimentally in several papers, in which expressions are usually derived for the equilibrium potential (E_e) of the $O_2 - H_2O_2$ system. This last question is discussed in [17]. The basis used for the discussion was the assumption, made in an earlier article, that at a dropping mercury electrode in alkaline solution the reaction

$$O_2 + H_2O + 2e \rightarrow HO_2^- + OH^-$$

proceeds under conditions close to equilibrium; the following equation was derived on the basis of this suggestion:

$$E_e = E^0 + \frac{RT}{2F} \ln \frac{p_{O_2}(1 + K_w/Ka_{OH^-})}{ca_{OH^-}}.$$

Here E^0 is approximately equal to the normal potential of the reaction in alkaline solution; p_{O_2} is the oxygen pressure; c is the stoichiometric peroxide concentration; a_{OH^-} is the activity of the ion; K is the dissociation constant of the peroxide (pK 11.65); K_w is the ionic product of water.

The equation is satisfied over fairly wide limits of variation in the parameters, and the value $E^0 = -0.045$ V is obtained; this does not differ greatly from the tabular value of $E^0 = -0.076$ V. The same authors [18] measured the decomposition rate of hydrogen peroxide at a mercury surface at pH 12-13. At low peroxide concentrations the results were found to be in approximate agreement with the hypothesis of linked electrochemical reactions; the rate increases sharply at $c \geq 0.15$ mole/liter, which evidently indicates an increasing contribution from the chain mechanism.

A unique point of view on the mechanism for establishment of potential in solutions of hydrogen peroxide was put forward by Bockris [8], on the basis of measurements of stationary potential E_{st} on smooth platinum and gold electrodes at pH values between 0 and 13.5. It was found that within wide limits of oxygen and hydrogen peroxide concentrations E_{st} does not depend on the concentrations provided that $[H_2O_2] > 10^{-6}$ mole/liter and satisfies the following equation (for a platinum electrode):

$$E_{st} = 0.835 - 0.059 \, pH. \tag{9}$$

It follows from this [8] that E_{st} cannot be a reversible potential for the $O_2 - H_2O$ or $O_2 - HO_2$ systems. Also unacceptable is the hypothesis that E_{st} is determined by one of the systems which can arise by oxidation of the metal with the peroxide:

$$Pt(OH)_2 + 2H^+ + 2e \rightleftarrows 2H_2O + Pt, \quad E^0 = 0.98 \text{ V};$$

$$PtO_2 + 2H^+ + 2e \rightleftarrows Pt(OH)_2, \quad E^0 = 1.1 \text{ V},$$

since the E values for them differ from the observed values. In addition it would be necessary to expect deviations from linearity in the E_{st}-pH curve in strongly acidic and alkaline regions. It was suggested that the initial event in the process is decomposition of the peroxide at the surface:

$$(H_2O_2)_{ads} \rightarrow 2OH_{ads}, \tag{10}$$

where Γ_{OH} = const if the amount of peroxide is sufficient to fill the surface with radicals. The electrode reaction:

$$OH_{ads} + e \rightarrow OH^- \tag{11}$$

proceeds until E becomes positive enough to stop it. Thus, E_{st} is regarded as the equilibrium potential for Reaction (11). Equation (9) can easily be derived if in addition it is assumed that Reaction (10) is irreversible.

It should be mentioned that some of the statements made by Bockris and Oldfield [8] are not consistent with the results from other researches. Thus, according to [19], the dependence of E_{st} on pH has a distinctly nonlinear character; in the alkaline region (even at pH 10.35) E_{st} clearly depends on $[H_2O_2]$. Serious arguments have even been put forward many times in favor of the view that the $O_2 - H_2O_2$ system is reversible or approximates such a state in alkaline medium [15, 17, 18, 20, 21]. This places under some doubt the main assumption by Bockris that the peroxide does not take part directly in the potential-determining step.

The decomposition mechanism of hydrogen peroxide at a metal surface was investigated in greatest detail by R. and H. Gerisher [19], who recorded the electrochemical characteristics of the process in parallel with measurements on the decomposition rate. It was noticed first of all that the condition of the platinum surface had a great effect on all the measured quantities and that the measurements were insufficiently reproducible owing to "aging" of the electrodes;

this fault is exhibited to a lesser degree by the electrodes with low catalytic activity which were employed. They were unable to interpret the effects in the alkaline region and restricted themselves mainly to the pH 1.8-7.6 region.

The earlier observations on the fact that a definite potential is established practically instantaneously in a hydrogen peroxide solution were confirmed; its value is, however, different on different electrodes and changes slowly with time owing to change in the surface condition. The term "stationary potential" (E_{st}) can only be applied provisionally on the understanding of a specific mean value. (The Gerishers preferred to call it the "rest potential".) It is significant that even in weakly alkaline solution E_{st} depends slightly on [H_2O_2] and that it is always more positive in the absence of peroxide.

The formation mechanism of E_{st} was established by analysis of sections of the polarization curves near E_{st}. Figure 1 shows curves for various pH values and Fig. 2 gives their semilogarithmic representation. The curves are closely satisfied by the equations of the retarded discharge theory; they give transfer coefficients ($\alpha = 0.20$, $\beta = 0.69$) which depend little on the pH value. The exchange current

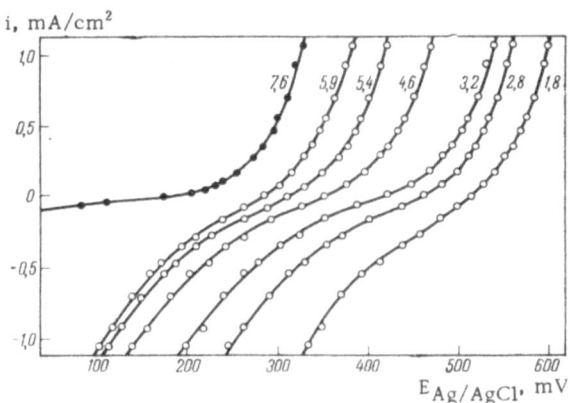

Fig. 1. Sections of i-E curves for hydrogen peroxide near E_{st} at various pH values (indicated on the curves).

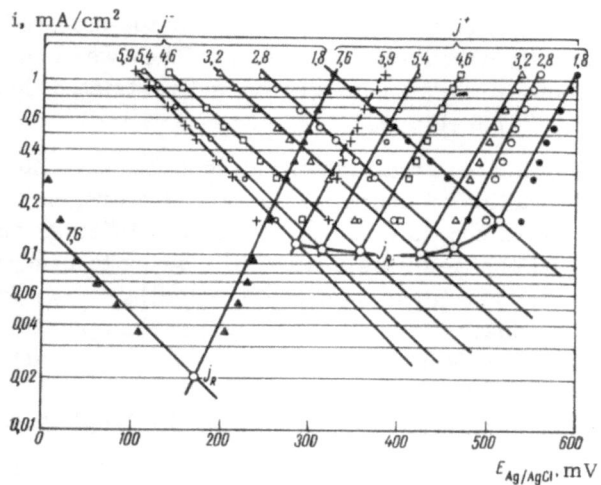

Fig. 2. The curves of Fig. 1 in semilogarithmic form (shown by points) [19]. Straight lines represent extrapolation for calculation of i_0 and E_{st}. The figures on the curves represent the pH values.

(i_0), which is proportional to $[H_2O_2]$, also hardly changes with pH. The fact that $\alpha + \beta$ is not equal to unity shows that E_{st} does not represent a reversible equilibrium potential of a single potential-determining reaction but a mixed potential, establishment of which involves at least two independent reactions. Since the exchange currents i^+ and i^- are proportional to $[H_2O_2]$, these can only be the reactions which initiate decomposition of the peroxide. In dicusssing their nature, the Gerishers [19] largely accept the Weiss scheme, i.e., consider Reactions (1a) and (2a) as the initial reactions but, in order to explain the nature of the relationship between the exchange currents and the pH, assume that the following analogous reactions take place in parallel:

$$H_2O_2 + H^+ + e \rightarrow OH + H_2O, \tag{1b}$$

$$H_2O_2 \rightarrow HO_2 + H^+ + e. \tag{2b}$$

The strongest argument in favor of the Weiss theory is the result from comparison of the exchange current and oxygen evolution rate (v)

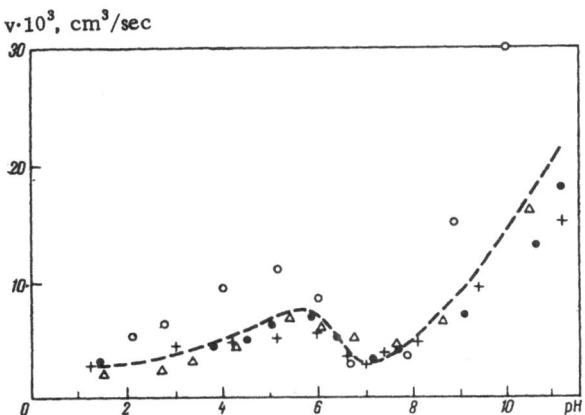

Fig. 3. Oxygen evolution rate (v) at various pH values
[19].

expressed in current units (i_v); in the conversion it is assumed that formation of one molecule of O_2 involves the transfer of two electrons in both directions. The ratio

$$Q = \frac{i_v}{|\alpha + \beta| i_0}$$

where the denominator was determined directly by a special method, was calculated for various pH values. If it is assumed that the chain of Reactions (3) and (4) is not involved in the process and that Reactions (1b) and (2b) are followed by Reactions (7) and (8), then with $\alpha + \beta = 0.8$-0.9 the calculated value of Q is 1.1-1.25. The following values were obtained in practice:

pH	0.9	2.7	4.1	5.0	6.1	6.6	7.5
Q_{mean} . .	1.35	1.35	1.55	1.50	1.70	2.45	5.0

In the acidic region the Q values only slightly exceed the calculated value, but they rapidly increase beginning at pH $\simeq 6.0$, which indicates an increasing contribution from the chain reaction. The postulated mechanism is also favored by the distinct resemblance of the v–pH and i_0–pH curves, of which the former is shown in Fig.3. Both curves have a characteristic minimum at pH 7-8, the origin of which is not clear. In [19] it is related to change in the degree of oxidation of the platinum surface.

Earlier attempts to demonstrate the presence of free radicals at the electrode surface by their induction of polymerization of acrylonitrile gave a negative result; from this it follows that the radicals are adsorbed and their activity is greatly reduced.

On the whole, convincing arguments are given in [19] in favor of the fact that (at least in acidic solution) the decomposition process is controlled mainly by the Weiss mechanism; consideration of a chain reaction also makes it possible to explain the complications observed in the alkaline region. However, the poor reproducibility of the measurements when one electrode is exchanged for another, the appreciable effect of aging, and also the purely qualitative agreement between the proposed scheme and experiment, show that it is impossible to give a comprehensive explanation of the observations without taking accompanying changes in the metal surface into account. Thus, the interpretation of the decomposition process given by Haber [13] does contain a kernel of truth, in spite of the fact that the effect of electrochemical factors was not examined.

3. REDUCTION OF PEROXIDES AT A MERCURY ELECTRODE

In this and the following section processes involving passage of current and described by i-E and E-t curves are examined. Here we are treating the mercury electrode separately, since the properties of this metal and the possibility of using it in the form of a dropping electrode make it possible to avoid (in a specific range of potentials) complications arising from variation in the surface conditions and to obtain reproducible results.

Investigation of the reduction of hydrogen peroxide at a dropping mercury electrode commenced early [22], but the question was later [23] examined in more detail. The i-E curves in aqueous solutions for the investigated pH range are characterized by the presence of a single wave with constant slope (b), equal to ~0.245 V, on the semilogarithmic curve. The half-wave potential is approximately $-0.79\,V$

(referred to the saturated calomel electrode) at pH values between 3 and 11, and $dE_{1/2}/dpH = b$ in the strongly alkaline region. All these features can be explained on the basis of the following mechanism. Only undissociated molecules are reduced (and, furthermore, irreversibly), and the first step postulated by Weiss is the slow step:

$$H_2O_2 + e \xrightarrow{\text{k}} OH + OH^-.$$

Subsequent steps are not considered, and it is sufficient to suppose that they are fast. The absence of a kinetic component of the current also in the region where pH → pK shows that the recombination rate is considerably higher than the diffusion rate. On these assumptions the following equation is derived for the wave:

$$E - \psi = \text{const} + \frac{RT}{\alpha F} \ln \frac{[H^+]}{[H^+]+K} + \frac{RT}{\alpha F} \ln \frac{i}{i_d - i}. \tag{12}$$

In general, Eq. (12) correctly reflects the form of the polarization curve up to pH 13.2 (with an accuracy of 10-15 mV); beyond this value the wave begins to split (see below).

The formation of a new wave for hydrogen peroxide in the presence of certain cations, e. g., Mg^{2+}, was first noticed by Rysselberghe et al. [24]. Later the same authors and simultaneously Chodkowski [25] observed a similar pattern in solutions of hydroxides MOH (where M = Li, Na, K, or NR_4) with concentrations greater than 0.25 mole/liter. According to data reported in [25], the new wave (which will subsequently be called the first wave) is displaced towards the positive side by approximately 0.5 V with respect to the "normal". With increase in the pH value the first wave increases, and the second wave decreases; on this basis it was suggested that the first wave belonged to the cation and the second wave to the undissociated molecule, which is extremely unlikely. The explanation given in [26] for the splitting is more plausible. A radical mechanism is proposed for the process and includes the following steps:

$$H_2O_2 + e \rightarrow OH + OH^-, \tag{I}$$

$$OH + e \rightarrow OH^-, \tag{II}$$

$$OH + OH \rightarrow H_2O_2, \tag{III}$$

$$OH + H_2O_2 \rightarrow HO_2 + H_2O, \tag{IV}$$

$$HO_2 + H_2O + 3e \rightarrow 3OH^-, \tag{V}$$

$$H_2O_2 + 2OH \rightarrow H_2O_3 + H_2O \tag{VI}$$

$$H_2O_3 + H_2O + 4e \rightarrow 4OH^- \tag{VII}$$

and so forth.

Reactions (I)-(V) have already been met repeatedly in discussing the various changes in hydrogen peroxide; all the following reactions are completely hypothetical and constructed by analogy with Reactions (IV) and (V) but involving two, three, and more OH radicals. It was then proposed that the electrode Reactions (I) and (II) lead to the appearance of the first wave and Reactions (V), (VII), etc. lead to the appearance of the second wave. Calculation shows that this mechanism does in fact explain certain characteristics of the waves, e.g., the effect of various factors on the value of R (the ratio of the limiting currents i_{1l}/i_{2l}); R decreases with increase in $[H_2O_2]$ and increases with increase in $[OH^-]$. However, other features (e. g. the specific action of Mg^{2+} ions and the various effects of the hydroxides of different metals) cannot be explained. The work has undoubtedly put forward a solution to the problem by illustrating with experimental data the complexity of the mechanism and the possible participation of radicals, but the specific mechanism proposed is not very convincing on account of the absence of direct or indirect evidence for existence of the radicals H_2O_3, H_3O_4, etc.

New data which characterize the nature of waves in strongly alkaline solutions (0.1-1.5 mole/liter NaOH) have recently been presented [27]. At the dropping mercury electrode i_{1l} is not controlled by diffusion, since it does not depend on the height of the mercury column and has a high temperature coefficient; its dependence on concentration is nonlinear. On a streaming electrode the situation is completely different; neither splitting of the wave nor the nonlinear relationship between i_l and c is observed. Furthermore surface-active substances, e.g., 0.04% gelatin, also eliminate splitting in the case of the dropping mercury electrode. Rius and Sacristan [27] conclude from this that the first wave is due to adsorption. Unfortunately this conclusion was not supported by direct measurements, and no attempt was made to establish the nature of the adsorbed particles. Bernard

and others [28, 29] made the earlier hypothesis somewhat more specific by assuming that adsorbed metal complexes of the peroxide are
reduced at the first wave; this explains the effect of the nature of the
metal. According to them the adsorption only takes place at attainment of the solubility product of the complex; this can explain the
existence of a concentration threshold for both peroxide and Mg^{2+}
ions, which is necessary for splitting of the wave. At more negative
potentials the HO_2^- anions, which are not adsorbed in the form of a
complex, are protonated, and H_2O_2 is reduced; this gives rise to the
appearance of the second wave. In this case, too, no calculations
are given which would support the hypothesis. Thus, the problem of
the reason for splitting of the wave has still not been finally resolved,
although it has been established that adsorption phenomena play an
important part; the methods of classical polarography are clearly
inadequate to give a final solution.

Compared with hydrogen peroxide the electrochemical behavior
of organic peroxides proper has been little investigated; although a
considerable number of polarographic papers have been published,
they are mostly concerned with special analytical procedures, and
certain general relationships governing the reduction are only derived
in a few of them in passing. Thus, in [30], it was found that the
length of the hydrocarbon chain and the composition of the aqueous
ethanolic solvent had an effect on the half-wave potential and slope of
the wave for C_4-C_9 alkyl hydroperoxides. The effect of the chain
length was investigated in a series of C_6-C_{18} peracids [31] and in a
series of tert-butyl peresters with C_9-C_{14} acid groups and dialkyl peroxides with C_9-C_{18} groups [32]; the solvent used in these two investigations was a 1:1 mixture of benzene and methanol. In the series
of alkyl hydroperoxides and peresters there is a distinct positive shift
of $E_{1/2}$ with increase in the chain length, and this cannot be explained
by the effect of the substituent on the electron density. Peroxides of
various types can be arranged in the following order corresponding to
decreasing ease of reduction (R = alkyl, Ac = acyl): Ac – O – OH >
Ac – O – O – Ac > Ac – O – O – R ≃ R – O – O – H > H – O – O– H > R
– O – O – R. In general this order corresponds to the expected inductive effect of the substituents, but there are deviations, which indicate that other more specific effects exist. No coulometric measurements were made in these papers, but in calculations using the

Ilkovič equation it was assumed that in all cases the wave is a two-electron wave.

In [33], which is mainly an analytical work, the most important result was determination of the effects of the solvent and electrolyte additions. On changing from water to nonaqueous solvents a large shift of the wave to the negative side was observed (Fig. 4), and this was particularly great in the case of peroxides containing a chain of five or more carbon atoms or a benzene ring. The conclusion can be drawn that the shift is due to lower adsorbability of the peroxides from the organic media and a correspondingly lower rate in the electrode reaction; this is the point of view which Mairanovskii successfully developed with respect to other classes of compounds. Measurement of the drop time before and after addition of the peroxide to the solution of electrolyte in fact confirmed this supposition in the case of peroxides of the type mentioned above. In the case of hydrogen peroxide, which is practically unadsorbed, the shift is considerably smaller:

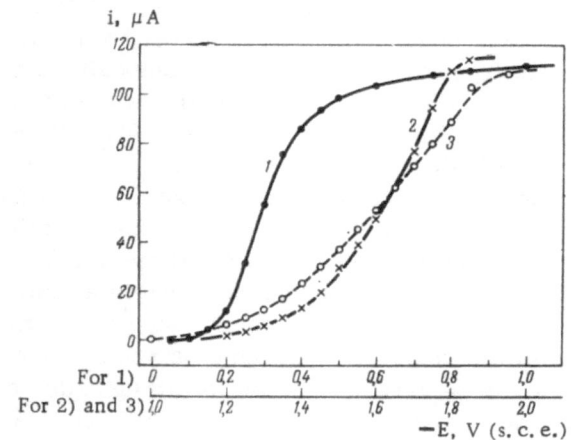

Fig. 4. Polarization curves for ethyl cumyl peroxide [33]. 1) In 20% ethanol beginning at 0 V; 2) in 95% ethanol, beginning at −1.0 V; 3) in benzene −methanol (1:1) mixture, beginning at −1.0 V (referred to saturated calomel electrode).

	$\Delta E_{1/2}$, V (water → 95% ethanol)	$\Delta E_{1/2}$, V (water → benzene + methanol)
H_2O_2	0.21	0.26
Hydroperoxides	0.56	0.85
Dialkyl peroxides	1.0	1.0

Cations (tetraalkylammonium salts) adsorbed on the mercury in this case give rise to a negative shift of the wave, i.e., the effect of retardation of the reaction predominates over the effect arising from change in ψ_1 potential. In accordance with the supposition made, this shift is much more clearly defined in the case of the aqueous solution.

Certain deliberations concerning the reduction mechanism are given in [34]. In a series of alkyl hydroperoxides, the effect of successive replacement of the hydrogen atoms at the carbon atom adjacent to the peroxide group by methyl groups on the half-wave potential was investigated. In aqueous buffer solution (pH 7.5) the following results were obtained:

Compound	$-E_{1/2}$, V (s. c. e.)
H—O—O—H	0.866
CH_3—O—O—H	0.791
CH_3CH_2—O—O—H	0.444
iso-C_3H_7—O—O—A	0.350
tert-C_4H_9—O—O—H	0.330

As in the above-mentioned cases, the direction and the large value of the $E_{1/2}$ shift cannot be explained by the inductive effect of the substituent. It was found (but the method was not indicated) that the limiting current has a diffusion character in the first three compounds and an adsorption character in the last two. On this basis the $E_{1/2}$ shift was related to change in adsorbability; this type of interpretation is acceptable in the general form, but the specific form in which it was expressed is clearly incorrect; the wave is regarded as a Brdička "prewave" caused by adsorption of the product, i.e., ethanol. In the meantime the pattern is not even outwardly reminiscent of that obtained by Brdička – i_l does not tend towards a limit with increase in the concentration of peroxide; in addition the Brdička theory is not at all applicable to irreversible electrode reactions. In the same paper it was found that, unlike other peroxides, the tert-butyl esters of substituted perbenzoic acids

$$O=C-O-O-C(CH_3)_3$$

give a small prewave, separated from the main wave by 0.75 V, on polarographic curves in aqueous dioxane mixtures; as shown by the form of the i-t curves recorded on a single drop, the current of the prewave is kinetic with a large dioxane content (80%), but when water is added it gradually becomes purely diffusional. The following hypothesis was put forward in [34]: the O − O bond, which in this case is slightly polar, becomes even more strongly polarized in the field of the double layer until the molecule undergoes heterolytic fission and the half bearing the positive charge readily captures electrons. Thus, heterolysis is the rate-determining stage. When dioxane is replaced by water the dielectric constant of the medium increases, heterolysis is facilitated, and kinetic limitation of the current is partly removed.

The papers considered above have the common feature that the conclusions are not based on quantitative measurements but are hypotheses. Some more soundly based conclusions relating to the reduction mechanism of hydroperoxides were made by the authors of the present review [35]. Representatives of the following two groups were taken as topics for investigation: an alkyl hydroperoxide containing the amyl radical (AHP) and the acyl hydroperoxide peracetic acid (PAA). Both groups possess weakly acidic properties; for alkyl hydroperoxides $pK_a \simeq 12$ and for peracids $pK_a \simeq 8$. The reduction was investigated mostly on mercury electrodes − dropping mercury and suspended drop electrodes. To determine the dependence of $E_{1/2}$ on the drop time (t_1) and on the ionic strength (μ_i), and also the dependence of the mean limiting current ($\bar{i}_l$) on t_1, the electrode with glass plate according to Mairanovskii was used. Typical $\bar{i}$-E curves are shown in Figs. 5 and 6; the form of the wave is normal in acidic solution, but in alkaline solution (apart from the main wave) there is a prewave followed by a current drop.

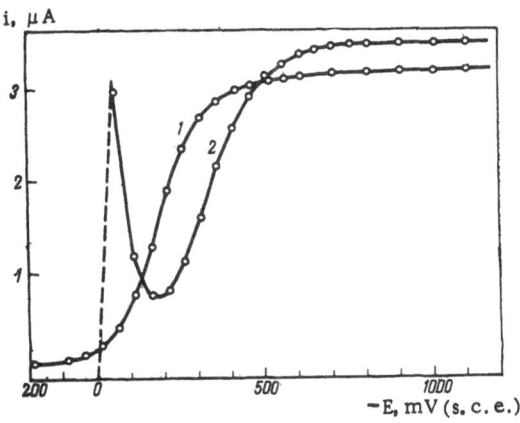

Fig. 5. Polarization curves for amyl hydroperoxide
[35]. 1) In 0.1 N perchloric acid; 2) in buffer solution at pH 12.2.

Turning to the processes which take place at the main wave, it
is necessary first of all to note its clearly defined irreversible nature, which follows even from the values of the reciprocal slope (b)
of the semilogarithmic curve; for the compound under consideration,
and also for the majority of peroxides investigated previously, b ≥
0.15 V. The effects arising from irreversibility are also confirmed
by experiment; thus, the equation for the dependence of $E_{1/2}$ on t_1 is
followed with great accuracy. The second manifestation of irreversibility (the dependence of $E_{1/2}$ on the ψ_1 potential) will be dealt with
below.

The main point of discussion in [35] is the nature of the particles
which react directly at the electrode, and the related question of the
nature of the current. Here it is more convenient to consider the
two types of hydroperoxides separately.

In the case of alkyl hydroperoxides (AHP) the semilogarithmic
curves are linear right up to pH 12.5, and b lies between 0.15 and
0.18 V; this shows that, at any pH value below 12.5, the same type of
particle is reduced with the same mechanism. The b values for AHP
(in water at 5°C) are given below:

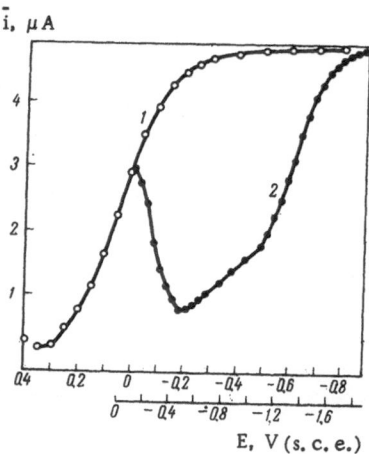

Fig. 6. Polarization curves of peracetic acid. 1) In 0.1 N perchloric
acid + 0.9 N sodium perchlorate; 2)
in glycocol buffer solution at pH
12.45 [35].

pH	1.0	9.0	10.2	10.6	11.8	12.7
b, V	0.15	0.15	0.15	0.16	0.185	0.31

The charge of the reacting particles was determined from the
dependence of $E_{1/2}$ on the ionic strength and thereby on the ψ_1 potential, using the widely-known Frumkin equations which describe the
reduction of particles bearing charge z. The experimental data in
both acidic and strongly alkaline solutions agreed best of all with the
supposition that z = 0. Thus, uncharged particles (specifically ROOH
particles) pass through the double layer. However, this still does
not make it possible to choose between the two concepts; according to
one [19] ROOH particles react directly on the electrode, and according to the other [8] radicals formed by dissociation of ROOH react at
the electrode. Data given by the Gerishers on the reduction mechanism of hydrogen peroxide indicate in favor of the first interpretation.

Since the undissociated acid reacts at the electrode, the process
should include a protonation stage in the pH region close to pH$_a$.

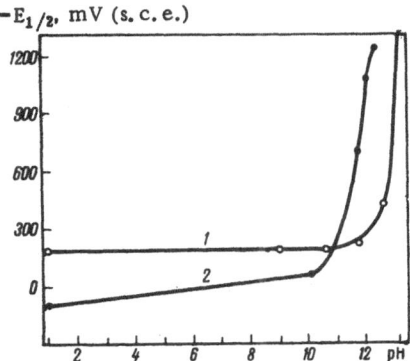

Fig. 7. Dependence of $E_{1/2}$ on pH [35].
1) Curve for amyl hydroperoxide, calculated from Eq. (12); 2) the same for peracetic acid; O represents the corresponding experimental points; ● represents the
points for peracetic acid.

However, the current does not show kinetic contribution in this region, which is also confirmed by the dependence of i_l on t_1. Indeed,
when the approximate equation

$$\bar{i}_l = At_1^p$$

was used at pH $\leq$ 13 p values of 0.22-0.26 were obtained (for cadmium and lead ions p = 0.22), and only in 1 N sodium hydroxide solution
does p reach a value (0.32) which indicates partial involvement of
kinetic control. Thus, the wave has a quasidiffusion character at pH
$\leq$ 13 (according to the terminology of Mairanovskii). Figure 7 shows
that it does in fact closely satisfy Eq.(12). The increase in p, beginning at pH $\simeq$ 13, indicates that the recombination rate here is already
insufficient to maintain the protolytic equilibrium at the electrode.

In the case of peracetic acid (PAA) constancy in b (0.20-0.22) is
only retained up to pH $\simeq$ 9; at pH values between 9 and 12.4 the wave
is very extended, and the semilogarithmic curve is curved or has an
inflection; beginning at pH 12.4 the curve is once again linear, but its

slope is somewhat greater. The b values for PAA (in water at 0°C) are given below:

pH	1.10	1.5	4.4	10.2	10.35	11.8	12.45	12.6
b, V	0.21	0.22	0.22	(0.36)	(0.58)	(0.63)	0.28	0.25

The dependence of $E_{1/2}$ on ψ_1 differs in acidic and strongly alkaline regions; the equation with $z = 0$ fits best in the former and the equation with $z = -1$ in the latter. Thus the undissociated molecule is reduced right up to pH values somewhat greater than pK_a, and the anion is reduced in fairly alkaline solution. It is natural to assume that in the intermediate region, where the form of the wave does not correspond to a simple irreversible reaction, both forms react simultaneously at the electrode. An equation for a similar case has been derived before [36]. Figure 8 shows that with appropriate selection of the parameters the experimental points lie satisfactorily on the calculated curves. There is one further argument in favor of the proposed interpretation. If the acid and anion are reduced simultaneously in the intermediate region, $\bar{i}_l$ should be controlled by their joint diffusion; in accordance with this, $p = 0.24$ not only in acid and strongly alkaline regions but also in the intermediate region (pH 9.5). The difference in behavior between the two types of hydroperoxides is not in principle determined mainly by difference in pK_a; the pH region where reduction of the acid is limited by the small protonation rate is only readily accessible and reduction of the anion only begins to appear with peracids.

Formation of a prewave (maximum current) in alkaline solutions is observed not only at a dropping electrode but also at solid electrodes (see below). Apart from this, other characteristics indicate the nonturbulent nature of the maximum. The current at the maximum ($\bar{i}_m$) is proportional to the concentration of peroxide (c_p), and although it is sensitive to the experimental conditions it does not always remain less than $\bar{i}_d$; it does not change from addition of maximum suppressors. The products of some chemical reactions involving the peroxide are clearly reduced at the electrode in the region of the prewave. The suggestion that this is catalytic decomposition to the aliphatic acid (or the respective alcohol) and oxygen can be rejected immediately; under identical conditions (with 0.1 mole/liter NaOH + 0.9

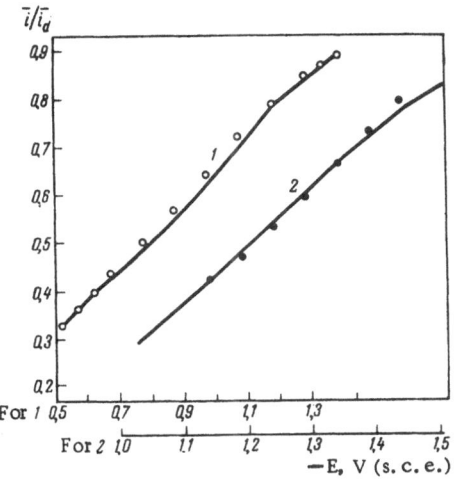

Fig. 8. Dependence of i/i_d on E for peracetic acid. The continuous lines represent calculated curves [36]. 1) pH 10.50; 2) pH 12.45. The circles represent experimental points [35].

mole/liter $NaClO_4$ as electrolyte) the maximum current lies at -0.10 V in the case of PAA (at -0.09 V in the case of AHP), while the oxygen wave begins at -0.15 V (Fig. 9). (The potentials are referred to the saturated calomel electrode.) Furthermore, the above-mentioned decomposition is a second-order reaction, which is inconsistent with the linear dependence of $\bar{i}_m$ on concentration. The most likely mechanism is the following:

$$Hg + HO_2^- \xrightarrow{k} HgO + OH^-, \tag{13}$$

$$HgO + H_2O + 2e \rightarrow Hg + 2OH^-, \tag{14}$$

i.e., oxidation of the mercury surface by the peroxide followed by cathodic reduction of the HgO. In fact the equilibrium potential of Reaction (14) is -0.08 V in 0.1 mole/liter sodium hydroxide solution, which agrees approximately with the prewave maximum for both peroxides. The suggestion was substantiated by chronopotentiometry on a hanging mercury drop. The electrode was kept in a solution of the

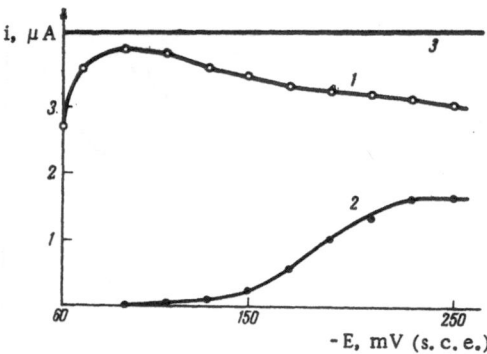

Fig. 9. i-E Curve in 0.1 N sodium hydroxide solution. 1) Beginning of peracetic acid wave; 2) beginning of oxygen wave; 3) limiting current for peracetic acid.

peroxide for a specific time (e.g., 10 sec) in open circuit, after which a current pulse of the required intensity i_0 was applied. Well-defined arrests were obtained at E = -0.08 V. If the electrode was also held under open circuit conditions in the electrolyte without the peroxide, there were no arrests, but they appeared if a potential of $+0.05$ V or more was applied to the electrode under potentiostatic conditions. In the last case the arrest, which also lies at -0.08 V, can only be attributed to reduction of HgO, formed by Reaction (14) taking place from right to left. The chronopotentiometric curves from the corresponding experiments are given in Fig. 10. By regarding the HgO film as equivalent to an adsorbed layer it is possible to calculate the surface concentration Γ. The Γ values obtained varied between $2.0 \cdot 10^{-10}$ and $2.2 \cdot 10^{-10}$ depending on the model used for the calculation, and this is an order of magnitude lower than the expected value of Γ_{∞}. The reason preventing the production of larger coverages is not yet clear.

It was possible to explain the form of the prewave, at least qualitatively, on the basis of the following considerations. In the potential region of the decrease of the prewave, the reduction of HgO takes place instantaneously, and the slow step can only be Reaction

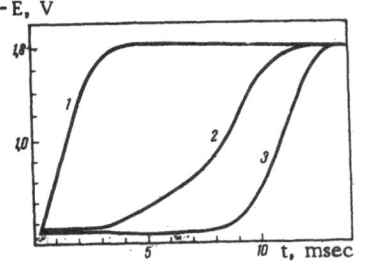

Fig. 10. Chronopotentiometric curves. 1) Solution of 0.1 N NaOH + 0.9 N NaClO$_4$, electrolyte, kept in open circuit for 10 sec; 2) electrolyte + 0.0015 mole/liter peracetic acid, kept in open circuit; 3) 0.1 N NaOH + 0.9 N NaClO$_4$, kept at E = +0.05 V.

(13); consequently, i is proportional to the surface concentration of peroxide anion, which decreases as the zero-charge point is approached.

4. REDUCTION OF PEROXIDES AT SOLID METAL ELECTRODES

As known, processes on electrodes with an unrenewed surface are to a large extent complicated by chemisorption of oxygen and hydrogen. This complication is particularly noticeable with peroxides since they are capable of oxidizing many metals by a purely chemical mechanism. The behavior of peroxides on solid electrodes has therefore become a subject for fairly extensive research; this is true about oxygen even to a larger extent. Many metals have been used, but the present review will only discuss data relating to the platinum electrode, which has been investigated in most detail. Nearly all that is known on the subject concerns hydrogen peroxide.

A stationary method is not very suitable for investigating processes on electrodes with a surface which changes continuously during polarization; the method usually employed until recently has been to record galvanostatic charging curves (chronopotentiometry) or i-E curves with saw-tooth or triangular pulses. Recently a series of interesting results have been obtained through the introduction of a new method, involving a rotating disk electrode with a ring, introduced by Frumkin et al., [37, 38].

Changes in the surface conditions of platinum during anodic and cathodic polarization, and also with other methods of treatment, have been investigated in a large number of papers [39-44], to which only reference can be made here. The most accurate results on the chemisorption of oxygen were obtained in [45, 46] owing to the use of an electrode which had been reduced in hydrogen and then completely degassed in vacuum at high temperatures. It was shown that oxygen was present on the surface in several forms with different adsorption and desorption potentials; the number of forms depended on the preceding electrochemical treatment. Cathodic polarization with rapid application of the pulse only removed half the oxygen, and oxygen thus remained on the surface even at hydrogen potentials. These observations explain many of the discrepancies between the results obtained in various investigations. The fact is that methods for electrode preparation are rarely the same, and description of an electrode by such a term as "reduced" (or "oxidized") does not represent a uniform characteristic — the surface condition and behavior of two such electrodes can be completely different.

There are two opposite points of view on the interpretation of the reduction mechanism (and also oxidation mechanism) of hydrogen peroxide on platinum; according to one, it is a purely electrochemical process, and according to the other, the deciding role is played by chemical stages (oxidation of the metal by the peroxide and catalytic decomposition of the peroxide at the electrode surface). This difference in approach has already been mentioned when discussing decomposition in the absence of current. At the present time the purely electrochemical theory has practically no supporters, but there is still a large amount of controversy about the details of the process. It is therefore necessary, albeit briefly, to indicate these different points of view.

Anson and King [47] noticed the accelerating effect of surface oxides on both steps in the reduction of oxygen and explained it by the fact that reduction of oxides leads to a slight platinizing effect on the electrode. The authors were inaccurate in that the first stage is not accelerated but retarded under the influence of oxides [48].

A similar point of view is held by the Linganes [49], who have carried out the detailed investigation of the reduction of hydrogen peroxide by chronopotentiometry. Two types of electrode were investigated. An "active" electrode was obtained by alternating anodic and cathodic treatment, repeated many times, and was distinguished by a considerable degree of surface oxidation; a "pure" electrode was kept for a day in 1 mole/liter sulfuric acid solution and, in the opinion of the authors, was completely free from oxides and from finely divided metal layer. The electrolyte was 0.5-1 mole/liter acid. Cathodic (to −0.20 V) and anodic pulses (to +1.30 V vs. s.c.e.) were applied in a specific order. The following results were obtained. On the pure surface there was no step for the first cathodic pulse, i.e., reduction did not occur. At the next anodic pulse there was a step at a potential which approximately coincided with the normal potential of the reaction:

$$Pt + 2H_2O = Pt(OH)_2 + 2H^+ + 2e, \quad E^0 = +0.98V$$

and approximately 0.4 V more positive than E^0 for the $O_2 - H_2O_2$ system. From this it was concluded that the anodic reaction involved oxidation of the platinum; it was followed by chemical oxidation of the peroxide by the oxide formed. A cathodic step was first observed after application of the cathodic pulse following the anodic pulse, and it increased to a certain limit with increase in the number of paired pulses. The Linganes [49] put forward the following explanation. During the anodic and the following cathodic run a finely divided platinum layer, which forms the activity carrier, is formed. During the next cathodic pulse this layer is oxidized chemically by the peroxide, and the cathode reaction amounts to reduction of the platinum oxides. While the activity of the electrode is small, the form of the E-t curve is controlled by the platinum oxidation rate and, when sufficient activity has been attained, by diffusion of the peroxide to the electrode. They also considered catalytic decomposition of the peroxide on finely divided platinum, which manifests itself in a decrease in the anodic transition time (τ_a) with increase in the activity of the electrode; this process is, however, not reflected in τ_c, since the decrease in current arising from the fall in hydrogen peroxide concentration is compensated by the oxygen reduction current. Thus the point of view most consistently expressed by the Linganes [49] suggests that platinum is mostly involved in the electrochemical steps, and decomposition of the peroxides and reduction of oxygen play a secondary role; however, this view is insufficiently well founded on the quantitative side.

Nekrasov and Muller [48, 50, 51] offered an important contribution to the solution of the problem by using the electrode with ring. The principal idea of the method is as follows: at the same time as the polarization curve (i_d-E_d) is recorded on the disk the limiting currents [anodic $(i_{c,a})$ and cathodic $(i_{c,c})$], obtained during application of a fairly high positive or negative potential respectively, are recorded on the ring. The currents at the ring arise from the fact that a definite proportion of the unreacted oxygen or peroxide at the disk is rejected onto the ring. By measuring $i_{c,a}$ and $i_{c,c}$ it is possible to calculate the rate of access of each of the two compounds to the ring; here use is made of the fact that in the case of the peroxide the number of electrons is equal to two both in reduction and in oxidation, and oxygen only makes its contribution to the cathodic limiting current. From the limiting current density of components A (i^A) its surface concentration on the disk (c_A^s) is calculated from the following equation:

$$i^A = N n_C F \frac{D_A c_A^s}{\delta_A},$$

where D_A is the diffusion coefficient; δ_A is the thickness of the diffusion layer on the rotating disk electrode. The proportionality factor N, which characterizes the proportion of compound (unreacted at the disk) which reaches the ring, is only determined by the geometry of the electrode and can be calculated. Together these data make it possible to characterize the process at the disk for any value of E_d.

The best defined results were obtained by this method during reduction of hydrogen peroxide in strongly alkaline medium (0.125 N potassium hydroxide). Figure 11 shows the characteristic features of the i_d-E_d curve — a maximum at 0.7 V and the gradual decrease of the current with a minimum at 0.15 V (vs. h. e. in the same solution). The black points in the figure represent the "corrected" values of the current, $i_d + i_d'$, where i_d' is the current calculated from the surface concentration of unreduced oxygen. The calculation gives the following picture. At the beginning of the wave (0.94 V) the amount of oxygen (corresponding to the complete decomposition of the peroxide) which diffuses to the electrode is recorded at the ring. In the region corresponding to the rise in the curve the oxygen formed is partially

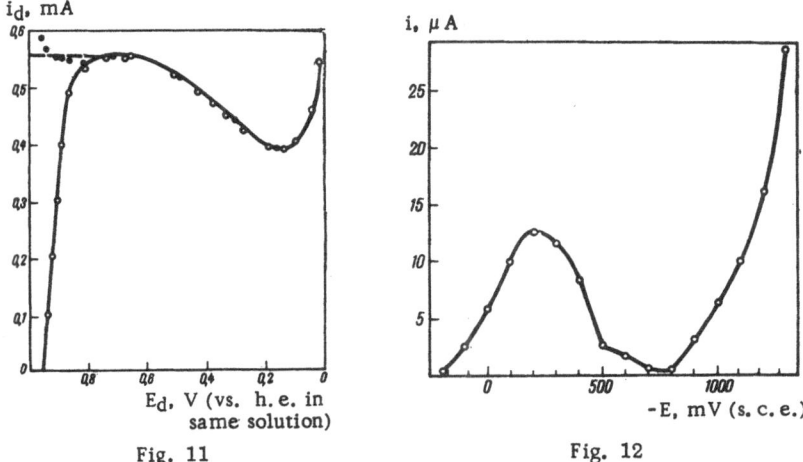

Fig. 11   Fig. 12

Fig. 11. i_d-E_d Curve for hydrogen peroxide in 0.125 N potassium hydroxide solution, recorded on rotating disk electrode of platinum [48]. Black points represent corrected values for current.

Fig. 12. i-E Curve for peracetic acid on rotating disk electrode of gold [35].
Electrolyte: 0.1 N NaOH + 0.9 N NaClO$_4$.

reduced at the disk, but the corrected current again corresponds to
complete cleavage of the peroxide. Beginning at the maximum no
oxygen at all is detected on the ring, since it reacts completely at the
disk; here increasing amounts of peroxide are recorded on the ring.
All this leads to the following conclusions. In the region of the first
wave catalytic decomposition of the peroxide takes place on the oxidized surface of the electrode, and its rate decreases with reduction
of the platinum oxide. Here, the electrochemical reaction involves
mainly oxygen, and direct reduction of the peroxide does not play a
substantial part. The latter begins to appear in the region corresponding to the decrease in the curve and completely controls the process at the next wave. In acidic solution the decrease in the current
is absent, but splitting of the wave is observed and calculation shows
that a similar explanation is possible.

Reduction of organic peroxides on solid electrodes has hardly
been investigated at all; there are only preliminary data for peracetic

acid on a rotating gold electrode [35]. Figure 12 shows the curve obtained in alkaline solution (0.1 N NaOH + 0.9 N NaClO$_4$); in general form it is reminiscent of the curve for hydrogen peroxide, but its maximum and minimum are shifted by approximately 0.15 V to the positive side. There is reason to suppose that the processes are similar in the two cases.

In summarizing, the following general pattern of electrochemical reduction in peroxides can be considered likely. At the potentials of the main wave, which normally has a diffusion or quasidiffusion character, the peroxide molecule (or its anion) takes part directly in the electrochemical steps; it is possible that both forms of particle participate at the same time. The reaction rate, and consequently also the half-wave potential, depend substantially on the adsorbability of the peroxide. Moreover, some peroxides are capable of entering into chemical reaction with the electrode material. In the case of mercury the oxide is formed, followed by electrochemical reduction, and this leads to the appearance of a wave with a decrease; in the case of platinum-group metals the peroxide decomposes at the surface, and the oxygen formed enters into the electrochemical reaction. These phenomena are only clearly observed in strongly alkaline solutions. Convincing evidence of homolytic fission of the peroxides, preceding the electron transfer, has not been found.

LITERATURE CITED

1.   F. Haber and R. Willstätter, Ber., 64:2844 (1931).
2.   F. Haber and J. Weiss, Naturwiss, 20:949 (1932).
3.   F. Haber and J. Weiss, Proc. Roy. Soc., A147:332 (1939).
4.   D. Lea, Trans. Faraday Soc., 45:81 (1949).
5.   J. Kolthoff and J. Jordan, J. Am. Chem. Soc., 74:4801 (1952).
6.   J. Kolthoff and K. Izutsu, J. Electroanalyt. Chem., 7:85 (1964).
7.   J. Baxendale, M. Evans, and G. Park, Trans. Faraday Soc., 42:155 (1946).
8.   J. O'Bockris and L. Oldfield, Trans. Faraday Soc., 51:249 (1955).
9.   P. Delahay, M. Pourbaix, and P. van Rysselberghe, Ind. Chim. Belge, 1b:396 (1951
10. M. Evans and N. Uri, Trans. Faraday Soc., 45:224 (1949).
11. A. Hickling and S. Hill, Trans. Faraday Soc., 46:557 (1950).
12. M. Evans, N. Hush, and N. Uri, Quart. Rev., 6:186 (1952).
13. F. Haber and S. Grinberg, Z. Physik. Chem., 34:513 (1900); Z. Elektrochem., 7:44 (1901).
14. J. Weiss, Naturwiss., 23:64 (1935).

15. A. Hickling and W. Wilson, J. Electrochem. Soc., 98:425 (1951).

16. G. Bianchi, F. Mazza, and T. Mussini, Electrochim. Acta, 7:458 (1962).

17. L. E. Yablokova and V. S. Bagotskii, Dokl. Akad. Nauk SSSR, 85:599 (1952).

18. V. S. Bagotskii and I. E. Yablokova, Dokl. Akad. Nauk SSSR, 95:1219 (1954).

19. R. Gerisher and H. Gerisher, Z. Physik. Chem. (N. F.), 6:178 (1956).

20. V. S. Bagotskii and D. L. Motov, Dokl. Akad. Nauk SSSR, 71:501 (1949).

21. V. I. Tikhomirova, V. I. Luk'yanycheva, and V. S. Bagotskii, Élektrokhimiya, 1:645 (1965).

22. Z. A. Iofa, Ya. B. Shimshelevich, and E. P. Andreeva, Zh. Fiz. Khim., 23:828 (1949).

23. V. S. Bagotskii and I. E. Yablokova, Zh. Fiz. Khim., 27:1663 (1953).

24. P. van Rysselberghe and C. Murdock, Compt. Rend. de la 2-d Réunion du CITCE, Milan, 1950.

25. J. Chodkowski, Roc. Chem., 27:309 (1953).

26. G. Murdock and P. van Rysselberghe, Proc. 6th Intern. Congr. of CITCE, 1955.

27. A. Rius and A. Sacristan, Electrochim. Acta, 6:155 (1962); 9:171 (1964).

28. M. Bernard and S. Hausswirth, C. R. Acad. Sci., 257:1282 (1963).

29. S. Hausswirth, B. Bregeon, and M. Bernard, Electrochim. Acta, 11:267 (1966).

30. D. Skoog and A. Lauwzecha, Analyt. Chem., 28:825 (1956).

31. W. Parker, C. Ricciuti, C. Ogg, and D. Swern, J. Am. Chem. Soc., 77:4037 (1955).

32. L. Silbert, L. Witnauer, D. Swern, and C. Ricciuti, J. Am. Chem. Soc., 81:2774 (1959).

33. M. F. Romantsev and É. S. Levin, Zh. Analiticheskoi Khim., 18:1109 (1963).

34. M. Schulze and K. Schwarz, Monatsber. Dt. Akad. Wiss. (Berlin), 6:515 (1964).

35. É. S. Levin and A. V. Yamshchikov, Élektrokhimiya, 4:54 (1968).

36. É. S. Levin, Élektrokhimiya, 2:955 (1966).

37. A. N. Frumkin and L. N. Nekrasov, Dokl. Akad. Nauk SSSR, 126:115 (1959).

38. A. N. Frumkin, L. Nekrasov, V. Levich, and Ju. Ivanov, J. Electroanalyt. Chem. Chem., 1:84 (1959).

39. F. Anson and J. Lingane, J. Am. Chem. Soc., 79:4901 (1957).

40. F. Anson, J. Am. Chem. Soc., 81:1554 (1959); Analyt. Chem., 33:934 (1961).

41. H. Laitinen and C. Enke, J. Electrochem. Soc., 107:773 (1960).

42. J. Lingane, J. Electroanalyt. Chem., 1:379 (1960).

43. S. Feldberg, C. Enke, and C. Bricker, J. Electrochem. Soc., 110:826 (1963).

44. V. I. Luk'yanycheva and V. S. Bagotskii, Dokl. Akad. Nauk SSSR, 155:160 (1964).

45. V. I. Tikhomirova, A. I. Oshe, V. S. Bagotskii, and V. I. Luk'yanycheva, Dokl. Akad. Nauk SSSR, 159:644 (1964).

46. V. I. Luk'yanycheva, V. I. Tikhomirova, and V. S. Bagotskii, Élektrokhimiya, 1:262 (1965).

47. F. Anson and D. King, Analyt. Chem., 34:362 (1962).

48. L. Myuller and L. N. Nekrasov, Dokl. Akad. Nauk SSSR, 154:437 (1964); 157:416 (1964); Zh. Fiz. Khim., 38:3028 (1964).

49. J. Lingane and P. Lingane, J. Electroanalyt. Chem., 5:411 (1963).
50. L. N. Nekrasov and L. Myuller, Dokl. Akad. Nauk SSSR, 149:1107 (1963).
51. L. N. Nekrasov, Élektrokhimiya, 2:438 (1966).